Plant and Vegetation

Volume 16

Series editor

M.J.A. Werger, Utrecht, The Netherlands

Plant and Vegetation is a new Springer series comprising a series of books that present current knowledge and new perspectives on world vegetation. Examining the ecology of plants and vegetation at all scales – from plant to landscape – and covering key issues such as globalization, invasive species, climate change and the dynamics of plant biodiversity, this book series draws together a wide range of material of interest to plant ecologists, vegetation scientists, and geographers around the world. The series provides a valuable resource for both graduate students and researchers in environmental and biological sciences, as well as for landscape planners and policy makers involved in land-use and restoration projects at local, regional and international levels.

More information about this series at http://www.springer.com/series/7549

Marcelino J. del Arco Aguilar
Octavio Rodríguez Delgado

Vegetation of the Canary Islands

Springer

Marcelino J. del Arco Aguilar
Department of Botany, Ecology and Plant Physiology
University of La Laguna
San Cristóbal de La Laguna, Tenerife, Spain

Octavio Rodríguez Delgado
Department of Botany, Ecology and Plant Physiology
University of La Laguna
San Cristóbal de La Laguna, Tenerife, Spain

ISSN 1875-1318 ISSN 1875-1326 (electronic)
Plant and Vegetation
ISBN 978-3-030-08411-0 ISBN 978-3-319-77255-4 (eBook)
https://doi.org/10.1007/978-3-319-77255-4

Softcover re-print of the Hardcover 1st edition 2018

Printed on acid-free paper

This Springer imprint is published by the registered company Springer International Publishing AG part of Springer Nature.
The registered company address is: Gewerbestrasse 11, 6330 Cham, Switzerland

Acknowledgments

We thank the editor of the *Plant and Vegetation* series Marinus J.A. Werger for his valuable help during the process of writing and editing this book. His suggestions, advice and corrections were important for improving its content. We are also grateful to Valeria Renaud (Publishing Editor) and Ineke Ravesloot (Assistant Editor) for their kind assistance as well as Sathiamoorthy Rajeswari (Project Manager at SPi Technologies India Pvt Ltd). Thanks also to Guido Jones Carter for his comments and preliminary proofreading.

Special thanks are due to all botanical colleagues at the University of La Laguna with whom we have shared laboratory and field work, and publications, during our careers.

Thanks to GRAFCAN for kind permission for including versions of our maps (Mapa de Vegetación de Canarias).

Introduction

The purpose of this book is to describe the vegetation of the Canary Islands, including relevant floristic aspects, for students and scholars of nature. The work is a synthesis of numerous publications and also provides data derived from the experience accumulated by the authors during many years of teaching and research in these subjects at the University of La Laguna. It tries to reflect the enormous floristic and phytocoenotic biodiversity of these islands, which constitute natural laboratories for studies of flora, vegetation, biogeography, geology, bioclimatology, ecology, and so on. To do this, in the preliminary chapters, we approach topics essential to a contextual understanding of the physical and biotic framework in which the vegetation units are described.

The "geographic framework" chapter situates the Canaries as oceanic and volcanic islands in the context of Macaronesia and provides geographical data of interest. The "climate" chapter defines the main factors determining the Canary climate and its variety, paying particular attention to the decisive influence of the trade winds in the generation of different climatic zones typical in the development of the distinct plant communities. In addition, the main types of weather are synthesized. The chapter on "bioclimate" provides selected key data, as considered in the world bioclimatic classification in the establishment of its various units. There is special reference to those data used for characterization of bioclimatic belts within the Mediterranean macrobioclimate. Correlations between potential climatophilous communities and bioclimatic belts in the Canary Islands are established, providing representative bioclimographs and examples of bioclimatic cartography.

The "biogeography" chapter begins with considerations about the plant colonization of the archipelago in a variable geological and paleo-climatic framework, since the appearance of the first subaerial edifices of the archipelago and its surroundings. The possible vectors of colonization and the main sources of flora are established. Finally, the biogeographic classification we follow is justified by means of geographic, bioclimatic, floristic, and phytocoenotic criteria, considering the Canaries as included within the Canarian-Madeiran subregion of the Mediterranean region. This chapter touches on different floristic aspects, which are further explained in the next chapter on "other floristic considerations." Insular plant (*s.l.*)

biodiversity is quantified and the importance of generic and specific endemics is highlighted, comparing them with those of the other Macaronesian archipelagos. Common and differential elements are exemplified.

The "vegetation" chapter constitutes the bulk of the work. Vegetation is characterized from a phytosociological point of view. First, we present an "overview" where the main potential communities growing from coast to summit are characterized, framing them within their altitudinal limits and bioclimatic belts. Physiognomy, structure, environmental aspects, floristic composition, origin and affinities, and distribution are considered in their descriptions. Their main substitutional communities are indicated and also, where appropriate, the other important edaphophilous communities growing in their climatophilous domains. In the section "current vegetation units," we follow a syntaxonomic nomination for the communities, although sometimes some physiognomic units are additionally included. The communities are later collected and ordered into a syntaxonomic catalog, in Appendix 1. Summarized descriptions of communities (physiognomy, structure, bioclimatic belts, environmental factors, dynamic status, character species, and distribution) are provided. Descriptions of special Canary communities are further detailed, especially those with a climatophilous character or greater representivity in the Canary landscape. For each of the four most representative phytosociological classes of climatophilous vegetation we have inserted a list of characteristic species of their communities. In the section "Potential vegetation", the climatophilous and edaphophilous vegetation series are summarized by mentioning the typical communities representing their final stages. The main permanent communities are also listed, ordered according to the environment in which they grow. Then, we briefly mention the most typical vegetation complexes and geosigmeta. The species are acompanied by the family to which they belong, their common name, biotype, and island distribution.

The chapter on "changes in the natural landscape through human influence" firstly considers the possible effect the aboriginal inhabitants exerted on the flora and vegetation of the islands, and later the more intense transformation after the conquest, from the end of the fifteenth century. It also considers the effect of climate change on island vegetation, particularly regarding possible elevational displacements of the vegetation, and habitat and diversity loss. The chapter on "conservation status of the Canary flora and vegetation" shows the efforts made to protect them. Red lists of threatened species are referred to, along with the legal framework for the protection of species, natural reserves, and habitats. Botanical gardens, seed and germplasm banks, and nurseries of utility in carrying out conservation and restoration tasks are mentioned. Attention is also drawn to the danger posed by the multitude of invasive plants spreading nowadays on the islands.

Finally, the volume is completed with several appendices. Appendix 1 presents the "syntaxonomic scheme," alphabetically arranged according to syntaxonomic classes, indicating the island distribution. In Appendix 2, a "brief history of botanical studies and explorations in the islands" is sketched. The main scientists concerned with the studies of flora and vegetation are chronologically mentioned, along with the current research centers in the Canaries carrying out studies on flora and vegetation. Appendix 3 contains "ethnobotanic notes" relating to phytonymy and phyto-toponymy, and about some special trees or notable specimens. Finally, Appendix 4 contains the most relevant literature used.

Tables, photos, and figures document and illustrate all the information provided.

Department of Botany, Ecology
and Plant Physiology
University of La Laguna
San Cristóbal de La Laguna, Tenerife, Spain

Marcelino J. del Arco Aguilar
Octavio Rodríguez Delgado

Contents

Chapter 1
Geographic Framework

The Canary Islands are an Atlantic oceanic archipelago off NW Africa located between 13°20′–18°10′W and 27°37′–29°25′N. It is formed by seven main inhabited islands: from E to W these are Lanzarote, Fuerteventura, Gran Canaria, Tenerife (Fig. 1.1), La Gomera, La Palma and El Hierro. To these are added the small islets and rocky components of the so-called "Chinijo" (=small) Archipelago, located in its Eastern sector to the North of Lanzarote; from biggest to smallest these are La Graciosa (with a stable population), Alegranza, Montaña Clara, Roque del Este and Roque del Oeste; the small islet of Lobos between Fuerteventura and Lanzarote, and a series of offshore rocks of small surface area. All the Canaries together have an area of about 7447 km^2. The tallest (3718 m; Mount Teide) and largest area island is Tenerife (2034 km^2) and the smallest El Hierro, 268.7 km^2. The oldest but lowest islands are the easternmost, Fuerteventura and Lanzarote (Table 1.1; Fig. 1.2).

Fig. 1.1 View of Tenerife from La Gomera. On clear days one can see several islands from each of the islands. On such days, from the peak of the Teide all seven islands are visible

M. J. del Arco Aguilar, O. Rodríguez Delgado, *Vegetation of the Canary Islands*,
Plant and Vegetation 16, https://doi.org/10.1007/978-3-319-77255-4_1

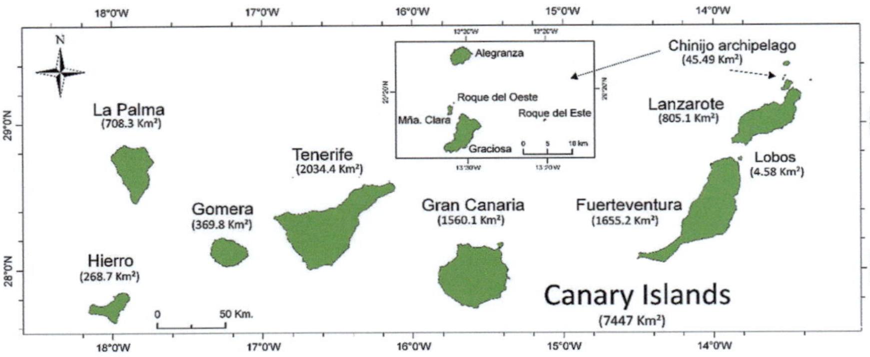

Fig. 1.2 Map of the Canary Islands

The Canaries are considered within a group of archipelagos known as Macaronesia (*macaron*: happy; *nesoi*: islands, a name derived from classical mythology used for the islands in the Atlantic ocean beyond the Pillars of Hercules in the Strait of Gibraltar). From North to South they are Azores, Madeira, Salvage islands, Canaries and Cape Verde. These are dispersed within a wide latitudinal and longitudinal range between 14–40°N and 13–32°W that stretches from mid-Atlantic waters west of Portugal to the coast of Senegal (Fig. 4.1). They all have a volcanic origin (Fig. 1.3), with variable ages for the oldest rocks of their sub-aerial edifices that span from Neogene to Pleistocene and volcanic activity still continues at some hot-spots. Azorean volcanism is linked to inter-plate activity near the mid-Atlantic ridge, and the volcanism of the other archipelagos is linked to a collision of the African plate with Europe.

There is a notable climatic variation across Macaronesia. It goes from a dominant Temperate climate in the Azores, to Mediterranean and Temperate in Madeira, Mediterranean in the Salvage islands and Canaries and to Tropical in Cape Verde.

From a floristic point of view the Macaronesian islands show some similarities, but also notable differences, and three clear groups can be distinguished: 1. Azores; 2. Madeira, Salvage islands and Canaries; 3. Cape Verde. These differences still stand when referring to vegetation.

The Canaries were the only Macaronesian archipelago with an aboriginal human population at the time of their European colonization. Its indigenous inhabitants are usually referred to generically as "Guanches", although there are particular names for the originally north-African inhabitants of each island. The

Fig. 1.3 All islands have a volcanic origin. The stratovolcano Teide gave rise to the central part of the island of Tenerife. It constitutes the Natural Monument of the Teide, right in the middle of the Teide National Park

archipelagos began to be colonized by the Portuguese during the fifteenth century, while the Canaries were progressively conquered by the Spanish Crown by the end of the fifteenth century. The suboceanic construction of the Canary archipelago seems to have originated in the middle of the Tertiary era by accumulation of volcanic materials over a Jurassic oceanic crust, coinciding with the collision of the African Plate on its way northeast towards Europe and Asia. Different theories have been put forward to explain its origin: in the 1960s the hot-spot theory was proposed and in the 1970s the propagating fracture and uplifted blocks theories; in the 1990s the bubbles model and laminar thermal anomaly, and finally in 2000 a unifying model theory based on a consensus among most of these theories was formulated. This hypothesis considers the existence below the Canaries of hot spots linked to a residual thermal plume that was active from the opening of the Atlantic, 200 Ma B.P.

Table 1.1 Geographic framework of the Canary Islands

Canary Islands 27°37′–29°25′ N 13°20′–18°10′ W	Geographic location within the archipelago	Biogeographic location within the archipelago	Area km^2	Maximum height m and site	Age Ma	Coast length Km	Geology	Population 01/01/2016 inhabitants number
Main islands								
Lanzarote	Eastern	Eastern Canary Subprovince	805.11	671 Peñas del Chache	15.5	191		145,084
Fuerteventura	Eastern	Eastern Canary Subprovince	1655.16	807 Pico de la Zarza	20.6 Basal complex outcrops (Betancuria): up to 65	304		107,521
Gran Canaria	Central	Western Canary Subprovince	1560.10	1949 Pico de las Nieves	14.5	258		845,195
Tenerife	Central	Western Canary Subprovince	2034.38	3718 Pico del Teide	Anaga: > 6.5 Teno: 6.7 Adeje: 11.6 Cañadas edifice: 2.15	342	Basaltic and salic differentiates emitted in several eruptive cycles from the Miocene to the present	891,111
La Gomera	Western	Western Canary Subprovince	369.76	1487 Garajonay	12 Basal complex outcrops (Vallehermoso): up to 20	100		20,940
La Palma	Western	Western Canary Subprovince	708.32	2423 Roque de los Muchachos	1.77 Basal complex outcrops (Caldera de Taburiente): up to 4	166		81,486
El Hierro	Western	Western Canary Subprovince	268.71	1501 Pico de Malpaso	1.12	110		10,587

Canary Islands 27°37′–29°25′ N 13°20′–18°10′ W	**Geographic location within the archipelago**	**Biogeographic location within the archipelago**	**Area km^2**	**Maximum height m and site**	**Age Ma**	**Coast length Km**	**Geology**	**Population 01/01/2016 inhabitants number**
Chinijo Archipelago								
Roque del Este	Eastern NE of Lanzarote	Eastern Canary Subprovince	0.065	84			Recent basaltic lava flow remains	
La Graciosa	Eastern N of Lanzarote	Eastern Canary Subprovince	29.05	266 Las Agujas			Recent volcanic cones and basaltic lava flows	Included within Lanzarote
Alegranza	Eastern N of Lanzarote	Eastern Canary Subprovince	10.3	289 La Caldera			Recent volcanic cones and basaltic lava flows	
Roque del Oeste	Eastern N of Lanzarote	Eastern Canary Subprovince	0.015	41			Dismantled volcanic cone	
Montaña Clara	Eastern N of Lanzarote	Eastern Canary Subprovince	1.48	256 La Mariana			Recent volcanic cones and basaltic lava flows	
Isla de Lobos	Eastern Between Fuerteventura and Lanzarote	Eastern Canary Subprovince	4.58	127 La Caldera o La Herradura			Recent basaltic lava flow remains	
					Islets and rocks	114		
Archipelago			**7.447**			**1.585**		**2,101,924**

TOURISM (2016).: Foreign 13,332,466. **National**: 1,648,648. **TOTAL**: 14,981,114

Biogeographic location: Rivas-Martínez 2011. Area and altitude: http://www.gobiernodecanarias.org/istac/jaxi-istac/menu.do?uripub=urn:uuid:fbc0bdc8-cacb-43b8-a5cb-a93f745dcff6. Geology: Rodríguez Losada, J.A. (2000), Carracedo & Troll 2016. Population census and tourism (ISTAC 11/09/2017), data for 2016

The islands' construction are thought to have gone through three stages:

In the first stage the oceanic crust fractures and blocks are lifted up, over which erupting lava flows and accumulates underwater. This began 70 Ma ago in the Fuerteventura edifice, and lasted for about 45 Ma, becoming highly active during the middle stages of the Alpine orogeny, that included the formation of the Alps and Atlas among many other mountains. Plutonic rocks and marine sediments were important components of this basal complex. Today it only outcrops on some islands (Fuerteventura, La Gomera and La Palma). The second stage, the Miocene-Pliocene sub-aerial construction, began about 20 Ma ago and underwent two cycles. The first or ancient series, Miocene – early Pliocene, gave rise to large volcanic edifices through the build-up of basaltic lava flows interspersed with pyroclastic deposits. The series started at the beginning of the period in the eastern and central islands and took place towards its end on the western islands of La Palma and La Gomera. These eruptions gave rise to the ancient massifs now recognizable on the islands (Fig. 1.4). This first cycle was followed by a long period of inactivity, which led to an active erosive dismantling process.

Fig. 1.4 Ancient massif of Famara in the North of Lanzarote

The second cycle, the third stage or recent series, is Plio-Pleistocene, and began with more diversified emissions where the salic series stand out, although basaltic emissions dominate in general. This cycle continues to the present day, including the so-called historical volcanism, documented for the islands of Lanzarote, Tenerife, La Palma and El Hierro.

The Canary Islands are therefore volcanic, oceanic islands that show an age progression towards the East in the oldest rocks of their sub-aerial edifices: Lanzarote (15.5 Ma), Fuerteventura (20.6 Ma), Gran Canaria (14.5 Ma), Tenerife (11.6 Ma), La Gomera (12 Ma), La Palma (1.77 Ma) and El Hierro (1.12 Ma) (Table 1.1). This is also clearly evident in the degree of erosion and dismantling of the island edifices. The latter are independent units, except for Fuerteventura and Lanzarote, and are deeply seated on the ocean floor with the depth increasing towards the west (up to 3000–4000 m. They have never been united to the African continent, from which they are separated by a minimum distance of about 96.5 km (from the SE of Fuerteventura to the African coast at about 35 km SW of Cape Juby) and a depth of about 1000 m in addition to a thick layer of sediment.

The islands have a very rugged relief, a consequence of the active volcanism that has formed them and the continuous action of erosive forces. Erosive dismantling is most notable in the older islands, although they have had some recent volcanic episodes. This is what happened in La Gomera, without recent volcanism, or in the south-western half of Gran Canaria, or in the ancient Anaga, Teno and Adeje massifs of Tenerife. Fuerteventura (807 m) shows a relief eventually softened by erosion and an advanced dismantling of its oldest massifs, Betancuria and Jandia. Also Lanzarote (671 m) shows dismantling in its ancient massifs of Famara and Los Ajaches, although it also has been rejuvenated with recent and historical eruptions; those of Timanfaya, between 1730 and 1736 produced approximately 1 km^3 of basaltic lavas. The islands of La Palma and El Hierro are younger, although they show massive erosive events such as the Caldera de Taburiente (Fig. 1.5) or the high and large escarpments and ramps derived from enormous landslides.

Fig. 1.5 The Caldera de Taburiente is an enormous erosive depression with very steep cliffs. The highest peak of its enormous circus of 8 km in diameter is Roque de los Muchachos (2426 m a.s.l.), the top of the island of La Palma

Among the most frequent volcanic materials are pyroclasts, known in the Canary Islands as "picón" or "rofe" (Lanzarote), which usually determine characteristic types of edaphophilous vegetation, and lavas which create more or less flat fields when they are fluid (pahoehoe; "lajiales") and fractured rock fields when they are more viscous (aa; "malpaíses"), and provide the conditions for permanent vegetation complexes. Basaltic materials predominate and the optimum natural vegetation in equilibrium with the macro-bioclimatic characteristics of the territory is the climatophilous vegetation developing on the soils derived from them. Among the intrusive materials are salic plugs and outcrops (Fig. 1.6), which tend to have some particular plants adapted to a higher degree of acidity. In some bioclimatic belts the salic territories have a particular vegetation discordant with the climatophilous vegetation of their adjacent basaltic territories, as happens on Tenerife in the salic lava flows of La Guancha-Icod where the pine forest descends in altitude almost to the coast.

As more typical elements of the volcanic landscape derived from simple eruptive activity, single volcanic cones, horseshoe-shaped cones, multiple volcanoes, dikes, volcanic domes, rocks and volcanic plugs stand out. Complex forms such as the old massifs have been generated by repeated eruptive activity. Currently these, such as Famara, Ajaches, Betancuria, Jandía, Tamadaba, Anaga, Teno, Adeje, harbour their own endemics generated during the time they were independent edifices or islands. Other complex forms are the dorsal ridges associated with the main structural axes of the islands and stratovolcanoes like Teide.

Among the most outstanding forms of erosion are highly developed deep ravines with a radial (*e.g.,* La Palma, La Gomera, Gran Canaria) or parallel arrangement (*e.g.,* S of La Palma, Tenerife), depending on the position of the main centers of volcanic activity of the islands or the alignments of their structural axes (Fig. 1.6). On the western islands, V-valleys with steep slopes abound and on the eastern, U-valleys and glacis. The ravines are associated with erosion calderas, alluvial deposits, and interfluves. Valleys, hillsides and interfluves are elements of the landscape that determine particular potential natural vegetation units or their main substitutional stages. Very abrupt terrains and cliffs are common, mainly in the central and western islands, and they harbor a rich, largely sheltered flora, displaced from its natural habitats by destruction, exploitation or pressure from introduced herbivores (Fig. 1.6). The steep slopes of ravines and cliffs constitute the main reserve of the endemic flora. Coastal cliffs are also the norm; they clearly prevail over beaches, especially in the western islands. They carry particular types of permanent vegetation linked to the presence of marine salt.

The beaches may consist of volcanic sands, generally black, in which case they usually lack apparent vegetation, or organic or mixed sand, "white beaches" that carry typical psammophilic communities (Fig. 1.6). The beaches of volcanic sand, pebbles or boulders are dominant in the western islands and north slopes of the central islands. Light-colored beaches are well developed on the eastern islands and southern Gran Canaria. On those islands dune fields with "newkas" have developed. The eolic sand fields ("jables") can be blown across the islands, as in the "Istmo de la Pared" (Fuerteventura) or "El Jable" (Lanzarote).

It is thought that the Canary Islands were colonized by Berber ethnic groups, perhaps brought by the Phoenicians or Romans, during the first millennium BC. These first inhabitants or aborigines, who did not use metals because of the lack of ores for producing utensils of iron or bronze, constituted mainly small pastoral populations operating a regime of seasonal altitudinal droving and also practised cereal and legume agriculture. With them came the medium-size herbivores (goats, sheep and pigs) to the islands, which was a problem for the insular plants that lacked adaptive defense systems. They began to open up territory for pasture and certain crops, with the help of fire. It is difficult to quantify the effects that the aboriginal population had on the vegetation of the islands, but it is evident that they began to transform the insular landscape.

The arrival of Europeans on the islands is documented before their definitive conquest, as in the case of Lancelotto Malocello, a Genoese who in 1312 settled in Lanzarote, where he remained for about 20 years. Subsequently, expeditions to the Canary Islands took place, such as those organized by King Alfonso IV of Portugal, with Majorcans and Castilians. The final conquest of the islands took place between 1402 (Lanzarote) and 1496 (Tenerife). In 1402 and later years the Norman Jean de Bethencourt together with Gadifer de la Salle, sent by King Enrique III of Castile, took the islands of Lanzarote, Fuerteventura and El Hierro (Bethencourt or Norman Conquest). La Gomera was conquered in 1450 by Castilian nobles who also took over the previously conquered islands (Castilian Conquest). Finally, between 1478 and 1496 the conquest of Gran Canaria, La Palma and Tenerife was undertaken by the Castilian Crown under the rule of Ferdinand and Isabella, and this led to the definitive assimilation and later colonization by the Spanish Crown.

Fig. 1.6 (**a**) Sand beaches developed mainly on the easternmost islands (Playa de Barlovento and Jandía Massif, Fuerteventura). (**b**) Cliffs are common and harbor a rich endemic flora (Teno cliffs, Tenerife). (**c**) Salic plugs spread over the island's geography (Roque El Cano, La Gomera). (**d**) Big ravines developed on the oldest islands (Ravine of Valle Gran Rey, La Gomera)

Therefore throughout the fifteenth century, and towards its end in a particularly extensive way, a progressive transformation and decline of the natural island landscapes of the archipelago took place. The initial villages had at first a preferential location at the middle altitudes, because of the benign climate, fertile soil for crops, presence of grazing land, water sources and a safe distance from the coasts that was subject to pirate and corsair attacks. This started the slow but inexorable deterioration of the thermo-sclerophyllous forests typical at those elevations. Soon, large areas were altered as a consequence of intensive sugar-cane cultivation. Sugar mills were constructed consuming a massive amount of laurel forest lumber for fuel throughout the sixteenth century, with its peak around 1550. This gradually yielded to the cultivation of vines and the wine trade during the seventeenth century. In the nineteenth century there was a recovery of sugar-cane cultivation for rum production. The pine forest was always exploited as a wood resource, but especially during the sixteenth and seventeenth centuries to obtain pitch, and in the nineteenth and twentieth centuries for the extraction of resin to obtain turpentine, and for shipbuilding, furniture, house building, etc. The impact on laurel and pine forests was enormous and orders had to be issued by the local authorities for their conservation; though these were not always followed. In addition, rural and pastoral livelihoods themselves consumed forest resources. Meanwhile, the expansion of crops continued for centuries, at first wheat, fruit and vines, but also crops coming from America took over, such as potatoes and maize. The cultivation of cacti of the genus *Opuntia* started at the beginning of nineteenth century for collecting their parasite, the cochineal beetle (*Dactylopius coccus*) for carmine dyes; but this started to decline at the middle of the nineteenth century because of the development of aniline dyes. These cacti have expanded particularly in the landscape of the lower zones. Later in the nineteenth century, monocultures such as those of bananas and tomatoes also occupied extensive territories in the lower regions, to the detriment of *Euphorbia* shrublands and thermo-sclerophyllous forest.

The displacement of the population nuclei to the coasts of the islands is a recent phenomenon. Initially the location of these urban areas was related to port and fishing activities, and they later expanded due to the growth of tourism, that also created new towns to accommodate visitors and to meet their service needs. The current population of the islands was estimated at 2,101,924 inhabitants in 2016 and the number of tourists who visited the islands during that year was 14,981,114. The pressure exerted by man on the islands is intense, focusing nowadays on the low and middle areas, the summits being more protected.

In spite of this great pressure and the demand for space by both resident and floating populations, the conservation policy followed has managed to include almost 50% of the surface of the archipelago into some category of protection, although it is worth mentioning that the distribution is not uniform in the different elevation zones, higher altitudes again being favoured.

Chapter 2
Canary Climate

The climate of the Canary Islands can be classified, according to the latitudinal situation and the oceanic character of the islands, as maritime subtropical, and, particularly by its seasonal regime with summer drought that lasts more than two months after the summer solstice, as Mediterranean. Within this general climatic framework there are important climatic variations mainly due to the elevation of the island blocks and topographic exposure. The trade-winds play a key role in delimiting the different climatic zones.

The islands are bathed by the Canary Current, a cool current forming part of the North Atlantic Gyre that descends along the western side of Africa towards the equator. Its waters are cold due to the ascending latitudinal trajectory of the Gulf Stream, its encounter with cold waters of the Labrador Current and overall by the supply of upwelling cold deep waters to the west of the African coast. Water temperatures in summer range between 18 and 20 °C and in winter between 15 and 18 °C. The Canary Current influences the temperatures of the insular coasts up to 300 m a.s.l., delaying the arrival of the lowest winter temperatures to the month of February and the highest of the summer to August. These cold waters cool the coastal areas, whose average relative humidity is 65%.

Three classic types of weather can be defined for the Canary Islands: Anticyclonic, Continental-Saharan and Disturbed.

2.1 Anticyclonic Weather

The Azores high-pressure nucleus is the main determinant of the Canary climate, conditioning what is known as anticyclonic weather. This anticyclone belongs to the belt of atmospheric high pressure known as the subtropical high. It tends to be located throughout the year at the latitude of the Azores, between 30°and 35°N.

M. J. del Arco Aguilar, O. Rodríguez Delgado, *Vegetation of the Canary Islands*, Plant and Vegetation 16, https://doi.org/10.1007/978-3-319-77255-4_2

A tropical circulation of winds is established as a consequence of a general dynamics associated with the Hadley cell between the equator and the subtropical high. This cell is a circulation of winds that comprises rising warm air at the equator (intertropical convergence zone), where it produces abundant rainfall, then flowing poleward at high atmospheric layers as hot and dry SW winds (anti trade-wind). Afterwards it descends in the subtropics, in our case being incorporated in the Azores high, and then flows at low atmospheric levels equatorward as a NE fresh and humid wind (trade-wind). This is because on reaching mid-latitudes it is cooled and it moistens in contact with the Canary Current. There is furthermore an aerial component which derives from the anti trade-wind that has descended before becoming part of the high pressure dynamics returning towards the equator as hot dry winds (NW trade-wind), generating a warm subsidence inversion layer above the NE trade-wind (Fig. 2.1).

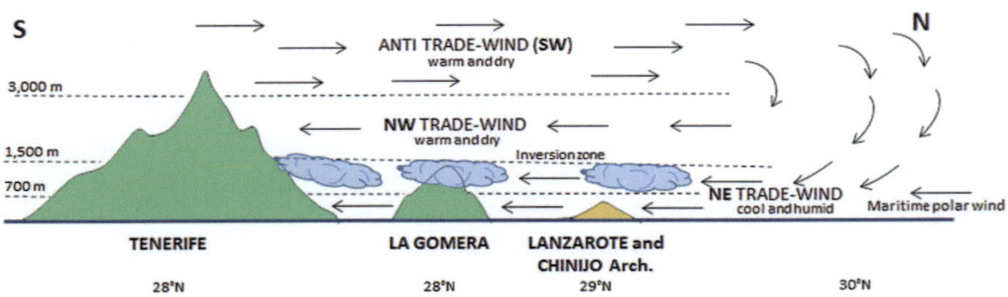

Fig. 2.1 Trade-winds dynamics around the Canary Islands, and the three types of islands (After Ceballos and Ortuño 1955; Huetz de Lemps 1969, modified)

On the islands with sufficient elevation a cloud layer is formed on slopes windward to NE trade-winds, which is capped on entering into contact with the subsidence inversion. This stratocumulus bank is known as the sea of clouds, due to its appearance from above. It not only develops on the island slopes (island component) but also over the ocean (oceanic component). The cloud roof averages 1500 m a.s.l. and there the thermal inversion (Von Ficker inversion) averages 5 °C (it can reach 10 °C). In any case the cloud roof undergoes considerable daily and seasonal fluctuations. In winter it can reach 1800 m a.s.l., while in summer it descends to about 1250 m or less, depending on the relief (Fig. 2.2). The cloud mass development also varies with topography. On long ramped slopes reaching high altitudes as within the Orotava valley in Tenerife (Fig. 2.3), the cloud reaches maximum in the evening, while it usually clears in the early hours of the morning. This is different at lower abrupt ridges like those of Anaga, where the maximum occurs at night and there is hardly any cloud at midday. The thickness of this fog bank varies from about 300 m in summer to about 500 m in winter, and also undergoes daily and seasonal fluctuations.

Fig. 2.2 Seasonal fluctuation of the sea of clouds. Summits of La Gomera and La Palma at their windward slopes without clouds in summer, above 1200 m a.s.l., as seen from the West of Tenerife

The beginning of the formation of the clouds is also highly variable, between 300 m and 1000 m a.s.l. In summer, NE trade-winds are more frequent, blowing 95% of the summer days. This fact is very important since it generates a cool moist atmosphere at the windward side that mitigates the high temperatures and dampens the summer water balance at mid-altitudes where laurel forest has developed. In winter, by displacement of the Azores anticyclone towards other territories, the frequency of trade-wind decreases to 55% of the days, but nevertheless at this time of the year the clouds reach their greater thickness and height as a consequence of the colder air and the effect of wedges of cold air incorporated below the trade-wind.

Fig. 2.3 Sea of clouds over La Orotava. Tenerife

Trade-wind clouds generate a particular climatic zone on the windward slopes of the islands, with a particular vegetation of laurel forest linked to the clouds, which has no equivalent on warmer and drier leeward slopes. The effect of the fogs is shown in the humidification of the environment, the generation of a protective cover that moderates evaporation, a greater water supply, and the buffering of altitudinal thermal variation through a lower adiabatic decrease. The increased water supply is mainly due to the capture of atmospheric humidity by the vegetation cover. This moisture interception has been called fog rain or horizontal rain. The amount of water captured depends on exposure to clouds, air temperature, air humidity percentage, shape of the capturing surface and wind speed. Accordingly water capture is not regular throughout the territory affected by fog. The abrupt ridges with maximum altitudes between 800 and 1200 m are the most efficient in capturing water because they are bathed in clouds in both winter and summer. This clouds accelerate downhill in their collapse towards leeward slopes where they eventually fade, sometimes forming spectacular cloud cascades (Fig. 2.4.).

2.2 Saharan Weather

The Harmattan or easterly wind causes a type of weather in which dry, and hot (in summer) African wind blows towards the islands, usually accompanied with atmospheric dust haze ("calima"). Such "tiempo sur" normally arrives as a S, SE or E wind. It coincides with the disappearance of the anticyclonic regime from the Canaries. In summer it is due to the presence of a weak storm over Morocco and

Fig. 2.4 Cascade of overflowing clouds towards the leeward slope. El Paso. La Palma

Mauritania with maintenance of the high altitude anticyclone over Northern Africa, while in winter it is caused by the presence of an anticyclone over Tunisia and the Mediterranean. Its influence on the islands is apparent from a strong temperature rise (less in winter when it can even be cool), a huge decrease in relative humidity, presence of dust in suspension and arrival of locusts (now prevented by fumigations at source). It usually lasts between 3 and 15 days, first affecting higher levels and gradually forcing the thermal inversion altitude to descend, and if it lasts long enough it eventually affects the coasts. These winds have a more direct influence on the southern slopes of the islands, accentuating their warm and dry character.

2.3 Disturbed Weather

There are certain disturbed atmospheric conditions that also condition the climate of the islands. These provide rainfall and take place mainly between late autumn and mid-spring:

– Some **ocean disturbances** are caused by low pressure nuclei that have separated from the south side of the polar front, with a NW – SE trajectory, affecting the islands in late winter and spring with greater or lesser power as winds from the

SW or NW quadrants. The lows provide varying rainfall depending on their particular characteristics.

- **SW storms** originating in undulations of the zonal flow from the west. They descend in latitude, at most to 25°N to the west of the islands. They are loaded with moisture and become warm, and soon they acquire a NE direction approaching the islands and surpassing them. If their intensity is weak they only produce showers on the southern slopes but if it is strong they cause abundant precipitations, pass over the summits and cause rains and violent winds on the northern slopes. They usually appear in winter and early spring and last for a few days.
- **Intrusions of polar air** coming from the NE arriving mainly in autumn and spring. These cold winds weakly affect the islands, disrupting the trade winds, and causing the humid lower air that normally circulates over the sea to rise. This generates precipitation and snowfall on the summits, where below zero temperatures can be reached.
- **Cold core lows** are a sporadic phenomenon in the Canary Islands. They arise through isolation of a cold air core within a ripple of the Jet Stream that becomes strangled, and moves southwards at high altitude as a cyclonic rotation nucleus. At low latitudes this collapses and causes intense local rainfall.

2.4 Main Climatic Parameters

2.4.1 Thermometric Distribution

The three factors that clearly affect the thermometric distribution in the islands are: relief, orientation and distance from the sea. Taking Tenerife as an example, this would be the distribution: The lower zones have the warmest average temperatures that range between 21° and 18 °C. Ascending in altitude and away from the coast the average temperatures fall, varying at mid-elevations between 18° and 11 °C. Finally, in the mountains the means are clearly lower, ranging between 11° and 3 °C on Teide (Fig. 2.5). The southern slopes are warmer than the northern. Winter is the coldest season of the year and January or February the coldest month. Summer is the warmest season and August is characteristically the warmest month. The altitudinal thermal gradient is modified on the northern slopes by the warm and dry subsidence stratum above the lower cool humid air of the NE trade-winds. This leads to an anomalous temperature rise in the upper layers of this lower stratum, which on average is about 5 °C. Above this discontinuity, the thermal gradient changes, gradually lowering the temperature. On southern slopes the thermal gradient is more continuous, although the temperature drop is slower between the coast and mid-elevations than between these and the summit, according to the humidity conditions.

Fig. 2.5 Frost is frequent from the end of fall till the middle of spring at the summits of La Palma and Tenerife. Snowfall happens especially in winter time. Hoar-frost on *Spartocytisus supranubius*

2.4.2 *Rainfall Regime*

Relief and orientation have a decisive influence on the rainfall distribution over the islands; this is particularly evident on the highest islands, which show notable variations between N and S slopes. The irregularity and hourly intensity of the rain are characteristic, but on average the rainfall reaches its maximum during the end of autumn and winter (from November to February) and its minimum in summer (July and August). On almost all the islands the percentage of rain falling in winter exceeds 50%. The coasts have low precipitation, ranging between 50 and 500 mm. On southern coasts, the period of aridity extends from mid-February to mid-October. The mid-elevations receive an average annual precipitation of 300–800 mm, the N slopes being rainier (with a maximum of about 1300 mm in the N of La Palma). Besides, the mid-altitudes of the north receive an additional water supply by fog rain, although that is very irregularly distributed. On the summits there is less rainfall, amounting to 500–650 mm. In general, it snows in winter above 1800 m as a result of cold polar air intrusions. Snowfall very occasionally descends to 1000 m.

2.4.3 *Average Relative Humidity*

There are notable variations in relative humidity depending on the distance from the sea and slope orientation. In general, leeward coasts have a relative humidity of about 65%, which decreases inland, being 55–60% at mid-altitudes. However on windward slopes up to 1500 m there is a relative humidity of 75–80%, being maximum in the areas bathed in trade-wind clouds. Above the temperature inversion zone, the relative humidity suddenly drops, and the summits show figures below 50%.

2.5 Island Type According to Altitude

Three types of islands can be established according to their height and the different climatic zones that determine the trade winds affecting them (Fig. 2.1). The first type is formed by rocky islets, the islands of the Chinijo Archipelago (Graciosa, Montaña Clara, Alegranza), and Lanzarote. Because of their low elevations, the NE trade-winds do not generate fogs on their slopes and they are therefore without the benefit of clouds, constituting essentially arid islands. This is not entirely true for Lanzarote that receives some mist at the summits of the Famara Massif (671 m). The second group is typically formed by La Gomera and El Hierro whose summits are normally immersed in the mists. We included Fuerteventura in this group, although only the highest peaks of the Jandía (807 m) peninsula receive mists, but they harbor impoverished dwarf relics of dry laurel forest. The third group consists of the higher islands of Gran Canaria, Tenerife and La Palma, on which mist or fog forms on the northern slopes. Tenerife presents the maximum variation of climatic zones on its slopes, and consequently the greater zonal distribution of vegetation types.

Chapter 3
Bioclimatology

This complex multidisciplinary science attempts to establish the mutual relations of living beings (biosphere) and atmospheric processes over long periods.

In this book, we deal with bioclimatology as an environmental science that deals with the relationship between climate and the distribution of plants and plant communities on the surface of the Earth, according to the methodology of the bioclimatic classification of the Earth by Rivas-Martínez.

The importance of temperature and precipitation in the distribution of plants and plant communities is well known. Both are basic climatic elements in the set of parameters and indices used by bioclimatology to establish physical frameworks (bioclimatic territories) in which living beings develop.

Solid bases for the characterization of the bioclimatic territories through their physical climatic parameters are deduced from knowledge of:

- distributional areas of plants and simple plant communities defined through classical phytosociology,
- vegetation series or sigmeta. These are plant communities linked by the process of succession within similar teselar territories, which are ecologically homogeneous territories and potentially able to sustain a single type of potential vegetation,
- geoseries or geosigmeta, (contacting series along an ecological gradient),
- geopermaseries (concatenation of permanent communities),

complemented with a statistical study of climatic threshold values that determine the vegetation boundaries. The spaces corresponding to the bioclimatic units (bioclimates, thermotypes, and ombrotypes) have been established and adjusted following the above procedure.

This global bioclimatic classification establishes five major macrobioclimates for the Earth: Tropical, Mediterranean, Temperate, Boreal and Polar. These, and their subordinate units or bioclimates, are inhabited by particular plant communities and formations. Within the bioclimates there are a certain number of variations in the seasonal rainfall rhythms: bioclimatic variants. In addition, thermal and ombric

M. J. del Arco Aguilar, O. Rodríguez Delgado, *Vegetation of the Canary Islands*, Plant and Vegetation 16, https://doi.org/10.1007/978-3-319-77255-4_3

territories are established according to certain thermal and pluviometric index ranges: thermotypes and ombrotypes, used to definine the bioclimatic belts.

According to this classification, the Canary Islands are included in the Mediterranean macrobioclimate. This is an extratropical macrobioclimate, whose main characteristic is the existence of a dry period (P < 2 T) lasting at least two months after the summer solstice. Of the eight bioclimates that it encompasses, four are recognized in the Canaries: hyperdesertic-oceanic, desertic-oceanic, xeric-oceanic and pluviseasonal-oceanic.

3.1 Bioclimatic Belts

A basic objective of bioclimatology is the establishment of the bioclimatic belts found within a territory. These are defined as each of the thermo-ombroclimatic spaces that occur along an altitudinal or latitudinal gradient. They are physical spaces, defined by climatic features, which contain certain plant communities. They must not be confused with vegetation belts, which are defined by the concordant physiognomy of various plant communities. The bioclimatic belts of a territory are delimited according to thermo-climatic factors (thermotypes -It, Itc, Tp) and ombro-climatic factors (ombrotypes -Io) (Tables 3.1 and 3.2).

Table 3.1 Thermotypes recognized in the Canary Islands and threshold values for Itc and Tp (Rivas-Martínez 2011)

Thermotype	**Index value**		**Horizon**	
	Itc	Tp[a]		Itc
1. Inframediterranean	580–450	> 2.400	Lower:	580–515
			Upper:	515–450
2. Thermomediterranean	450–350	> 2.100	Lower:	450–400
			Upper:	400–350
3. Mesomediterranean	350–220	> 1.500	Lower:	350–285
			Upper:	285–220
				Itc/Tp
4. Supramediterranean	< 220	> 900	Lower:	220–150/1500–1200
			Upper:	150–(120)/1200–900
5. Oromediterranean	–	450–900	Lower:	–/900–675
			Upper:	–/675–450

[a]In territories outside the eutropical latitude zone, when Ic ≥ 21 (continental) or when It or Itc < 120, the thermotype is determined from Tp values

Table 3.2 Ombrotypes recognized in the Canary Islands and threshold values for Io (Rivas-Martínez 2011)

Ombrotype	Index value	Horizon	
	Io		Io
1. Hyperarid	0.2–0.4	Lower	0.2–0.3
		Upper	0.3–0.4
2. Arid	0.4–1.0	Lower	0.4–0.7
		Upper	0.7–1.0
3. Semiarid	1.0–2.0	Lower	1.0–1.5
		Upper	1.5–2.0
4. Dry	2.0–3.6	Lower	2.0–2.8
		Upper	2.8–3.6
5. Subhumid	3.6–6.0	Lower	3.6–4.8
		Upper	4.8–6.0
6. Humid	6.0–12.0	Lower	6.0–9.0
		Upper	9.0–12.0

A few climatic values and indices are enough to establish the thermotypes and ombrotypes:

Compensated Thermicity Index (Itc) is used to determine thermotypes, territorial entities characterized by thermal values. It is an index that balances the average annual temperature with cold intensity, a limiting factor for many plants and plant communities. Positive temperature index can be used alternatively to determine thermotypes in some cases as seen below.

$$\mathbf{Itc} = \mathrm{It} \pm \mathrm{C}$$

It (thermicity index) = (T + M + m) × 10
T = mean annual temperature
M = mean maximum temperature of the coldest month of the year
m = mean minimum temperature of the coldest month of the year
C = compensation value

When the continentality index (**Ic** = difference between mean temperatures of the warmest and coldest months of the year) is <8 (oceanic) or >18 (continental), a compensation value (**C**) is subtracted or added to **It** to obtain **Itc**. This value is used in the territories outside the eutropical latitudinal zone of the Earth (north of 23°N and south of 23°S) to compensate for the extra winter cold of highly continental territories or the extra winter warmth in highly oceanic ones, so that the resulting compensated thermicity index (**Itc**) is comparable all around the Earth. In the

Canaries, only Ic values <8 require compensation. Such compensation values are obtained from: $C = (8.0 - Ic) \times 10$.

Positive Temperature (Tp) is the value in tenths of degrees resulting from the sum of the monthly means above 0 °C. In cold climates, when $Itc < 120$ or when $Ic > 21$, the positive temperature shows greater accuracy than Itc for determining thermotypes.

Ombrothermic Index (Io) is a pluvio-thermal coefficient used to establish bioclimates, subdivisions of the macrobioclimates, and ombrotypes, territorial entities characterized by their rainfall values balanced with their corresponding thermal values. The Io is defined as follows: $Io = (Pp/Tp) \times 10$.

Positive Rainfall (Pp) is the annual rainfall in mm, taking into account only the months with mean temperature higher than 0 °C. Since this is the case for all the available thermopluviometric weather stations in the archipelago, except on Mount Teide approximately above 3000 m (N facing) and 3200 m (SE facing) where subzero mean temperatures are recorded in winter time, Pp generally has the same value as P (mean annual rainfall).

Apart from the bioclimatic values and indexes used for characterizing bioclimatic belts one must consider the presence or absence of trade-wind clouds mainly over windward territories (Table 3.3). These clouds are an inherent part of the climate of the Canary Islands, but their action is often not included in the data used for defining bioclimatic belts. The presence or absence of clouds under the same biocli-

Table 3.3 Bioclimatic belts recognized in the Canary Islands

	HA	A	SA	D	SH	H
Inframediterranean	X	X	X X	X X	X X	
Thermomediterranean			X X	X X	X X	X X
Mesomediterranean				X X	X X	X X
Supramediterranean			X	X	X	
Oromediterranean		X	X	X		

Ombrotypes: HA: Hyperarid, A: Arid, SA: Semiarid, D: Dry, SH: Subhumid, H: Humid. Grey areas: with trade-wind clouds (c).

matic combinations (thermotype + ombrotype) determines very different conditions that imply different climatophilous vegetation series in them. We have characterized belts under the potential area of trade-wind clouds as follows:

– With trade-wind clouds (c), which refers to windward areas directly affected by clouds all year round, although with diverse intensity;
– With overflowing trade-wind clouds (oc), which refers to leeward slopes under the influence of clouds overflowing from the windward slopes;
– With trade-wind clouds except in summer (ces), as normally occurs on windward slopes above about 1250 m a.s.l.;
– Without trade-wind clouds (wc).

The delimitation of bioclimatic belts establishes a good physical characterization of the territory and a mutual relation with climatophilous vegetation series (Table 3.4). This is used for designing potential climatophilous vegetation maps (Fig. 3.1) and combining other characteristics such as geology, soils, etc., as a solid basis for preparing more complete potential natural vegetation maps.

3.2 Bioclimatic Belts in the Canary Islands

Many bioclimatic belts have been recognized (Tables 3.3 and 3.4), and they can be linked with climatophilous vegetation series (Figs. 3.2, 3.3, 3.4, 3.5, 3.6 and 3.7; Table 3.4). Common and scientific names of vegetation series are shown for each island (Table 3.5).

Table 3.4 Bioclimatic combinations and climatophilous vegetation series in the Canary Islands

	Bioclimatic combinations	Climatophilous vegetation series (common names)	C.C.	R.C.	B
1	Hyperarid Inframediterranean	Sweet spurge shrubland			1
2	Arid Inframediterranean	Sweet spurge shrubland			2
3	Upper-arid Upper-Inframediterranean	Cardón shrubland			
4	Lower-semiarid Inframediterranean	Cardón shrubland			3
5	Semiarid Inframediterranean (oc)	Thermo-sclerophyllous woodland, humid faciation		T	
6	Upper-semiarid Inframediterranean	Thermo-sclerophyllous woodland			
7	Upper-semiarid Inframediterranean (c)	Thermo-sclerophyllous woodland, humid faciation		T	
8	Lower-dry Inframediterranean (wc)	Thermo-sclerophyllous woodland			
9	Lower-dry Inframediterranean (c)	Dry laurel forest		P,T	5
10	Upper-dry Inframediterranean (wc)	Thermo-sclerophyllous woodland			
11	Upper-dry Inframediterranean (c)	Dry laurel forest			
12	Subhumid Inframediterranean (wc)	Thermo-sclerophyllous woodland		P	
13	Subhumid Inframediterranean (c)	Humid laurel forest		P	
14	Lower-semiarid Thermomediterranean	Thermo-sclerophyllous woodland			
15	Upper-semiarid Thermomediterranean (wc)	Thermo-sclerophyllous woodland			4
16	Upper-semiarid Thermomediterranean (c)	Dry laurel forest			
17	Upper-semiarid Thermomediterranean (oc)	Dry laurel forest		G,T	
18	Lower-dry Thermomediterranean (wc)	Thermo-sclerophyllous woodland			
19	Lower-dry Thermomediterranean (c)	Dry laurel forest			
20	Lower-dry Thermomediterranean (oc)	Dry laurel forest		G,T	6
21	Upper-dry Thermomediterranean (wc)	Canary pine forest			
22	Upper-dry Thermomediterranean (oc)	Canary pine forest		G, T	
23	Upper-dry Thermomediterranean (c)	Dry laurel forest			
24	Subhumid Thermomediterranean (wc)	Canary pine forest			12
25	Subhumid Thermomediterranean (c)	Humid laurel forest			7
26	Subhumid Thermomediterranean (oc)	Dry laurel forest		G	
27	Subhumid Thermomediterranean (ces)	Cold laurel forest		P	
28	Humid Thermomediterranean (wc)	Canary pine forest		P	
29	Humid Thermomediterranean (c)	Humid laurel forest		P	
30	Humid Thermomediterranean (ces)	Cold laurel forest		P	
31	Lower-dry lower-Mesomediterranean	Canary pine forest			11
32	Lower-dry lower-Mesomediterranean (c)	Humid laurel forest		G,T	
33	Lower-dry lower-Mesomediterranean (ces)	Cold laurel forest		T	
34	Lower-dry lower-Mesomediterranean (oc)	Canary pine forest, humid faciation		T	
35	Upper-dry lower-Mesomediterranean	Canary pine forest			
36	Upper-dry lower-Mesomediterranean (c)	Humid laurel forest		G,T	
37	Upper-dry lower-Mesomediterranean (ces)	Cold laurel forest		T	
38	Upper-dry lower-Mesomediterranean (oc)	Canary pine forest, humid faciation		T	
39	Subhumid Lower-Mesomediterranean (wc)	Canary pine forest			
40	Subhumid Lower-Mesomediterranean (c)	Humid laurel forest			8
41	Subhumid Lower-Mesomediterranean (ces)	Cold laurel forest			9
42	Subhumid Lower-Mesomediterranean (oc)	Canary pine forest, humid faciation		G,T	10
43	Humid Lower-Mesomediterranean (wc)	Canary pine forest		P	
44	Humid Lower-Mesomediterranean (c)	Humid laurel forest		P	
45	Humid Lower-Mesomediterranean (ces)	Cold laurel forest		P	
46	Humid Lower-Mesomediterranean (oc)	Canary pine forest, humid faciation		P	
47	Dry upper-Mesomediterranean	Canary pine forest			
48	Subhumid Upper-Mesomediterranean	Canary pine forest			
49	Humid Upper-Mesomediterranean (wc)	Canary pine forest		P	
50	Humid Upper-Mesomediterranean (ces)	Cold laurel forest		P	
51	Upper-semiarid Supramediterranean	Summit broom shrubland		T	
52	Dry Supramediterranean	Summit broom shrubland			
53	Subhumid Supramediterranean	Summit broom shrubland		T,P	13
54	Humid Supramediterranean	Summit broom shrubland		P	14
55	Arid Oromediterranean	Teide violet community		T	
56	Semiarid Oromediterranean	Teide violet community		T	
57	Dry Oromediterranean	Summit broom shrubland		T	

Abbreviations: **C.C.**: common bioclimatic combination (grey); **R.C.**: rare bioclimatic combination restricted to some islands (**P**: La Palma, **G**: La Gomera, **T**: Tenerife). **(c)** with trade-wind clouds, (ces) with trade-wind clouds except in summer, **(oc)** with overflowing trade-wind clouds, **(wc)** without trade-wind clouds. **B**: Number of bioclimograph in Fig. 3.8.

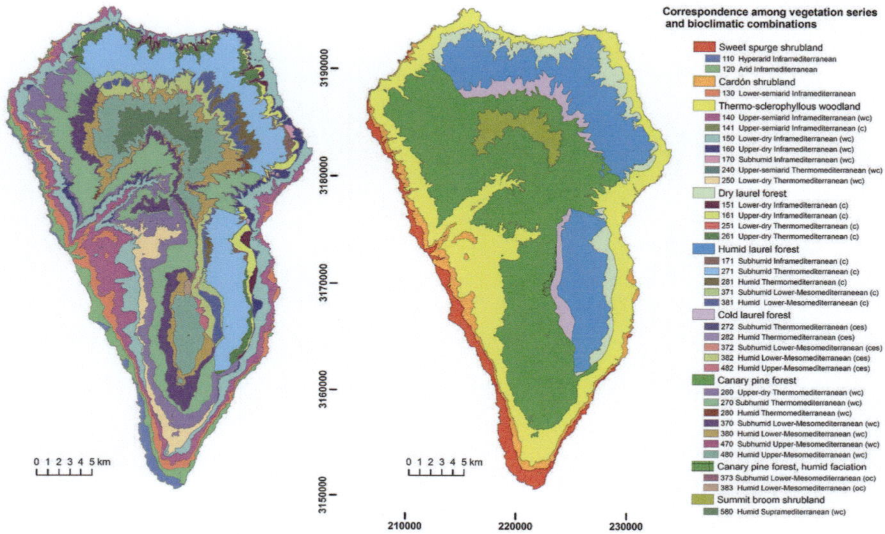

Fig. 3.1 La Palma. Map showing the correspondence between bioclimatic belts and climatophilous vegetation series, as detailed in the legend and in Table 3.4 (Based on Garzón et al. 2014, corr)

Fig. 3.2 *Asterisco intermedii-Euphorbio balsamiferae S.* growing within the territory of the arid Inframediterranean bioclimatic belt. Malpaís de La Corona. Lanzarote

Fig. 3.3 *Periploco laevigatae-Euphorbio canariensis S.* growing within the territory of the lower-semiarid Inframediterranean bioclimatic belt. Valle Jimenez. Tenerife

Fig. 3.4 *Brachypodio arbusculae-Juniperetum canariensis S.* growing within the territory of the upper-semiarid Thermomediterranean bioclimatic belt. Agulo-Vallehermoso. La Gomera

Fig. 3.5 *Ixantho viscosi-Lauro novocanariensis* communities growing within the territory of the subhumid-humid Thermo-Mesomediterranean bioclimatic belt. Laurel forest of Los Tilos. La Palma

Fig. 3.6 *Loto hillebrandii-Pino canariensis S.* growing within the territory of the subhumid Mesomediterranean (wc) bioclimatic belt. Caldera de Taburiente. La Palma

Fig. 3.7 *Spartocytiso supranubii S.* growing within the territory of the dry Supramediterranean bioclimatic belt. Las Cañadas. Tenerife

Table 3.5 Common names and scientific names of climatophilous vegetation series

Community common name	Vegetation series name
Sweet spurge shrubland	*Asterisco intermedii-Euphorbio balsamiferae S.* (L)
	Lycio intricati-Euphorbio balsamiferae S. (F)
	Euphorbio balsamiferae S. (C)
	Ceropegio fuscae-Euphorbio balsamiferae S. (T)
	Neochamaeleo pulverulentae-Euphorbio balsamiferae S. (G)
	Echio breviramis-Euphorbio balsamiferae S. (P)
	Rubio fruticosae-Euphorbio balsamiferae S. (H)
Cardón shrubland	*Asterisco intermedii-Euphorbio balsamiferae S. aeonietosum lacerotensis* (L)
	Kleinio neriifoliae-Euphorbio canariensis S. (F)
	Aeonio percarnei-Euphorbio canariensis S. (C)
	Periploco laevigatae-Euphorbio canariensis S. (T)
	Euphorbio berthelotii-canariensis S. (G)
	Echio breviramis-Euphorbio canariensis S. (P)
	Aeonio valverdensis-Euphorbio canariensis S. (H)

(continued)

Table 3.5 (continued)

Community common name	Vegetation series name
Thermo-sclerophyllous woodland	*Convolvulo lopezsocasii-Oleo cerasiformis S.* (L)
	Micromerio rupestris-Oleo cerasiformis S. (F)
	Pistacio lentisci-Oleo cerasiformis S. (C)
	Junipero canariensis-Oleo cerasiformis S. (T)
	Brachypodio arbusculae-Junipero canariensis S. (G)
	Rhamno crenulatae-Junipero canariensis S. (P)
	Rubio fruticosae-Junipero canariensis S. (H)
Thermo-sclerophyllous woodland, humid faciation	*Junipero canariensis-Oleo cerasiformis S. ericetosum arboreae* (T)
	Brachypodio arbusculae-Junipero canariensis S. ericetosum arboreae (G)
	Rhamno crenulatae-Junipero canariensis S. ericetosum arboreae (P)
	Rubio fruticosae-Junipero canariensis S. ericetosum arboreae (H)
Dry laurel forest	*Echio handiense-Visneo mocanerae S.* (F)
	Visneo mocanerae-Arbuto canariensis S. (C,T,G,P,H)
Humid laurel forest	*Lauro novocanariensis-Perseo indicae S.* (C,T,G,P,H)
Cold laurel forest	*Pericallido murrayii-Morello fayae S.* (C,T,P,H)
	Violo rivinianae-Morello fayae S. (G)
Canary pine forest	*Pino canariensis S.* (C)
	Sideritido solutae-Pino canariensis S. (T)
	Cisto gomerae-Pino canariensis S. (G)
	Loto hillebrandii-Pino canariensis S. (P)
	Bystropogono ferrensis-Pino canariensis S. (H)
Canary pine forest, humid faciation	*Pino canariensis S.* (C) *ericetosum arboreae*
	Sideritido solutae-Pino canariensis S. (T) *ericetosum arboreae*
	Cisto gomerae-Pino canariensis S. (G)
	Loto hillebrandii-Pino canariensis S. (P) *ericetosum arboreae*
	Bystropogono ferrensis-Pino canariensis S. (H) *ericetosum arboreae*
Summit broom shrubland	*Descurainio bourgeauanae-Spartocytiso supranubii S.* (T)
	Spartocytiso supranubii S. (T)
	Genisto benehoavensis-Adenocarpo spartioidis S. (P)
Teide violet community	*Violo cheiranthifoliae S.* (T)

3.3 Bioclimographs

They are ombrothermic diagrams showing the variations of temperature and rainfall along the year, complemented with several other climatic, bioclimatic and phytocenotic features. They clearly show the humid (P curve above T curve) and dry (P curve below T curve) periods of the year and summarize the climatic and bioclimatic characteristics of a locality and its correspondence with a climatophilous vegetation series. Figure 3.8 shows bioclimographs covering typical localities in a representative set of climatophilous vegetation series for the archipelago.

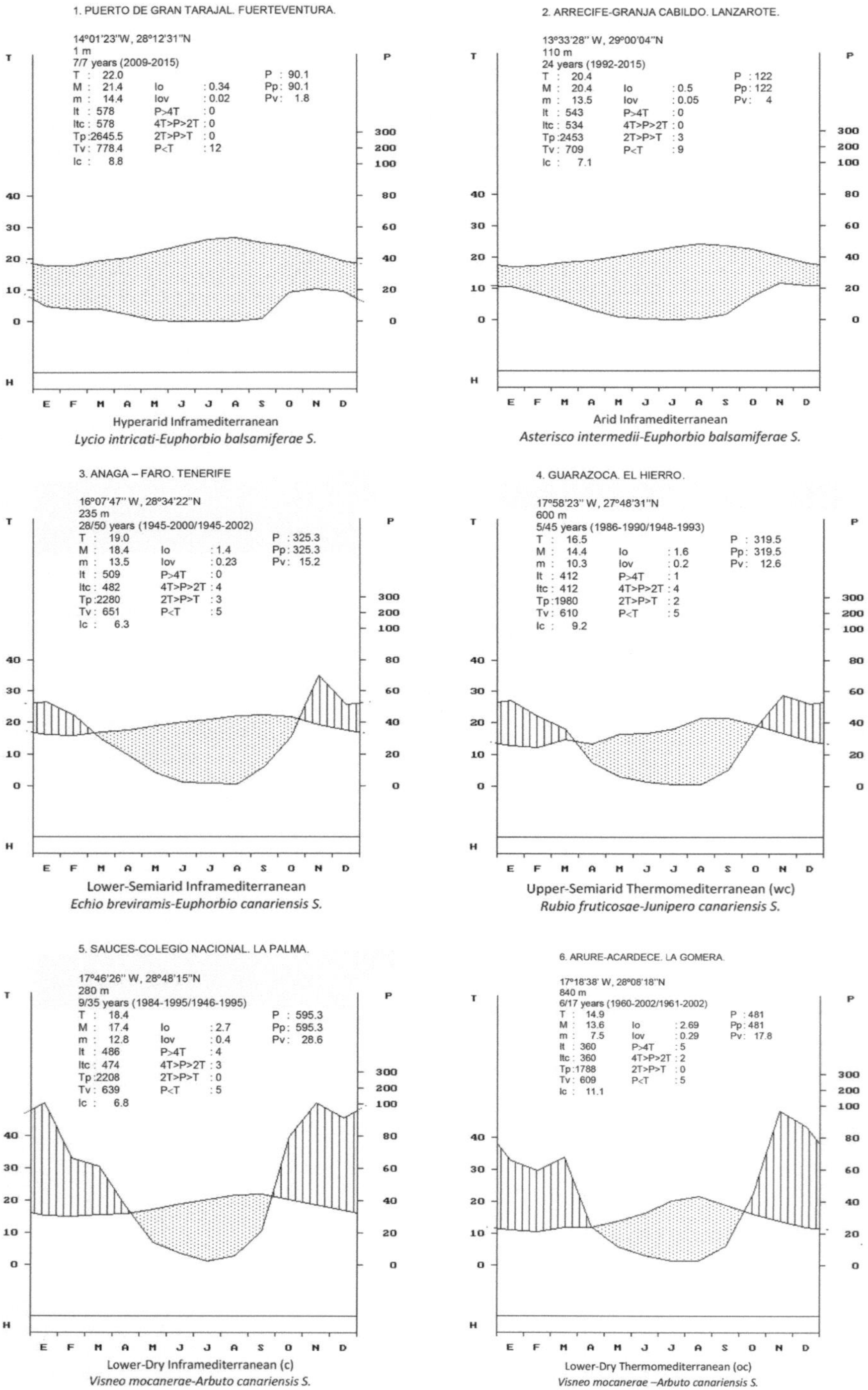

Fig. 3.8 Bioclimographs of selected meteorological stations in the archipelago showing parameters and indexes for the usual bioclimatic belts and their relation with climatophilous vegetation series. T = Mean annual temperature in °C; M = Mean maximum temperature of the coldest month; m = Mean minimum temperature of the coldest month; It = Thermicity index; Itc = Compensated

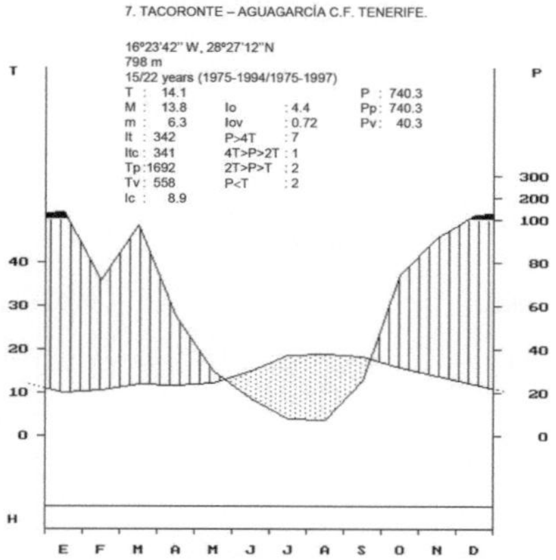

Subhumid Thermomediterranean (c)
Lauro novocanariensis-Perseo indicae S.

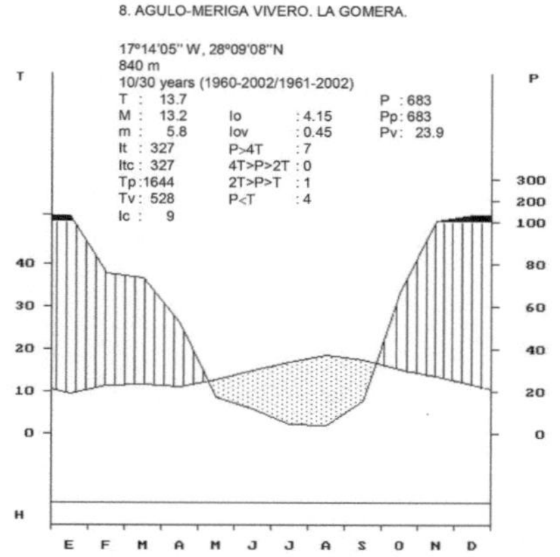

Subhumid Lower-Mesomediterranean (c)
Lauro novocanariensis-Perseo indicae S.

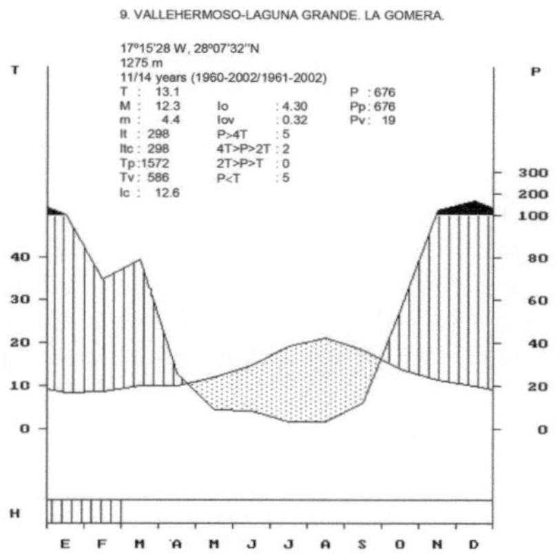

Subhumid Lower-Mesomediterranean (ces)
Violo rivinianae-Morello fayae S.

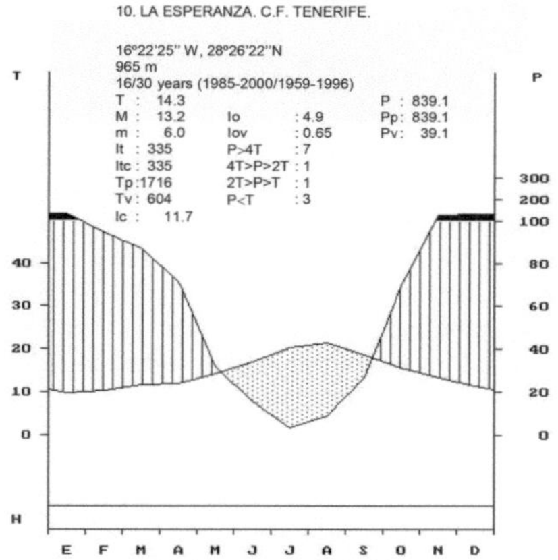

Subhumid Mesomediterranean (oc)
Sideritido solutae-Pino canariensis S.
ericetosum arboreae

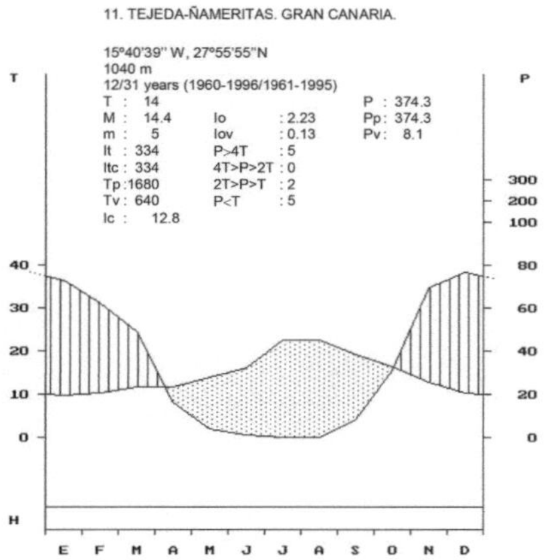

Dry Lower-Mesomediterranean
Pino canariensis S.

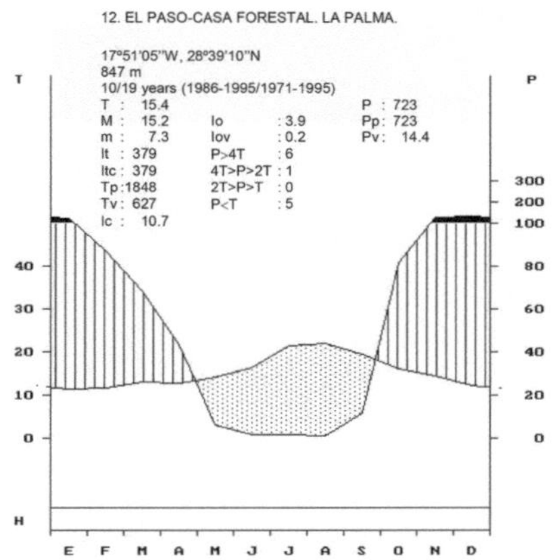

Subhumid Thermomediterranean (wc)
Loto hillebrandii-Pino canariensis S.

Fig. 3.8 (continued) thermicity index; Tp = Positive temperature; Tv = Summer temperature; Ic = Continentality index. Io = Ombrothermic index; Iov = Summer ombrothermic index; P > 4 T = Months in which the rainfall value (in mm) is greater than four times the temperature value (in °C); 4 T > P > 2 T = Months in which the rainfall value is between twice and four times

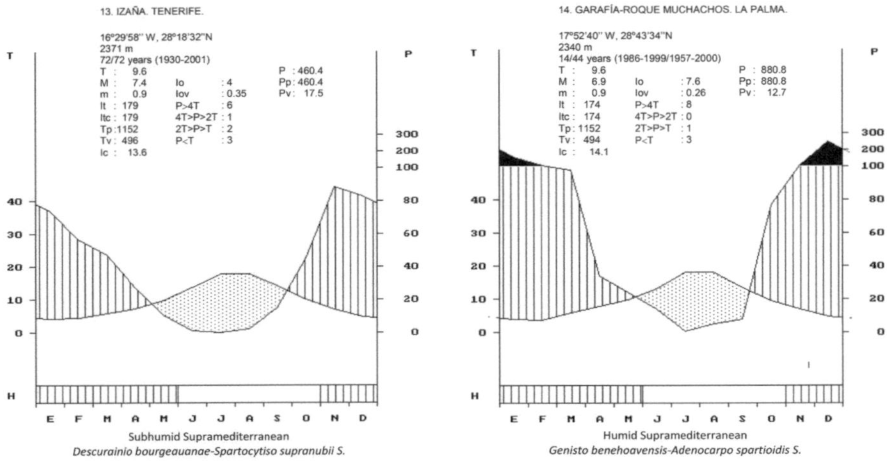

Fig. 3.8 (continued) the temperature value; 2 T > P > T = Months in which the rainfall value is greater than the temperature value but smaller than two times this value; P < T = Months in which the rainfall value is smaller than the temperature value. P = Annual rainfall in mm; Pp = Positive rainfall; Pv = Summer rainfall. H: Frost period [blank = frost-free period (mi > 0 °C); black = frosts certain (mi ≤ 0 °C); hatched = probable frosts (m′i ≤ 0 °C); mi = monthly mean daily minimum temperature; m′i: monthly mean absolute minimum temperature]. (c): With trade-wind clouds; (oc): With overflowing trade-wind clouds; (ces): With trade-wind clouds except in summer; (wc): Without trade-wind clouds

Chapter 4
Biogeography

4.1 Colonization

4.1.1 Islands and Continental Territories

The birth of the Canary Islands coincides with the halting of the northeast-bound African Plate by its impact against Europe and Asia and with the resultant Alpine orogeny. The rotation of the African Plate and distension in the Central Atlantic ridge released an effusion of magma that built up the Canary Archipelago. The present subaerial island edifices arose in the Miocene-Pleistocene periods, and has taken about 20 Ma. They are independent units, except for Lanzarote and Fuerteventura that were united (as the paleo-island "Mahan"), with their bases seated on an oceanic crust that increases in depth towards the W. These easternmost islands are separated from the mainland by depths of about 1000 m, apart from a thick layer of sediments, whereas between the western islands there are depths of up to 4000 m.

For the purposes of island colonization, not only the currently emerged islands and mainland around the Canary Islands should be considered, but also ancient volcanic islands today submerged, constituting seamounts or more specifically "guyots", some near the ocean surface. If we take into account the life-cycle of volcanic islands under a sufficient time-lapse, tentative stepping-stone corridors become apparent in the geographic area of Macaronesia. These islands show decreasing ages from NE to SW due to the displacement of hot spots as the African plate rotated counterclockwise towards the Northeast. These paths show curved parallel hot-spot tracks in the archipelagos of Madeira and the Canaries (including the

M. J. del Arco Aguilar, O. Rodríguez Delgado, *Vegetation of the Canary Islands*, Plant and Vegetation 16, https://doi.org/10.1007/978-3-319-77255-4_4

Salvages Islands), which over time would enable the transit of biota from mainland areas to the islands and between the islands of these Archipelagos.

These series of seamounts and present islands represent the different phases of the life-cycle of volcanic islands, from the initial stage of birth to their senescence as guyots. They comprise a temporary strip of volcanic activity that dates back to the opening of the Atlantic basin. The Canary plume begins in the NE at the Essaouira Seamount (68 Ma) and ends at La Palma and El Hierro (4 Ma, active up to the present). The Madeira corridor begins with the Ormonde Seamount (67–65 Ma) and ends with Madeira itself (5–0 Ma). Within the Canary Islands Seamount Province there are older seamounts with ages reaching 142 Ma, within the late Jurassic, which emerged when Africa and America were separated by only 900 km. These seamounts are located to the SW of the present-day Canaries and show an almost orthogonal alignment to the Canary hot-spot track. Their genesis and growth are explained by other geological processes, including passive mantle upwelling beneath the mid-ocean ridge. This may have produced recurrent melting anomalies and seamounts from the Late Jurassic to the Recent period, or multiple and successive mantle plumes, among other explanations (Fig. 4.1).

Therefore, within the colonization framework of a dynamic insular environment, these volcanic edifices date back to maximum ages in the Late Jurassic (Mesozoic). They originated as the Atlantic basin opened, with limited life-spans, variable altitudes, and shifting surface areas according to their ontogenic stage and the changing sea-level. This latter depended on the greater or lesser power of the world's ice caps, which also affected their relative distances. They appeared and disappeared, and may have acted as stepping-stones in the process of island colonization, complementary to that via continental corridors. Out of these many island territories only the present islands survive, which in the case of the Canaries reach a maximum of only 20.5 Ma in age (Miocene), in Fuerteventura.

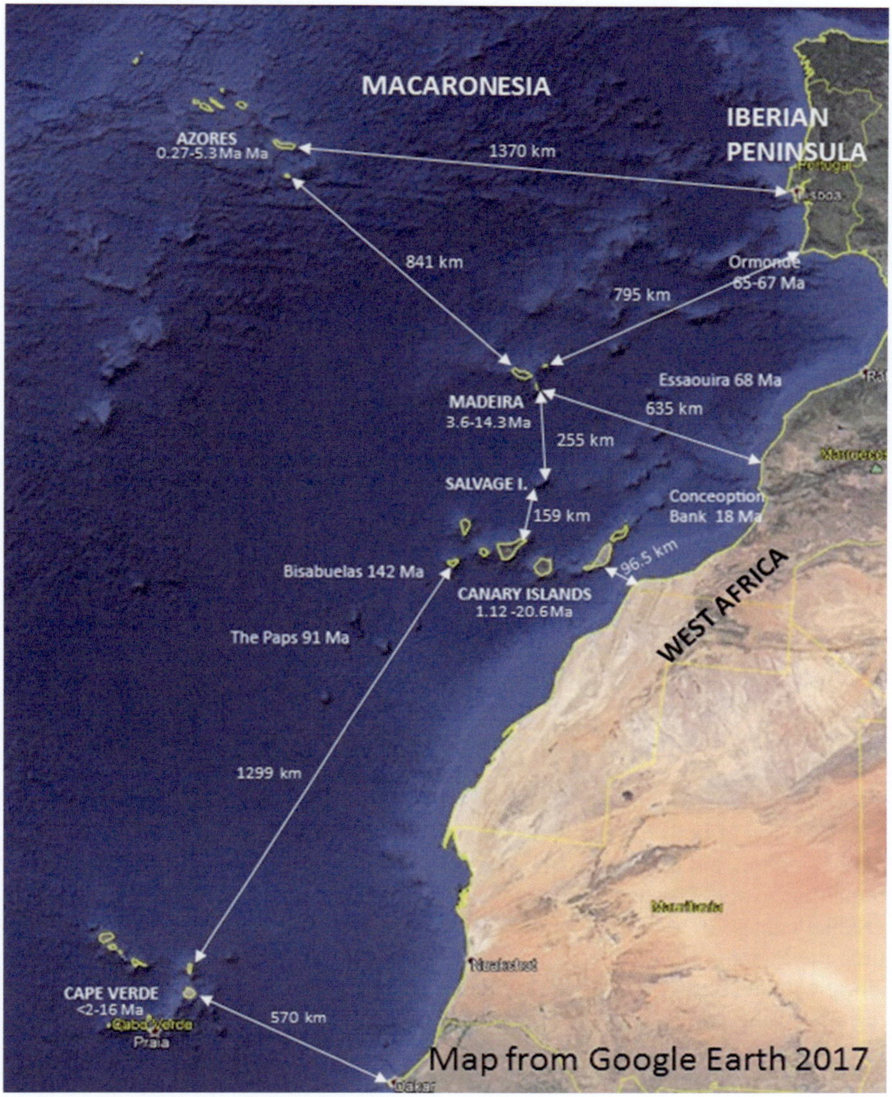

Fig. 4.1 Some geographic data of Macaronesia: shortest distances between archipelagos, geological age of subaerial edifices, and age of some seamounts within the Madeiran and Canary seamount provinces

4.1.2 Geological and Paleoclimatic Framework

In the Paleogene Period (65–25 Ma), the earliest paleo-islands of the Madeira and Canary hot-spot tracks were ready to be colonized. Both North Africa and Southern Europe were characterized by a warm humid tropical climate, with monsoon summer rains conditioned by the warm east to west circum-equatorial global marine current. This connected the Atlantic Ocean with India across the Tethys Sea. From the Late Cretaceous to Late Miocene, a Paleotropical flora spread on both sides of this sea, forming moist forests with a predominance of Lauraceae. A good part of the flora of the current Macaronesian laurisilva has its origin in this area, including its present-day four genera (*Apollonias, Laurus, Ocotea* and *Persea*) and numerous genera of ferns such as *Culcita, Davallia, Diplazium, Dryopteris, Trichomanes,* and *Woodwardia*.

The subaerial construction of the present Canary Islands, whose earliest date is in the Miocene, coincides with major paleo-bioclimatic events. A gradual cooling began in the Oligocene at the global level, culminating in the Quaternary glacial period. The present climatic zones were gradually delimited by a progressive reduction in area of the warm equatorial and tropical zone. The warm equatorial sea current in the Tethys Sea became closed off during the Miocene because of the collision of the Anatolian and Arabian plates, thus beginning a cooling of the Central-European climate. Marine transgressions covered much of Mediterranean Africa, also forming numerous lakes in the Central and Western Sahara. At the end of the Miocene, in the Messinian (6 Ma), the Mediterranean was closed by tectonic elevation of the arc of Gibraltar and became a saltpan, but soon later (5.3 Ma) it was refilled.

Already increasing in the late Miocene, aridity spread and intensified during the Pliocene and in both Europe and North Africa. In the Sahara there were cycles of desertification, but the "hamadas" were covered by temporary lakes and seas. In the transition to the Pliocene (3.2–2.6 Ma), in the area of the Canaries there was a gradual change to an arid climate alternating with numerous rainy periods. At this time the Panama Isthmus closed and the northwestward warm Gulf Stream began, along with the southwestward cool Canary Current (associated with its accompanying prevailing winds now called trade-wind). The onset of the Mediterranean climate coincides with the increased formation of ice in the Arctic and Antarctic regions. The Pleistocene was characterized in Europe by a six-phase glacial period interrupted by warm or warm-temperate interglacial phases. These affected the African continent with rainfall phases corresponding to European glacial phases and dry warm interglacial phases, sometimes arid, corresponding to the interglacial phases. After the glacial period, the African floras became poorer and the lakes and seas in the Sahara disappeared (5000 years ago). Then the desertification process became more intense in N Africa and its current state was progressively reached.

The vegetation dynamics on the European and African continents since the Islands emerged provide the key to their main sources of flora, as we will consider later.

4.1.3 Dispersal and Colonization

The colonization of the Canary Islands has been possible through the dispersal process, whereby the diaspores (any part of a plant that spreads and is capable of generating new individuals in a new territory) are transferred and may lead to the establishment of new colonies, completing the migration process. This process is slow and takes place through the dissemination of vegetable diaspores by external agents such as water, wind, animals and man, for which the plants may have evolved particular adaptive syndromes.

Anemochory is wind-borne dispersal. In this way light diaspores like spores, small seeds, etc. are transferred, but also seeds and fruits with special devices such as wings, hairy pappus, etc. (Fig. 4.2).

Fig. 4.2 Dispersal by anemochory. (**a**) *Kleinia neriifolia* – fruits with a hairy pappus. (**b**) *Aeonium urbicum* – small seeds

Zoochory is dissemination by animals. It is seen especially in the colonization of islands through birds: ornithochory (Fig. 4.3). Birds can transport diaspores both outside their bodies – epizoochory, and inside – endozoochory (Fig. 4.4). For the latter to be effective, the resistance of the diaspore to digestion should be greater than its retention time in the body of the bird. The transport of diaspores by insects, entomospory, is important for small diaspores, such as fungal spores.

Fig. 4.3 Birds involved in transporting seeds and fruits. (**a**) *Corvus corax canariensis*, the Canarian raven. (**b**) *Pyrrhocorax pyrrhocorax barbarus*, a red-billed chough endemic to La Palma. (**c**) *Fringilla coelebs palmae*, La Palma chaffinch, typical in laurel forest. (**d**) *Columba bollii*, an endemic pigeon of laurel forest. (**e**) *Fringilla teydea*, the Tenerife blue chaffinch of pine forest and summit broom shrubland

Fig. 4.4 Dispersal by endozoochory. (**a**) *Arbutus canariensis*. (**b**) *Laurus novocanariensis*. (**c**) *Juniperus cedrus*

Hydrochory is dissemination by water. It requires buoyancy of the diaspores, to be transported passively or blown by the wind (anemo-hydrochory) (Fig. 4.5). The Canary Stream or Current, an eastern southward branch of the Gulf Stream, which was well established by the end of the Pliocene and still bathes the Islands, has undoubtedly carried diaspores there. "Vegetation islands", clumps of trees and soil detached from the banks of rivers and swept out to sea, can float onward to distant places and may have carried diaspores to the Canary Islands during various climatic phases, such as the torrential Quaternary rainfall phases. Transport of diaspores has even been possible via tsunamis caused by massive landslides that have occurred in the erosive dismantling of oceanic islands. This is documented for some insect groups, and could equally have occurred in plants *s.l.*

Fig. 4.5 Dispersal by hydrochory. (**a**) *Tetraena (Zygophyllum) fontanesii*. (**b**) *Patellifolia webbiana*. Both with floating fruits

Anthropochory, dispersal by mankind, began with the arrival of aboriginal populations from North Africa from at least 2500 B.P. onwards. It is proved that Phoenicians and Romans visited the easternmost islands during the first millennium B.P. until the 2nd A.D. century and could have contribute to dispersal. At the beginning of the fourteenth century the Genoese Lancelotto Malocelo stayed for 20 years in Lanzarote, and expeditions of Majorcans and Castilians commanded by Alfonso IV of Portugal took place in that century. But plant dissemination by humans was greatly intensified after the European conquest during the fifteenth century and subsequent settlements of colonists in the Canaries, making the current percentages of introduced vascular plants on the Islands very high (for some authors even more than 50%) (Fig. 4.6).

It has been estimated that the ancestors of the present Canarian endemic flora reached the islands by the following means: 33.9% by endozoocory, 18.8% by epizoochory, 25.8% by anemochory, 4.3% by hydrochory and 17.2% through diaspores with indeterminate dispersal but with anemochory and endozoochory as the most probable pathways. This means that 52.7% of the diaspores were zoochorous, which shows the importance of zoochory in island colonization by long-distance dispersal, especially considering that the percentage may be higher when considering possible zoochorous plants within the indeterminate diaspores.

The colonization of the Canary Islands has proceeded while they have been emerging from the ocean, with a flow of diaspores between them and the continents in varying directions. Within the Canaries hot-spot track it seems logical that the general direction of colonization should be established from E to W, as the islands emerged, firstly between mainland and the older Paleocenic islands and then with those that were progressively being formed. This would correspond to a model of colonization by successive hops between islands and the continents, eventually reaching the oldest volcanic edifices of the present-day Islands. This colonization route would also have been combined with other leaps from the islands of the Madeiran hot-spot track. Such a scenario would imply the possible arrival on the islands of the European tertiary flora at a time well before that indicated by the

Fig. 4.6 Anthropochory. (**a**) *Achyranthes aspera* – tropical. (**b**) *Aloe vera* – Arabia and East Asia. (**c**) *Ageratina adenophora* – Mexico. (**d**) *Asclepias curassavica* – Tropical America. (**e**) *Alternanthera caracasana* – Central and South America. (**f**) *Aptenia cordifolia* – South Africa

direct colonization route from the African continent. The latter is an important established route to the oldest volcanic edifices of the present islands, from their initial appearance. From them, plants were progressively transported onwards to the subsequently emerging ones, carrying African, paleotropical Tethyan flora (displaced southwards by Quaternary glacial processes), and later modern Mediterranean diaspores.

Recent biogeographic studies on the relationship between Macaronesia and the Americas point to the possibility of colonization of early island territories by Mesozoic flora at the beginning of the opening of the Atlantic basin. Seamounts lying to the SW of the Canaries, which reach up to 142 Ma in age, give credence to this hypothesis intended to explain the origin of endemic Macaronesian clades from Mesozoic ancestors and their relations with American, African and Asian clades. According to this theory the ancestors of these clades colonized the islands and their descendants gradually passed to other new islands as the previous ones disappeared. The existence of these ancient seamounts raises the possibility of colonization bridges Africa-Macaronesia-America very early on, which could explain hitherto little understood floristic relations with distant territories. This is exemplified for vascular plants in the following genera: *Clethra*, also distributed throughout America and Asia, with endemic *C. arborea* present on Madeira (extinct in the Canaries?) that has *C. alnifolia* of Eastern North America as sister group; *Morella*, which is also distributed throughout the Americas, Africa, Madagascar and East Asia, with its endemic species *M. rivas-martinezii* (Canaries) and *M. faya* (Azores, Madeira and Canaries) showing unresolved relations with an American sister group of 9 species and another in Africa and Madagascar with 11 species; and in other genera such as *Persea-Apollonias, Ilex, Sedum, Ocotea, Dryopteris, Cneorum, Ruta, Daucus, Descurainia* and *Fagonia*. These genera form part of a macro-continental flora with a more or less continuous past distribution. With the movements of the continents and geo-climatic shifts this flora has been subject to disjunctions and area loss in part or the whole area of the present islands and continental region. This recent theory of such early colonization leads to rethinking the processes of dispersal to islands and their chronology, lending greater importance to the hypothesis of vicariance for some groups of organisms. It is usually conceived of as colonization by chance (chance dispersal). In this model, at the beginning of the Atlantic opening, the geomorphological dynamics of rift-valley formation, crustal thinning, and associated volcanism resulting in the spread of organisms over a contiguous region of oceanic and mainland islands, were later disrupted by the formation and widening of the Atlantic. The current geological evidence suggests the possible existence of extensive subaerial archipelagos linked to the Central Atlantic ridge, shortening the distances between islands and generating an ancestral Macaronesian range that fostered divergence from mainland relatives. This model is based on primitive continental islands supplying the later oceanic ones, which shorten the distances to the new oceanic islands that emerge later, for which the described processes of colonization by jumps between islands are still valid.

Whatever the routes followed in colonization, there is a need for long-distance colonization to overcome the oceanic barrier. The speed of plant colonization has depended on the distance from its insular or continental sources, on the possibility of action of dispersal means, and on the accessibility of habitats. In turn, the current flora of the Islands depends on the chorological history of plants, their migration abilities, adaptation to the environments they reach, and capability of diversification. Such processes are clearly linked to their genetic diversity, reproductive capacity, and possibilities of hybridization, among others.

In general, the species richness of islands depends on four highly interrelated factors: (1) their surface area, on which the extinction rate depends; (2) the physiographic diversity, which determines the number of available habitats and therefore the extinction rate; (3) the distance to the source of supply, influencing the rate of immigration; (3) and age, which govern the level of endemism reached and the robustness of recombination. These factors have to be considered together. Thus, for example, Fuerteventura, the island closest to Africa and second in size and age (1655 km^2; 20.6 Ma), is now poorer in species than the western islands, which are smaller and younger (*e.g.,* La Palma – 708 Km^2; 1.8 Ma). This is largely due to its present low physiographic diversity. The island with the highest number of plant species is Tenerife. The present consideration of the number of wild vascular species on each island distorts the conclusions to be drawn in this respect, since introductions have been massive in recent times and the islands of greater physiographic diversity have probably undergone a greater degree of naturalization.

The diversification of plants arriving on the islands has depended to a large extent on (1) intrinsic potential regarding their genetic capacity, (2) the processes of hybridization and constitution of singameons, (3) the genetic and geographic isolation reached, and (4) human exploitation of different habitats. The original immigrants have sometimes remained more or less unchanged up to now, but in general their populations underwent natural processes of differentiation and speciation.

The processes that seem to have had most influence on the evolution and speciation of Canary spermatophytes are those of divergent selection, adaptive radiation, and genetic drift. The first two are more evident within each island, especially when the number of colonizable niches is greater. Genetic drift between the different islands is related to the "foundation effect" and within each island by the bottleneck effect after populations become isolated. Adaptive radiation is a process in which there is a rapid diversification of taxa over time from a pre-adaptive ancestor as a consequence of adaptation to different types of habitats under a strong selective pressure. This has notable examples on the islands, as for example in the genera *Echium* and *Argyranthemum.*

The arrival of diaspores to the islands since their formation has been continuous, but the current floristic composition depends on the performance of the above biological mechanisms when faced with the action of ecological barriers or filters (geographic, edaphic, climatic and biotic) that oppose or halt migrations.

4.1.4 Sources of Flora

Through time, the Islands have been subject to the influence of varied floras, depending on the paleo-bioclimatic changes in their original surroundings. Subsequently, plants have remained, evolved or disappeared. According to the theory of founder events, it has been thought that the primitive colonization of the islands occurred by single chance-colonization events, possibly with low genetic variation. However this fails to explain in any way the actual high genetic diversity found. Nevertheless, the Surfing Singameon Hypothesis considers that there were later secondary contacts in the processes of colonization between different genotypes from both the continent and other islands, which would have widened genetic diversity through processes of hybridization and backcrossing. This would explain the high genetic diversity of the present Canary flora and the diversification of many of its endemic lineages. The following chrono-sequence shows taxa from various floristic sources that have influenced the past and present floristic composition in the Canaries. The number of terrestrial vascular plants (ferns *s.l.*, gymnosperms and angiosperms), according to the catalog assembled by Acebes et al. (2010), amounts to 2091 species. Out of these, 1389 (66.4%) and 702 (33.6%) are respectively considered autochthonous (= native) and introduced. Of the natives, 539 species are endemic, which represent 38.8% of the native vascular plants.

4.1.4.1 Mesozoic Macro-Continental

This flora developed before the separation of the American and European plates. As we have seen, genera like *Clethra, Morella, Persea-Apollonias, Ilex, Sedum, Ocotea, Dryopteris, Cneorum, Ruta, Daucus, Descurainia* and *Fagonia* belong to it (Fig. 4.7). Some of them became important later, in the late Cenozoic flora of the Tethys Sea basin. The stepping-stones theory via oceanic islands along the Macaronesian hot-spot tracks, and also the colonization theory of continental and volcanic islands during the rift-valley formation of the Atlantic basin, can explain the early arrival of this type of flora to the present older easternmost Canary islands, perhaps in the Miocene, long before the later arrival promoted by Pleistocene glacial fluctuations.

Fig. 4.7 *Morella faya*. Of Mesozoic macro-continental origin

4.1.4.2 Rand-Flora

This ancient Paleoafrican floristic source is strictly Gondwanan and belongs to arid to dry territories. Indeed, the Rand Flora is essentially xerophilous and today has its greatest exponent in marginal regions such as South Africa where it remains optimal, Madagascar, East Africa, South Arabia, Macaronesia, etc., and some African mountain areas. In its former area, the Rand Flora spread from the Cape Region to North Africa in areas where precipitation was insufficient for the development of humid tropical or subtropical forests. The Post-Tertiary periods of drought that caused massive extinctions in the African flora and led to the desertification of the Sahara were also responsible for its retreat. A sample of Rand Flora currently survives in the Canary Islands, especially in the Infra- and Thermomediterranean thermotypes in arid, semi-arid and dry arid regions. *Aeonium, Camptoloma canariensis, Campylanthus salsoloides, Canarina canariensis, Caralluma burchardii, Ceropegia, Dracaena, Euphorbia balsamífera, Euphorbia canariensis, Kleinia neriifolia, Periploca laevigata,* among other autochtonous plants show many signs of such an origin (Fig. 4.8).

Fig. 4.8 *Ceropegia dichotoma*. Of Rand-Flora origin

4.1.4.3 Tethyan-Tertiary (Mesogean)

These belong to the flora of the Tethyan-Tertiary region of the Pliocene, which bordered the N coasts of the Tethys Sea. Due to the cooling that had already begun in the Oligocene and the succession of glacial and interglacial phases in the Quaternary, this flora was forced to undergo successive latitudinal advances and retreats. Many of the subtropical and tropical floristic components of the European flora of those times did not withstand these displacements, mainly due to the geographical barrier of the Mediterranean Sea, which largely prevented their escape to the South. The occasional land bridges through it were likely between Gibraltar and North Africa, between Italy, Sicily and Tunisia, and as today between Sinai and Egypt (across the present Suez Canal). This Mesogean contingent that had already arrived impoverished into the African continent became even more so after their leap to the Canaries. So, the present representation of this flora on the Canaries does not express the ancient richness of the Tethyan-Tertiary vegetation, although among such typical Tethyan-Tertiary floristic elements there coexist some older (Mesozoic) floristic elements proper of the macro-continental flora of the late Cretaceous, existing before the separation of the American and European tectonic plates. The Mesogean flora still holds on to its refuge in the Islands, mainly in the cloudy environments of the laurisilva of the dry to humid Infra- to Mesomediterranean belts. Trees like *Apollonias spp, Laurus azorica, Persea indica, Ocotea foetens, Pleiomeris canariensis* (*Myrsine canariensis*), *Maytenus canariensis, Prunus lusitanica, Picconia excelsa, Viburnum tinus, Visnea mocanera, Ilex canariensis* or ferns like *Culcita macrocarpa, Davallia canariensis, Dryopteris spp., Diplazium caudatum, Hymenophyllum spp., Trichomanes spp.,* found as fossil remains in the Mediterranean, have been used as testimonies to attribute their origin in this Tertiary-Tethyan flora (Fig. 4.9).

Fig. 4.9 *Prunus lusitanica*. Of Tethyan-Tertiary (Mesogean) origin

4.1.4.4 Mediterranean Biogeographic Region

In the islands these plants are of a more recent origin. After the Pleistocene glaciations, three large floristic regions originated from the Mesogean source: Mediterranean, Saharo-North-Arabian and Iranian-Turanian. Through the NW

African corridor, a number of plants arrived in the Canary Islands from the Mediterranean region, which settled on the islands or became diversified, giving rise to the Mediterranean floristic element of the Canaries. This is the most abundant within non-endemic native plants (*e.g., Pistacia lentiscus, Phyllirea angustifolia, Tamarix canariensis*). In addition, endemics of Mediterranean affinity are the largest proportion within the endemics (*e.g., Olea cerasiformis, Echium spp., Sonchus* (subg. *Dendrosonchus*) *spp., Micromeria spp., Sideritis spp., Teline spp., Erysimum spp., Reseda scoparia*) (Fig. 4.10).

Fig. 4.10 *Tamarix canariensis*. Originating from the Mediterranean biogeographic region

4.1.4.5 Saharan-North-Arabian Biogeographic Region

Many Canary plants have this origin, especially those of the low arid zones of the islands, among native non-endemic species (*e.g., Launaea arborescens, Pteranthus dichotomus, Tetraena (Zygophyllum) fontanesii, Astydamia latifolia, Cenchrus ciliaris, Tetrapogon villosus, Chenoleoides tomentosa*), and among the endemics *Forsskaolea angustifolia, Salsola divaricata*, etc. (Fig. 4.11).

Fig. 4.11 *Cenchrus ciliaris*. Originating from the Saharan-North-Arabian biogeographic region

4.1.4.6 Anthropic Taxa

More than a third of the Canarian flora is considered to be introduced by man. The maritime traffic from the Mediterranean to the islands and America has contributed greatly to this, since the first expeditions of the European conquest. Numerous American plants like *Opuntia spp., Ageratina spp., Agave spp.*, *Acacia spp.*, etc. occur widespread in the Islands. In addition, to date about 700 vascular plants from these and other continents and diverse islands of the world have been introduced.

4.2 Biogeographic Classification

A basic objective of biogeography is the establishment of a typology for the characterization of territories, mainly based on the distribution areas of taxa (organisms) and syntaxa (communities). Additionally, information from other sciences (Geography, Geology, Earth Sciences, Bioclimatology, etc.) is considered. This typology can be established for various chronological periods, including the past in the paleobiogeographic, and the present in the current biogeographic typology.

Several hierarchical ranks are recognized, the main ones being: Kingdom, Region, Province, Sector, District and Tesela. Each of these units can in turn be divided into subunits of lower rank. Tesela is the basic unit of the system and is defined as a homogeneous territory that, according to its ecological characteristics, can harbor a single type of potential vegetation.

The main criteria for the recognition of biogeographic a areas can be grouped into:

- Geographic (formation of continents, orogeny, insularity, etc.).
- Bioclimatic (macrobioclimates, bioclimatic belts, etc.).
- Floristic or taxonomic (affinities and mainly chorology). Degree of endemicity has been the traditional criterion.
- Phytocoenotic (phytocoenotic affinities, association area, vegetation series, geoseries, permanent communities, altitudinal zonation, etc.).

The use of these criteria provides arguments to characterize and differentiate biogeographic ranks. The traditional floristic criterion of endemicity in combination with taxonomic range (species, genus, family, etc.) weigh heavily in the differentiation of biogeographic ranks. Table 4.1 shows some of these criteria.

Table 4.1 Main criteria for recognition of biogeographic ranks

Ranks	Floristic criteria	Geographic, phytocoenotic and bioclimatic criteria
Tesela		A single type of potential vegetation (1 series)
District	Endemic **subspecies**	Traditional use of territory
		Several associations
Sector	Endemic **species**	Wide territory with geographic entity
		Own associations
		Peculiar zonation (generally because existence of own permanent and subserial communities)
Province	Endemic **genera**	Large territory
	Many endemics (typifying phytogeographic subelement – endemics of province or sector)	Characteristic climactic dominions (potential areaof an association), series, geoseries, permanent communities and typical altitudinal zonation of vegetation
Region	Endemic **genera** and **families**	Very large territory
	Flora or floristic constituent (endemics of region or higher rank) very characteristic	Particular climactic dominions (potential area of an alliance) and territories, thus its own series, geoseries and bioclimatic belts
Kingdom	Taxonomic and ecosystematic features	
	Flora and fauna origin. Continental drift. Paleoclimate. Climate. etc.	
	Endemic orders and families	

The biogeographic framework of the Canaries has been diverse throughout history. Here are the main trends, followed by some of the arguments put forward to support them. From the beginning of these classifications the concept of Macaronesia as a biogeographic unit has been in use, although its geographic limits have been diverse. At present, as we have seen in the chapter on the geographic framework, Macaronesia is a territory defined geographically by the archipelagos of the Azores, Madeira, Salvages, Canaries and Cape Verde, but from a biogeographic perspective this is not always the case. Table 4.2 shows chronologically the main trends in this regard.

Table 4.2 Chronological biogeographic frameworks of the Canary Islands

J.F.SCHOUW (1825) Includes the Canaries in the *Sempervivorum* province, within the *Labiatarum* and *Caryophyllearum* kingdom (=Mediterranean region), together with the Azores, Madeira (with doubts) and the NW African coast.
WEBB, around 1850 refers to Macaronesia from a biogeographical point of view to highlight the floristic affinities of Madeira, the Salvages and the Canaries.
GRISEBACH (1872) considers the islands of the middle Atlantic under the denomination "oceanic islands" together with Ascension, Saint Helena and Tristan da Cunha, without deciding about their unitary character or position in this region.
ENGLER (1879) defines the biogeographic region of Macaronesia, which includes the Azores, Madeira and the Canary Islands. Cape Verde only appears on the map attached to the descriptive text. He justifies the classification in explaining that the territory is a group of islands that gathers Mediterranean, Indian, African and American floristic elements and which lodge relict vegetation of the tertiary laurisilva of the S of Europe. In a later work, ENGLER considers Cape Verde within Macaronesia, without allusions to the unity of the group.
STRASBURGER (1986) based on MEUSEL, JAEGER and WEINERT (1965) considers the Canary Islands together with the Azores and Madeira as the Macaronesian subregion of the Mediterranean-Macaronesian region.
SUNDING (1979) considers the Canaries within the Macaronesian region, which also includes the Azores, Madeira, Cape Verde and a part of the African west coast near the Canary Islands that he calls "Moroccan enclave".
KUNKEL (1980) follows the criterion of Sunding but he adds a small sector of the NW of the Iberian Peninsula to the Macaronesian territory.
LOBIN (1982) proposes the abandonment of the term Macaronesia from a biogeographical point of view. He includes the Canary Islands together with Madeira and the African (Moroccan) enclave in the Mediterranean region, within the Canary-Madeira subregion.
TAKHTAJAN (1986) considers the Canary Islands as part of the Macaronesian region, to which he assigns the same territory as Sunding (1979). The region is based on the existence of numerous endemics of generic and specific rank (in addition to some endemic sections). The majority of generic endemics are mainly confined to Madeira and the Canary Islands. He highlights the existence of relict endemisms, either taxonomically isolated or with close relatives in distant geographical regions. He indicates the presence of an ancient African flora and laurel forests related to the Miocene and Pliocene forests of Europe and the Caucasus.
RIVAS MARTINEZ (2009) in a comprehesive study of the Macaronesian archipelagos places the Canary Islands in the Canarian-Madeiran subregion, within the Mediterranean region, together with Madeira and the Salvages. This classification is the one followed in this book.

We adopt the bioclimatic classification of Rivas-Martínez (2009), which breaks up the single biogeographic territory of the Azores, Madeira, Canaries, Salvages, Cape Verde and African Macaronesian Enclave, and classifies it into two kingdoms and four floristic regions, according to the scheme in Fig. 4.12.

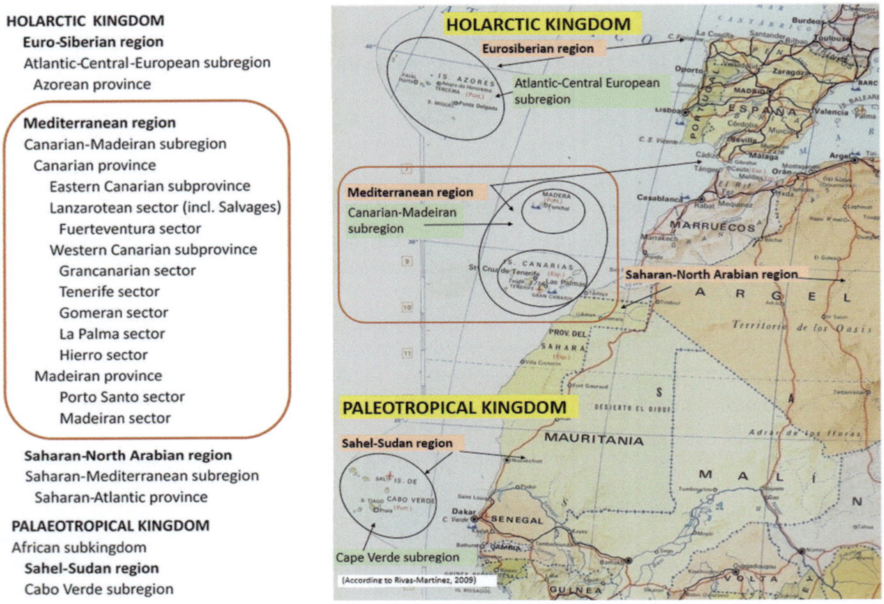

Fig. 4.12 Biogeographic characterization of Macaronesia, and particularly the Canarian-Madeiran subregion of the Mediterranean region

Supporting this classification, particularly for the Canary Islands, the following arguments based on the main criteria for biogeographic classification are shown as follows:

4.2.1 Geographic Criteria

The Canary, Salvage and Madeira Islands (hereinafter referred to as the Canarian-Madeiran subregion) are three closely related oceanic archipelagos located between 27°38′-33°07′N and 13°20′-17°-15′W, within the Subtropical latitudinal band, which is a differential feature compared with the Azores (36°43′-39°56′N and 29°46′-31°30′W), located in the Eutemperate latitudinal band, and Cape Verde (14°45′-17°10′N and 22°40′-25°20′W) in the Eutropical latitudinal band. Following the shortest route, the longest sea-bound distance between islands in the Canarian-Madeiran subregion is established at 255 km between the islet Bugio (Madeira) and

Selvagem Grande (Salvage Islands), much less than the minimum distance between the Azores (Santa Maria) and the island of Madeira (841 km), and the Canaries (El Hierro) and Cape Verde (Sal) of 1299 km. The small archipelago of the Salvages is 159 km from the Anaga peninsula (Tenerife, Canary Islands) and biogeographically more related to the Canaries than to Madeira (Fig. 4.1). This proximity between the three Macaronesian central archipelagos lends them great bioclimatic, floristic and phytocoenotic similarity and contributes to separation in these respects from the more distant archipelagos of the Azores and Cape Verde, as explained below.

4.2.2 Bioclimatic Criteria

The Canary Islands, Salvages and Madeira share a Mediterranean macrobioclimate which, among other characteristics, is differentiated by at least two months drought ($P<2T$) after the summer solstice. Madeira also partially shows a Temperate macrobioclimate (in its sub-Mediterranean variant), detectable above 200–300 m on its north slope and above 700–800 m on the south slope. A Temperate macrobioclimate, among other defining characteristics, presents summer rains and in the case of a dry summer period it never reaches a two month duration. The Azores clearly shows a Temperate macrobioclimate and the Mediterranean macrobioclimate is only shown in some southern coastal locations of its central and eastern islands. Cape Verde shows a Tropical macrobioclimate (Fig. 4.13).

4.2.3 Floristic Criteria

The archipelagos of Madeira, the Salvages and Canaries, due to their geographic location, as we have seen, show a strong Mediterranean influence in their native flora, both endemic and non-endemic, with its greatest exponent in the Canaries. This separates them from the Azores which show a strong floristic influence from the Atlantic-European Temperate flora, and Cape Verde with a strong floristic influence from the Tropical African flora.

The Canarian-Madeiran subregion (Madeira, the Salvages and Canaries) is the Macaronesian territory with the largest number of endemic vascular plants, of both generic and specific rank, the Canaries being the main contributor due to greater diversity and endemicity.

The Macaronesian archipelagos as a whole encompass a total of 45 endemic genera (Table 4.3). Of these, 39 are exclusive to the Canarian-Madeiran subregion (they include 29 endemic genera exclusive to one archipelago and 10 shared by at least two archipelagos). Of the remaining 6 genera, 4 are shared with the Azores, 1 is exclusive to the Azores and another to Cape Verde.

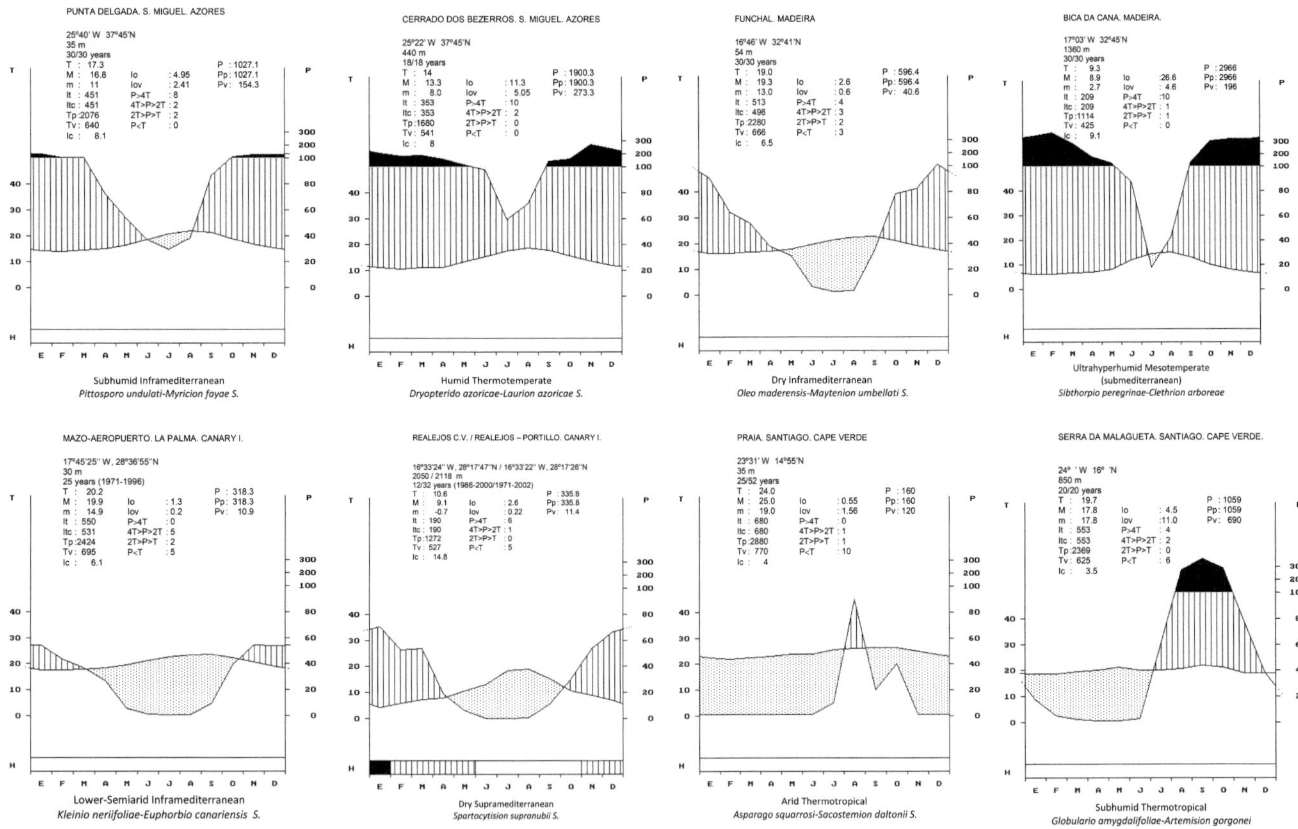

Fig. 4.13 Representative bioclimographs of the macrobioclimates found in Macaronesia

Table 4.3 Number of endemic genera of the Macaronesian archipelagos and the Canarian-Madeiran biogeographic subregion

	Endemic genera	
	Macaronesian archipelagos	**Canarian-Madeiran Subregion**
Azores	1	
Madeira	5	5
Salvages		
Canaries	24	24
Cape Verde	1	
Single archipelago endemics	**31**	**29**
Azores-Madeira-Canaries	4	
Madeira-Canaries	8	8
Salvages-Canaries	2	2
Shared by at least two archipelagos	**14**	**10**
Total	**45**	**39**

The Canarian-Madeiran subregion brings together a total of 672 endemic species and 118 endemic subspecies exclusive to one archipelago, plus 50 endemic species and 10 endemic subspecies shared by more than one archipelago. Thus, a grand total of 722 species and 128 endemic subspecies (Tables 4.4, 4.5 and 4.6; Figs. 4.14 and 4.15).

Fig. 4.14 *Ranunculus cortusifolius* is the single endemic Spermatophyte species with the widest distribution in Macaronesia. It is present in three of the biggest archipelagos (Azores, Madeira, and Canaries)

Table 4.4 Number of endemic species and subspecies of the Canarian-Madeiran biogeographic subregion

	Species	**Subspecies**
	Single archipelago endemics	
Madeira	128	14
Salvages	5	2
Canaries	539	102
	672	**118**
	Shared endemics	
Madeira-Canaries	35	9
Madeira-Salvages	5	
Madeira-Salvages-Canaries	6	1
Salvages-Canaries	4	
	50	**10**
Total	**722**	**128**

Table 4.5 Shared endemic species and subspecies within the Macaronesian archipelagos

	Azores	**Madeira**	**Salvages**	**Canaries**	**Cape Verde**
Adiantum reniforme subsp. *pusillum*	.	X	.	X	.
Aichryson villosum	X	X	.	.	.
Apollonias barbujana	.	X	.	X	.
Ammi majus subsp.*procerum*	.	?	.	X	.
Arum italicum subsp. *canariense*	.	X	.	X	.
Asparagus scoparius	.	X	?	X	X
Asplenium aethiopicum subsp. *braitwaitii*	.	X	.	X	X
Asplenium anceps	X	X	.	X	.
Autonoe madeirensis	.	X	X	.	.
Bupleurum salicifolium subsp. *salicifolium*	.	X	.	X	.
Carex peregrina	X	X	.	.	.
Carex viridula subsp. *cedercreutzii*	X	X	.	.	.
Carlina salicifolia	.	X	.	X	.
Cedronella canariensis	.	X	.	X	.
Cerastium vagans	X	X	.	.	.
Ceterach aureum	.	.	.	X	X
Cheilanthes pulchella	.	?	.	X	.
Chenopodium coronopus	.	?	X	X	.
Crepis divaricata	.	X	X	.	.
Dactylis smithii subsp. *hylodes*	.	X	.	X	X
Diphasiastrum madeirense	X	X	.	.	.
Dracaena draco subsp. *draco*	.	X	.	X	.
Dracunculus canariensis	.	?†	.	X	.

(continued)

Table 4.5 (continued)

	Azores	Madeira	Salvages	Canaries	Cape Verde
Drusa glandulosa	.	X	.	X	.
Dryopteris oligodonta	.	.	.	X	X
Elaphoglossum semicylindricum	X	X	.	.	.
Erysimum arbuscula	.	X	?	.	.
Erysimum bicolor	.	X	.	X	.
Euphorbia mellifera	.	X	.	X	.
Fumaria montana	.	X	X	X	.
Galium geminiflorum	.	X	X	X	.
Globularia salicina	.	X	.	X	.
Heberdenia excelsa	.	X	.	X	.
Huperzia dentata	X	X	.	.	.
Huperzia suberecta	X	X	.	.	.
Hypericum canariense	.	X	.	X	.
Hypericum glandulossum	.	X	.	X	.
Hypericum grandifolium	.	X	.	X	.
Ilex canariensis	.	X	.	X	.
Jasminum odoratissimum	.	X	.	X	.
Juniperus cedrus	.	X	.	X	.
Juniperus turbinata subsp. *canariensis*	.	X	.	X	.
Laurus novocanariensis	.	X	.	X	.
Lavandula pinnata	.	X	.	X	.
Limonium papillatum	.	.	X	X	.
Lolium canariense	.	X	.	X	X
Lolium lowei	.	X	.	X	.
Lotus glaucus	.	X	X	?	.
Lotus lancerottensis	.	X	.	X	.
Luzula elegans	.	X	.	X	.
Melica canariensis	.	X	X	X	.
Monizia edulis	.	X	?†	.	.
Notholaena marantae subsp. *subcordata*	.	X	.	X	X
Ocotea foetens	.	X	.	X	.
Patellifolia procumbens	.	X	X	X	X
Pelletiera wildpretii	.	.	X	X	.
Periploca laevigata	.	.	X	X	.
Persea indica	.	X	.	X	.
Phyllis nobla	.	X	.	X	.
Picconia excelsa	.	X	.	X	.
Plantago arborescens subsp. *maderensis*	.	X	.	X	.
Prunus lusitánica subsp. *hixa*	.	X	.	X	.
Ranunculus cortusifolius	X	X	.	X	.
Rhamnus glandulosa	.	X	.	X	.
Romulea columnae subsp. *grandiscapa*	.	X	.	X	.

(continued)

Table 4.5 (continued)

	Azores	Madeira	Salvages	Canaries	Cape Verde
Rubia fruticosa subsp. *fruticosa*	.	X	X	X	.
Rubus bollei	.	X	.	X	.
Rumex bucephalophorus subsp. *canariensis*	X	X	.	X	.
Rumex maderensis	.	X	.	X	.
Salix canariensis	.	X	.	X	.
Schizogyne sericea	.	.	X	X	.
Semele androgyna	.	X	.	X	.
Senecio incrassatus	.	X	X	X	.
Smilax canariensis	.	X	.	X	.
Tamus edulis	.	X	.	X	.
Teline stenopetala	.	.	.	X	X
Tolpis succulenta	X	X	.	.	.
Teucrium heterophyllum	.	X	.	X	.
Urtica morifolia	.	X	.	X	.
Urtica portosanctana	.	X	X	.	.
Visnea mocanera	.	X	.	X	.
Wahlembergia lobelioides subsp *lobelioides*	.	X	X	X	X

X: recorded, ?: uncertain, †: extinct

Table 4.6 Summary of the number of shared endemic species and subspecies within the Macaronesian archipelagos

	Shared endemics for archipelagos		
	spp	subsp	Total
AM_9	8	1	9
AMC_3	2	1	3
MS_3	3		3
M(S)_2	2		2
MC_41	33	8	41
(M)C_3	2	1	3
SC_4	4		4
MSC_5	4	1	5
MS(C)_1	1		1
(M)SC_1	1		1
MSCV_2	1	1	2
M(S)CV_1	1		1
MCV_4	1	3	4
CV_3	3		3
Summary for groups of archipelagos			
AM_9	8	1	9
AMC_3	2	1	3
MSC_60	50	10	60
MSCV_7	3	4	7
CV_3	3		3

Archipelagos (A: Azores, M: Madeira, C: Canaries, S: Salvages; V: Cape Verde. Brackets = uncertain record. Accompanying figure: number of taxa

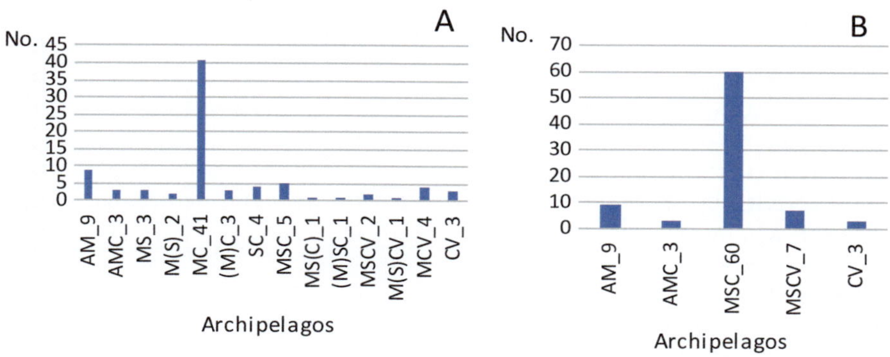

Fig. 4.15 Shared endemic species and subspecies within the Macaronesian archipelagos. (**a**) detailed; (**b**) abbreviated). Archipelagos (A: Azores, M: Madeira, C: Canaries, S: Salvages; V: Cape Verde. Brackets = uncertain record. Accompanying figure: number of taxa)

The **Canary Islands** is the archipelago with the most endemicity in the area. It has 24 endemic genera (*Allagopapppus, Atalanthus, Babcockia, Bethencourtia, Chrysoprenanthes, Gonospermum, Lactucosonchus, Sventenia, Vieraea, Rutheopsis, Tinguarra, Todaroa, Parolinia, Ceballosia, Dicheranthus, Neochamaelea, Ixanthus, Navaea, Pleiomeris, Spartocytisus, Bencomia, Dendriopoterium, Kunkeliella* and *Gesnouinia*) (Table 4.7; Figs. 4.16.1 and 4.16.2), an unstable figure that has even decreased to 15 genera or increased depending on phylogenetic evidence, nowadays mainly provided by molecular studies. In addition, there are 8 endemic genera shared with Madeira (*Bystropogon, Heberdenia, Marcetella, Phyllis, Isoplexis, Normania, Visnea* and *Semele*), which highlights their close floristic affinity, 4 endemic genera shared both with Azores and Madeira (*Pericallis, Aichryson, Cedronella* and *Picconia*) from the laurisilva area, 1 endemic genus shared with Madeira and Salvages (*Argyranthemum*) and two endemic genera shared with Salvages (*Schizogyne* and *Monanthes*) (Table 4.8; Figs. 4.17 and 4.18). The Canary archipelago has 2091 wild species of vascular plants from which 539 are endemic. Besides, 102 endemic subspecies can be added. It shares 35 endemic species (two of which with uncertain records) and 9 endemic subspecies with Madeira, 2 species and one endemic subspecies with Azores and Madeira, from the humid laurel forest, 4 endemic species with Salvages, 6 species (of which one species has uncertain records in Madeira and one has uncertain records in the Canaries) and one endemic subspecies with Madeira and Salvages, 2 endemic species (one of uncertain record in Salvages) and one endemic subspecies with Madeira, Salvages and Cape Verde, and finally one species and 3 endemic subspecies with Madeira and Cape Verde (Tables 4.4, 4.5 and 4.6; Fig. 4.15).

Table 4.7 Vascular plant genera endemic to the Canary Islands

	Islands							No.	B
RUTHEOPSIS (Apiaceae)	L	F						1	H
TINGUARRA (Apiaceae)[1]				T	G	H	P	1	H, Ch
TODAROA (Apiaceae)[1]			C	T	G	H	P	1	Ch
ALLAGOPAPPUS (Asteraceae)			C	T	G	H		2	N
ATALANTHUS (Asteraceae)[2]	L?	F	C	T	G		P	6	N
BABCOCKIA (Asteraceae)[2]			C					1	H
BETHENCOURTIA (Asteraceae)				T	G		P	3	N
CHRYSOPRENANTHES (Asteraceae)[2]			C					1	N
GONOSPERMUM (Asteraceae)[3]				T	G	H	P	7	Ch-N
LACTUCOSONCHUS (Asteraceae)[2]					G		P	2	H/G?
SVENTENIA (Asteraceae)[2]			C					1	Ch
VIERAEA (Asteraceae)				T				1	Ch
CEBALLOSIA[4] (Boraginaceae)	L	F	C	T	G	H	P	1	N
PAROLINIA (Brassicaceae)	L	F	C	T	G			6	N
DICHERANTHUS (Caryophyllaceae)				T	G			1	Ch
SPARTOCYTISUS[5] (Fabaceae)			C	T		H	P	2	N
IXANTHUS (Gentianaceae)			C	T	G	H	P	1	N
NAVAEA (Malvaceae)[6]				T				1	N
PLEIOMERIS (Primulaceae)			C	T	G		P	1	M
BENCOMIA (Rosaceae)			C	T		H	P	4	N
DENDRIOPOTERIUM (Rosaceae)			C					2	N
NEOCHAMAELEA[7] (Rutaceae)			C	T	G	H	P	1	N
KUNKELIELLA (Santalaceae)			C	T				4	N
GESNOUINIA (Urticaceae)			C	T	G	H	P	1	N

Islands: **L** = Lanzarote; **F** = Fuerteventura; **C** = Gran Canaria; **T** = Tenerife; **G** = La Gomera; **H** = El Hierro; **P** = La Palma. **No.** = number of species. **B** = Biotype according to Raunkiaer (**H** = hemicryptophyte, G= geophyte, Ch = chamaephyte, **N** = nanophanerophyte, **M**= macrophanerophyte *s.l.*). Chorology and species number according to Acebes *et al.* 2010, except for inclusion of *Bethencourtia, Navaea* and *Gesnouinia* and exclusion of *Plocama*

1: *Tinguarra* is considered within the non-endemic genus *Athamanta* (Euro+Med Plantbase, 2017) (EMP 2017). Downie et al. 2010 and EMP 2017 retain *Todaroa.* **2:** *Atalanthus*, *Babcockia, Chrysoprenanthes, Lactucosonchus* and *Sventenia* are within *Sonchus* according to EMP 2017, but Bramwell (2001) considers *Chrysoprenanthes* as an endemic genus. **3:** *Gonospermum* now includes the earlier monotypic genus *Lugoa;* furthermore it includes the previously Canary endemic species of *Tanacetum.* **4:** *Ceballosia* is accepted in EMP 2017; The Plant List (2017) (TPL 2017) includes it within *Tournefortia.* **5:** *Spartocytisus* is retain for *S. supranubius* in TPL 2017 but it is within *Cytisus* in EMP 2017 as Cubas et al, 2002 consider. **6:** EMP 2017 subordinates *Navaea* to *Malva.* Escobar et al. 2009 retain *Navaea.* **7:** *Neochamaelea* is accepted in Euro+Med Plantbase (2017); The Plant List (2017) includes it within *Cneorum.* Note: *Greenovia,* before considered as endemic genus, is now considered a Canarian endemic section of the genus *Aeonium* (Bañares 2015)

Fig. 4.16.1 Vascular plant genera endemic to the Canary Islands. *E.g.*, (**a**) *Todaroa montana*. (**b**) *Allagopappus canariensis*. (**c**) *Vieraea laevigata*. (**d**) *Bethencourtia palmensis*. (**e**) *Dicheranthus plocamoides*. (**f**) *Ceballosia fruticosa*

Fig. 4.16.2 Vascular plant genera endemic to the Canary Islands. *E.g.*, (**a**) *Parolinia ornata.* (**b**) *Ixanthus viscosus*. (**c**) *Pleiomeris canariensis*. (**d**) *Bencomia caudata*. (**e**) *Kunkeliella subsucculenta*. (**f**) *Gesnouinia arborea*

Table 4.8 Vascular plant genera endemic to Macaronesia and shared by at least two archipelagos

		ARCHIPELAGO					
	Mac	**A**	**M**	**S**	**C**	**Ce**	**B**
ARGYRANTHEMUM (Asteraceae)	24		3	1	20	20	N
PERICALLIS (Asteraceae)	13	1	2		10	10	H-Ch
SCHIZOGYNE (Asteraceae)	2			1	2	1	N
AICHRYSON (Crassulaceae)	14	1	2		11	11	Th-Ch
MONANTHES (Crassulaceae)	10			1	9	9	Th-Ch
BYSTROPOGON (Lamiaceae)	7		2		5	5	N
CEDRONELLA (Lamiaceae)	1	1	1		1		N
SEMELE (Liliaceae)	2		1		2	1	L
PICCONIA (Oleaceae)	2	1	1		1		M
VISNEA (Pentaphylacaceae)	1		1		1		M
ISOPLEXIS (Plantaginaceae)	4		1		3	3	Ch-N
HEBERDENIA (Primulaceae)	1		1		1		M
MARCETELLA (Rosaceae)	2		1		1	1	N
PHYLLIS (Rubiaceae)	2		1		2	1	Ch-N
NORMANIA (Solanaceae)	2		1		1	1	N

ARCHIPELAGO: A = Azores, **M** = Madeira, **S**: Salvages, **C**: Canaries). **Mac** = in Macaronesia. **Ce**: Canary endemics. **B** = Biotype according to Raunkiaer (**H** = hemicryptophyte, **Ch** = chamaephyte, **N** = nanophanerophyte, **M**: macrophanerophyte *s.l.*), **L**= liana). **Figures**: number of species. Chorology and species number according to Silva et al. (2008), Jardim et al. (2008), Acebes et al. (2010)

Fig 4.17 Vascular plant genera endemic to Macaronesia and shared with the Canaries. *E.g.,* (**a**) *Argyranthemum teneriffae.* (**b**) *Bystropogon canariensis.* (**c**) *Visnea mocanera.* (**d**) *Marcetella moquiniana*

Fig. 4.18 Vascular plant genera of laurel forests, endemic to Macaronesia and shared with the Canaries. *E.g.*, (**a**) *Pericallis echinata*. (**b**) *Aichryson laxum*. (**c**) *Cedronella canariensis*. (**d**) *Picconia excelsa*

Madeira has 5 exclusive endemic genera of vascular plants (*Chamaemeles, Melanoselinum, Monizia, Muschia* and *Sinapidendron* (Table 4.9), and the already mentioned 8 endemic genera shared with the Canaries, 4 endemic genera shared with the Azores and Canaries and 1 genus shared with the Salvages and the Canaries (Fig. 4.17). It has 1155 wild species of which 128 are endemic. Besides these, 14 endemic subspecies can be added. It shares 8 species and 1 endemic subspecies with the Azores, 2 species and 1 endemic subspecies with the Azores and Canaries, from the humid environments of laurel forest, 5 endemic species with the Salvages (two of them with uncertain records in the Salvages), 35 species (1 of them with uncertain records in Madeira) and 9 endemic subspecies (1 of them with uncertain records in Madeira) with the Canaries, 6 species (1 with uncertain records in the Canaries and another with uncertain records in Madeira) and 1 endemic subspecies with the Salvages and Canaries, and finally 3 species and 4 endemic subspecies whose distribution extends to the Canaries and Cape Verde (Tables 4.4, 4.5 and 4.6; Fig. 4.15).

Table 4.9 Vascular plant genera endemic to Madeira

	Islands			No.	B
MELANOSELINUM (Apiaceae)	M			1	H-Ch?
MONIZIA[a] (Apiaceae)	M	D	P	1	H
SINAPIDENDRON (Brassicaceae)	M			6	Ch-N
MUSCHIA (Campanulaceae)	M	D		3	Ch-N
CHAMAEMELES (Rosaceae)	M			1	N

Islands: **M** = Madeira; **D** = Desertas; **P** = Porto Santo. **No.** = number of species. **B** = Biotype according to Raunkiaer (**H** = hemicryptophyte; **Ch** = chamaephyte; **N** = nanophanerophyte). Chorology and species number according to Borges et al. (2008)

[a]*Monizia* has been recorded for Salvages, although now considered extinct there (?)

Chorology and species number according to Silva et al. (2008), Jardim et al. (2008)

The **Salvages** archipelago does not present any endemic genus. It seems that the endemic genus of Madeira *Monizia* was present there (there is an uncertain record of *M. edulis* for the Salvages, but it is currently considered extinct). The Macaronesian genera in the Salvages are *Argyranthemum* (endemic to Madeira, Salvages, and Canaries), and *Schizogyne* and *Monanthes* (endemic to Salvages and Canaries) (Fig. 4.17). There are 104 wild species of vascular plants, of which 5 are exclusive endemisms (in addition there are two endemic subspecies). It shares 5 endemic species with Madeira (two with doubtful records in the Salvages), 4 endemic species with the Canaries, 6 endemic species (one with uncertain records in Madeira and one with uncertain records in the Canaries) and 1 endemic subspecies with Madeira and the Canaries, and 3 endemic species with Madeira, Canaries and Cape Verde (one of them of uncertain record in the Salvages). Sixteen out of 19 specific and subspecific Macaronesian endemics existing in the Salvages and shared with some other Macaronesian archipelago are restricted to the Canarian-Madeiran subregion and only 3 extend to Cape Verde (Tables 4.4, 4.5 and 4.6; Fig. 4.15).

4.2.4 Phytocoenotic Criteria

The Macaronesian archipelagos harbor diverse types of potential climactic vegetation in accordance with island geography, island ages, distances from diaspore source, different types of climate (and bioclimate), and different floristic origins. For many scientists, the survival of laurisilva in the three large archipelagos of the North, a Tethyan-Tertiary relict vegetation, has had great weight in the biogeographic characterization. It should be mentioned that the laurisilva of the Azores has a great floristic influence in the Atlantic-European Temperate flora and is included in its own phytosociological class, *Lauro azoricae-Juniperetea brevifolii*. The laurisilva of Madeira and the Canary Islands, with a more Mediterranean influence, is included in the class *Pruno hixae-Lauretea novocanariensis*, which shows the following common arboreal elements between the Canaries and Madeira: *Apollonias barbujana, Erica arborea, Heberdenia excelsa, Ilex canariensis, Laurus novocanariensis, Ocotea foetens, Persea indica, Picconia excelsa, Prunus lusitanica subsp. hixa,* and *Salix canariensis* (water-course banks), and *Sideroxylon mirmulans* and *S. canariensis* as geo-vicariants. There are also some geo-vicariants with the Azores laurisilva: *Ilex perado* subsp. *azorica, Laurus azorica, Picconia azorica* and *Prunus azorica*.

The presence of tropical African flora in the two southerly archipelagos, mainly linked to the Rand-flora, has also weighted the biogeographic classification, but on the other hand it has almost no influence on the two more northern archipelagos. For the rest, there is no similarity of the other climatophilous communities of the Azores and Cape Verde with those of the Canarian-Madeiran subregion.

The archipelagos of the Canarian-Madeiran subregion, although different in some respects as just mentioned, are those that show greater phytocoenotic similarity (Table 4.10).

There are communities of *Kleinio neriifolia-Euphorbietea canariensis*, a class endemic to the Canarian province in the Canaries and potentially in the small highly disturbed Salvages archipelago. There, the isolated presence of *Asparagus scoparius, A. nesiotes, Euphorbia anacoreta, Periploca laevigata, Rubia fruticosa* and *Schizogyne sericea* leads us to presume the earlier presence of the above communities.

Rhamno crenulatae-Oleetea cerasiformis, a class endemic to the Canarian-Madeiran subregion, is present in the Canaries as the alliance *Mayteno canariensis-Juniperion canariensis.* This includes the climatophilous thermo-sclerophyllous forests. In Madeira it appears as the alliance *Mayteno umbellatae-Oleion maderensis*, which includes the *Euphorbia piscatoria* shrubland and other thermo-sclerophyllous thickets.

Pruno hixae-Lauretea novocanariensis, an endemic class in the Canarian subregion, comprises the alliance *Visneo macanerae-Apollonion barbujanae* which is common to the Canaries and Madeira, apart from *Ixantho viscosae-Laurion novocanariensis* in the Canaries and *Sibthorpio peregrinae-Clethrion arboreae* and *Polysticho falcinelli-Ericion arboreae* in Madeira.

The class *Chamaecytiso-Pinetea canariensis* is endemic to the Canarian province.

The geographical proximity between the central Macaronesian archipelagos, their Mediterranean macrobioclimate, their large number of endemic genera and species, the great influence of the Mediterranean flora on their native flora, and their phytocoenotic similarities, all together constitute a powerful biogeographic argument that supports the separate subregion rank given to the Canaries, Salvages and Madeira within the Mediterranean biogeographic region.

Table 4.10 Distribution of climatophilous vegetation classes and alliances within the Canarian-Madeiran subregion, and their origin

Phytocoenotic criteria			
Main climatophilous vegetation units			
Canaries	**Salvages**	**Madeira**	**Origin**
Succulent *Euphorbia* scrub and shrubland *Kleinio neriifoliae-Euphorbietea canariensis* *Kleinio neriifoliae-Euphorbion canariensis* (arid and lower-semiarid Infra-Thermomed.)	**Succulent Euphorbia scrub and shrubland** *Kleinio neriifoliae-Euphorbietea canariensis* *Kleinio neriifoliae-Euphorbion canariensis ?*		Tropical African Paleo-African Rand Flora
Thermo-sclerophyllous forest *Rhamno crenulatae-Oleetea cerasiformis* *Mayteno canariensis-Juniperion canariensis* (upper-semiarid – lower dry Infra-Thermomed.)		**Thermo-sclerophyllous forest** *Rhamno crenulatae-Oleeteea cerasiformis* *Mayteno umbellatae-Oleion maderensis* (upper-semiarid Inframed. -climatophilous; dry thermomed. -edaphophilous)	Mediterranean
Laurel forest *Pruno hixae-Lauretea novocanariensis* *Visneo mocanerae-Apollonion barbujanae* (upper-semiarid – dry Infra-Thermomed.(c)) *Ixantho viscosae-Laurion novocanariensis* (subhumid-humid Thermomed. (c) and dry-humid Mesomed. (c))		**Laurel forest** *Pruno hixae-Lauretea novocanariensis* *Visneo mocanerae-Apollonion barbujanae* (subhumid-humid Thermomed.) *Sibthorpio peregrinae-Clethrion arboreae* (humid-hyperhumid Thermo-Mesotemp.[a]) *Polysticho falcinelli-Ericion arboreae* (ultrahyperhumid Supratemp.)	Tethyan-Tertiary Macrocontinental Late Mesozoic flora

(continued)

Table 4.10 (continued)

Phytocoenotic criteria			
Main climatophilous vegetation units			
Canaries	**Salvages**	**Madeira**	**Origin**
Pine forest and summit shrubland and scrubland *Chamaecytiso-Pinetea canariensis* *Cisto symphytifolii-Pinion canariensis* (upper-dry – humid Mesomed.) *Spartocytision supranubii* (upper-semiarid – humid Supramed. and dry Oromed.) *Violion cheiranthifolii* (arid-semiarid Oromed.)			Tethyan-Tertiary (*Pinus canariensis*). Mediterranean

[a]med.: Mediterranean, temp.: Temperate, (c): with trade-wind clouds

Chapter 5
Other Floristic Considerations

5.1 Non-vascular Flora

The Canarian non-vascular flora (*s.l.*) is estimated at 5508 species of which 163 are considered endemic. It includes algae, fungi, lichens and bryophytes, highly diverse organisms belonging to five kingdoms, *Fungi, Bacteria, Protozoa, Chromista* and *Plantae*, living in terrestrial and marine environments. They notably contribute to the biodiversity of the islands as shown in Table 5.1, which includes data from several recent catalogs.

Table 5.1 Biodiversity figures for the Canaries

	Terrestrial[2]		Marine[1]		Total species	Total endemics
	Species	Endemics	Species	Endemics		
Algae *s.l.*[3]	?	–	1542	9	1542	9
Fungi *s.l.*	1893	107	19	1	1912	108
Lichens and lichenicolous fungi	1548	40	3	0	1551	40
Bryophytes	503	6	0	0	503	6
Vascular plants	2091	539	3	0	2094	539
Total	**6035**	**692**	**1567**	**10**	**7602**	**702**

From Moro et al. (2003)[1], Arechavaleta et al. (2010)[2] and Afonso-Carrillo (2014)[3]

Algae are diverse organisms, ranging from unicellular to complex multicellular forms, many meters long, currently included in four different Kingdoms (*Bacteria, Protozoa, Chromista* and *Plantae*) (Fig. 5.1). They mainly live in aquatic environments, but also colonize other moist surroundings. There are many species linked to fresh water, but in the Canaries these have not been cataloged. The most studied and

M. J. del Arco Aguilar, O. Rodríguez Delgado, *Vegetation of the Canary Islands*,
Plant and Vegetation 16, https://doi.org/10.1007/978-3-319-77255-4_5

Fig. 5.1 Algae. *Cystoseira*

diverse algae are those that grow in the sea, cataloged at 1542 species (Table 5.2). They are commonly known in the islands as “mujos”, “orchillas” and “confites”. Along the coastal landscape the red algae (*Rhodophyta*) are the most numerous, but the brown (*Phaeophyceae*) are dominant due to their greater size. The abundance of green algae (*Chlorophyta*) in the most polluted environments is striking. The endemic species of the Warm Temperate East Atlantic Region (32%) dominate among the Canary algae, followed by those of Tropical and Subtropical origin (31%), then Temperate (17%), Cosmopolitan (16%) and Disjuncts (4%). Algal communities on the island coasts follow defined distribution patterns, generally according to littoral zones, whose composition and richness is determined by the greater or lesser exposure to the waves. Intertidal pools constitute special environments, with their particular flora and vegetation zonation, depending on the salinity and temperature of the water that varies according to the duration of emersion.

Table 5.2 Marine Algae *s.l.* of the Canary Islands

Marine *Algae s.l.*	**Species**		**Endemics**	
Kingdom Bacteria(Phylum Cyanobacteria)		72		–
Kingdom Protozoa		1		–
Kingdom Chromista		890		1
Class Phaeophyceae	141		1	
Other Chromista	749		–	
Kingdom Plantae		556		13
Phylum Rhodophyta	425		11	
Phylum Chlorophyta	131		2	
Total		**1542**		**14**

From Moro et al. (2003) and Afonso-Carrillo (2014)

Fungi *s.l.* are very diverse heterotrophic organisms, ranging from unicellular forms to large macroscopic bodies (typical mushrooms) (Fig. 5.2). Currently they are included in three different Kingdoms (*Protozoa*, *Chromista* and *Fungi*). They are most developed in the terrestrial environment and only a small number are able to live in the sea. They live on decomposing organic matter (saprophytes), parasitizing other living organisms (parasites), or associated with the roots of some plants (mycorrhizal). According to their structure and morphology they receive very different names, especially the group of mushrooms, which is the most striking and best popularly known. There are 1912 fungi cataloged for the Canaries, the vast majority being terrestrial, since only 16 are marine. The most numerous groups are Basidiomycota and Ascomycota, with 1133 and 396 species respectively (Table 5.3). Their generally easy dispersal by spores allows them to occupy wide areas. According to Beltran Tejera (2011) the Holarctic (64%) are dominant, followed by the Tropical (24%) and Cosmopolitan (9%), and only a few are Endemic (5.6%). The greatest diversity is reached in the laurel forest and pine forest, being scarcer in the xeric communities of the lower belts and in the Canary high mountains. As a curiosity, lignicolous fungi are particularly abundant in laurel forest while the terricolous are in the pine forest.

Table 5.3 Fungi *s.l.* of the Canary Islands

	Terrestrial		**Marine**
Fungi s.l.	**Species**	**Endemics**	**Species**
Mycetozoa	172		
Oomycota	15		
Blastocladiomycota	1		
Chytridiomycota	1		
Glomeromycota	2		
Zygomycota	2		
Ascomycota	380	52	16
Basidiomycota	1133	15	
Fungi Imperfecti	187	40	3
Total	**1893**	**107**	**19**

From Moro et al. (2003) and Archavaleta et al. (2010)

Fig. 5.2 Fungi. *Auricularia*

Lichens and **lichenicolous fungi** are organisms resulting from the symbiosis of algae and fungi, which are now included within the latter (Fig. 5.3). They are highly resistant to inhospitable environments. They colonize and thrive on recent volcanic lava flows, tree trunks, bare rocks, etc., generally facing North. They have diverse biological forms, ranging from the simplest powdery appearance (leprose), to the most complex arbuscular (fruticose) structure. Only a few, mostly those that are used by man, have common names, such as "orchillas" (*Roccella* spp.), "ajicanes" (*Ramalina* spp.), "barbas de pino" (*Usnea* spp.). There are 1551 species recorded for the islands, all terrestrial except three that live in the intertidal zone (Table 5.4). Most of them are widely distributed, the Mediterranean species predominating, followed by the Tropical and Subtropical, and very few are Endemic (only 40, 2.6%). Some species characterize communities that are widely represented over the insular territory, such as *Sterocauletum vesuviani* on recent lava flows.

Table 5.4 Lichens and lichenicolous fungi of the Canary Islands

Lichens and lichenicolous fungi	Species	Endemics
Ascomycota	1491	38
Basidiomycota	16	
Fungi imperfecti	44	2
Total	**1551**	**40**

From Arechavaleta et al. (2010), modified

Fig. 5.3 Lichens. *Ramalina*

The **bryophytes** are terrestrial laminar or arbuscular-like cryptogams commonly known as "mosses", "liverworts" and "hornworts", which belong to the Kingdom *Plantae* (Fig. 5.4). They mostly live in humid environments on soil, rocks or tree bark, which can form carpets at favorable places. The largest group is mosses, with 70% of the total. Five hundred and three bryophytes are cataloged for the islands (Table 5.5). Due to their easy dispersion by spores, they are generally widespread. Most of the Canary bryophytes have a wide Holarctic distribution, in which Mediterranean and Oceanic elements stand out, followed by Temperate and Boreal, and they show very few Endemics (only 6, 1.2%). The moist environment of laurel forest is the most favorable for their development, and there they show their greatest abundance. However, they appear also in the rest of the island habitats, from the most arid coastal belts of *Euphorbia* scrub to the broom shrublands of the high mountains.

Table 5.5 Bryophytes of the Canary Islands

Bryophytes	Species	Endemics
Bryophyta	356	5
Marchantiophyta	141	1
Anthocerotophyta	6	0
Total	**503**	**6**

From Arechavaleta et al. (2010)

Fig. 5.4 Bryophytes. *Targionia*

5.2 Vascular Flora

This group, belonging to the Kingdom *Plantae*, is made up of the ferns and spermatophytes. The catalog of wild vascular plants of the Canary Islands consists of 2094 species, grouped in 781 genera and 173 families. Of these, 64 are ferns and 2030 spermatophytes (3 are marine). One thousand three hundred and ninety two are native (1342 spermatophytes and 50 ferns) and 702 (688 spermatophytes and 14 ferns) are introduced (33.5%). Of the total, 539 are endemic (25.7% of the total, and 38.7% of the native species) (Table 5.6).

Table 5.6 Vascular plants of the Canary Islands

	Terrestrial		Marine	Total	
Vascular plants	**Species**	**Endemics**	**Species**	**Native**	**Introduced**
Pteridophyta	64	2		50	14
Spermatophyta	2027	537	3	1342	688
Total	**2091**	**539**	**3**	**1392**	**702**

From Moro et al. (2003) and Archavaleta et al. (2010)

The number of species on the islands is not a product of chance, and in general tends to depend on the island's surface area, physiographic diversity (in particular), age and distance from the continent. Tenerife is the most diverse island and hosts 1468 terrestrial species, Gran Canaria 1308, La Palma 904, La Gomera 903, Fuerteventura 748, El Hierro 704 and Lanzarote 701 (Table 5.7). But these figures do not suffice to fully quantify the taxonomic diversity; that is greater, because some species have diversified in endemic subspecies or varieties.

Table 5.7 Insular distribution of terrestrial vascular plants

Terrestrial vascular plants	El Hierro		La Palma		La Gomera		Tenerife		Gran Canaria		Fuerteventura		Lanzarote		Canaries	
	Spp No.	End.	Spp No.	End.	Spp No.	End.	Spp No.	End.	Spp No.	End.	Spp No.	End.	Spp No.	End.	Spp No.	End.
Pteridophyta	34	1	42	1	41	1	53	1	47	2	16		15		64	2
Spermatophyta	670	118	862	176	862	176	1415	296	1261	213	732	78	686	69	2027	537
Total	**704**	**119**	**904**	**177**	**903**	**177**	**1468**	**297**	**1308**	**215**	**748**	**78**	**701**	**69**	**2091**	**539**

Spp No.: species number; End.: endemics
From Arrechavaleta et al. (2010)

Asteraceae (295), Poaceae (212), Fabaceae (179), Lamiaceae (89), Brassicaceae (84) and Crassulaceae (68) are the families with the greatest number of species. Asteraceae (131), Lamiaceae (52), Crassulaceae (51), Fabaceae (46), Brassicaceae (31) and Boraginaceae (24) are those with most endemic species (Table 5.8).

Table 5.8 Important families of vascular plants according to species number and endemic species

Vascular plants	**No.**
Families	173
Genera	781
Species	2094
Best represented families	**Species No.**
Asteraceae	295
Poaceae	212
Fabaceae	179
Lamiaceae	89
Brassicaceae	84
Crassulaceae	68
Families with most endemic species	
Asteraceae	131
Lamiaceae	52
Crassulaceae	51
Fabaceae	46
Brassicaceae	31
Boraginaceae	24

From Moro et al. (2003) and Archavaleta et al. (2010)

As already mentioned in the chapter on biogeography, the autochthonous vascular plants of the Canary islands come from diverse sources and with a different chronology: the Mesozoic macro-continental, Rand-Flora, Tethyan-Tertiary (Mesogean), Mediterranean biogeographic regions, and the Saharan-North-Arabian biogeographic region. In the current floristic composition, the Mediterranean floristic element is of great importance, both for non-endemic and endemic species the Mediterranean affinity is dominant. Considering that one-third of the current vascular plants were introduced after the conquest, approximately 40% of the primitive composition was endemic.

The number of genera endemic to the Canaries is very high, although it varies according to taxonomic revisions. Table 5.7 shows 24 endemic genera: *Allagopappus*, *Atalanthus*, *Babcockia, Bencomia, Bethencourtia*, *Ceballosia*, *Chrysoprenanthes*, *Dendriopoterium*, *Dicheranthus*, *Gesnouinia*, *Gonospermum*, *Ixanthus*, *Kunkeliella*, *Lactucosonchus, Navaea, Neochamaelea*, *Parolinia*, *Pleiomeris*, *Rutheopsis*, *Spartocytisus*, *Sventenia*, *Tinguarra*, *Todaroa* and *Vieraea/Vieria.* There are another 15 Macaronesian endemic genera, which in addition to the Canaries occur in some other archipelago. Of these, 11 are strictly from the Canarian-Madeiran subregion and only 4 extend to the Azores. None reach Cape Verde (Table 5.8): *Aichryson*, *Argyranthemum*, *Bystropogon*, *Cedronella*, *Heberdenia*, *Isoplexis*, *Marcetella*, *Monanthes*, *Normania*, *Pericallis*, *Phyllis*, *Picconia*, *Schizogyne*, *Semele* and *Visnea*. In addition, 26 non-endemic genera deserve to be highlighted for having greatly diversified or having at least four endemic species: *Aeonium* (28), *Sideritis* (24), *Echium* (23), *Sonchus* (18), *Cheirolophus* (16), *Limonium* (16), *Lotus* (15), *Micromeria* (14), *Crambe* (13), *Helianthemum* (11), *Convolvulus* (10), *Euphorbia* (9), *Cistus* (8), *Polycarpaea* (8), *Silene* (8), *Teline* (8), *Descurainia* (7), *Tolpis* (7), *Pimpinella* (5), *Vicia* (5), *Carduus* (4), *Carlina* (4), *Lavandula* (4), *Ononis* (4) and *Pterocephalus* (4) (Table 5.9) (Figs. 5.5.1 and 5.5.2).

Table 5.9 Non-endemic Macaronesian vascular plant genera showing at least four Canary endemic species

	No.	Ce	B
PIMPINELLA (Apiaceae)	5	5	Ch-N
CARDUUS (Asteraceae)	6	4	H
CARLINA (Asteraceae)	5	4	N
CHEIROLOPHUS (Asteraceae)	16	16	N
SONCHUS (Asteraceae)	23	18	H-Ch-N
TOLPIS (Asteraceae)	9	7	Ch
ECHIUM (Boraginaceae)	25	23	Th-Ch-N
CRAMBE (Brassicaceae)	13	13	Ch-N
DESCURAINIA (Brassicaceae)	7	7	Ch-N
POLYCARPAEA (Caryophyllaceae)	9	8	Ch
SILENE (Caryophyllaceae)	26	8	Th-Ch
CISTUS (Cistaceae)	10	8	N
HELIANTHEMUM (Cistaceae)	15	11	Ch
CONVOLVULUS (Convolvulaceae)	15	10	Ch-N-Li
AEONIUM (Crassulaceae)	32	32	Ch-N
PTEROCEPHALUS (Caprifoliaceae)	4	4	Ch-N
EUPHORBIA (Euphorbiaceae)	33	9	N-M
LOTUS (Fabaceae)	23	15	Ch-N
ONONIS (Fabaceae)	16	4	Th-Ch-N
TELINE (Fabaceae)	9	8	N
VICIA (Fabaceae)	19	5	Th?
LAVANDULA (Lamiaceae)	7	4	Ch-N
MICROMERIA (Lamiaceae)	15	14	Ch-N
SIDERITIS (Lamiaceae)	24	24	Ch-N
LIMONIUM (Plumbaginaceae)	22	16	Ch-N

No = Number of species recorded for the Canaries. **Ce** = number of Canary endemic species. **B** = Biotype according to Raunkiaer (Th = therophyte; H = hemicryptophyte; Ch = chamaephyte; N = nanophanerophyte; M = microphanerophyte). Exclusions: *Ceropegia* 2 endemic species (before 6). *Viola* now 3 end. sp. (probably there are 4). *Asparagus* 3 end. sp., apart from 2 end. Subsp. Distribution and species number according Acebes et al. 2010, excluding *Senecio* which diminishes after segregating *Bethencourtia*

Fig. 5.5.1 Non-endemic genera with more than 14 endemic species in the Canary Islands. *E.g.*, (**a**) *Aeonium canariense*. (**b**) *Cheirolophus metlesicsii*. (**c**) *Sonchus congestus*. (**d**) *Echium leucophaeum*

Fig. 5.5.2 Non-endemic genera with more than 14 endemic species in the Canary Islands. *E.g.*, (**a**) *Lotus pyranthus*. (**b**) *Micromeria pineolens*. (**c**) *Sideritis canariensis*. (**d**) *Limonium bourgeaui*

5.3 Molecular Studies

Insular speciation in Macaronesia has been a very active process and there have been numerous molecular studies on the subject in recent years. Through these, the origin in time and geography of the colonizing ancestors, insular diversification and the possible migratory routes from and towards other islands and continents have been traced. It has been estimated that many widely diversified genera derive from a single colonizing ancestor and are the result of active speciation through processes of adaptive radiation and vicariance. Adaptive radiation has been most effective in some taxa *e.g.*, *Aeonium* aliance (incl. *Aichryson, Monanthes* and *Greenovia* – now *Aeonium), Echium, Sonchus, Sideritis* sect. *Marrubiastrum, Argyranthemum, Micromeria,* and vicariance in others like *Crambe* sect. *Dendrocrambe, Cheirolophus*, etc. The hypothesis of the single ancestor contrasts with the "Surfing Syngameon Hypothesis". This postulates that secondary contacts with continental genotypes, or those of surrounding island regions, generated syngameons with high levels of genetic variation that played a key role in the successful colonization of the various regions of the islands, promoting gene flow cessation, drift, and/or eventual selection processes leading to species radiation. The relictual character of the Tethyan-Tertiary flora classically attributed to many groups of Canarian woody plants has been questioned, based on taxonomic, morphologic, cytologic or palynologic studies. It has been concluded that many of them are actually of recent origin in the islands. Notable examples are found in the following cases. The endemic Macaronesian species of the genus *Echium* are derived from a single herbaceous ancestor of Mediterranean origin that colonized the Canaries and from there it was dispersed to Madeira and Cape Verde. In the Canaries it acquired a woody biotype and diversified its forms (arbuscular-candelabriform, monocarpic), producing two reversals towards herbaceous biotypes (therophyte and hemicryptophyte). The eight endemic Macaronesian species (Canarian-Madeiran subregion) of the *Bencomia* alliance (*Bencomia*, *Dendriopoterium* and *Marcetella*), have significantly increased their woodiness and size in the islands. They are considered to be derived from a Mediterranean woody (*Sarcopoterium*) or subfrutescent ancestor (*Sanguisorba ancistroides*), which in turn derives from herbaceous ancestors (*Sanguisorba minor*). Molecular studies of the clade of Macaronesian *Crassulaceae* (*Aeonium, Greenovia, Monanthes* and *Aichryson*) show that it originated from continental herbaceous ancestors. It shows that *Greenovia* is nested within *Aeonium*, a widely diversified genus in the Canaries that, like *Aichryson* and *Monanthes*, has also developed secondary woodiness and are not considered relictual. From the Canary Islands, *Aeonium* has dispersed to Madeira and Cape Verde, and curiously later on it has also colonized Africa from the islands. *Argyranthemum*, with a woody biotype, also has Mediterranean herbaceous ancestors and its Madeira taxa are considered basal for the genus. The Macaronesian taxa of the *Sonchus* alliance, which includes the genera *Babcockia, Sventenia, Chrysoprenanthes, Lactucosonchus, Sonchus* subg. *Dendrosonchus* and *Taeckholmia*, also show secondary insular woodiness and derive from a continental herbaceous ancestor. Secondary woodiness is very common; it occurs in at least 34 genera (15 families) of the Canarian flora, *e.g.*, *Crambe, Ixanthus, Lobularia, Lotus, Parolinia, Sideritis*, etc., in addition to those named above.

Cheirolophus originated in the Mid-Late Miocene in the Mediterranean and began its colonization of Macaronesia on the Canaries in the early Pleistocene with a very high speciation rate, giving rise to 16 endemic species in the Canaries. This was mainly by allopatric speciation.

There are also studies that show the arrival of flora from the African continent in older periods. Such is the case of taxa of the genus *Euphorbia* sect. *Aphyllis* subsect. *Macaronesiae* (some Macaronesian dendroid succulent spurges) that arrived in the Canaries, specifically on the paleo-islands of Tenerife (Roque del Conde, Teno and Anaga) and La Gomera that acted as repositories and as stepping stones. The disjunction between the African (subsect. *Africanae*) and the Macaronesian groups is about 6 Ma (prior to the formation of the Sahara) and is in accordance with a climate-driven vicariance process that has also occurred in other groups like *Canarina, Plocama, Campylanthus, Camptoloma* and *Hypericum*.

Molecular studies also relate *Dracaena draco* to East African *Dracaena* species, such as *D. ellembeckiana*.

The Tethyan-Tertiary relict hypothesis for the origin of *Persea indica-Apollonias barbujana* also has been questioned, postulating a long-distance dispersal from

North America to Macaronesia, since molecular analyzes have placed them as a sister to Neotropical *Persea* clades from which they diverged at a much later time than the segregation between North America and Eurasia. This North American connection explained by long-distance dispersal could be similar to that postulated by molecular phylogenetic studies for *Pericallis*, a genus closely related to the North American *Packera*. However, other molecular studies have concluded that the genus *Bystropogon* for which a South American affinity with *Minthostachys* was postulated, has its closest relatives in Oldword species of the genus *Clinopodium*.

Monophyly has been established for genera such as *Aeonium, Argyranthemum, Crambe, Pericallis, Sideritis* sect. *Marrubiastrum, Micromeria*, and polyphyly for others such as *Lavatera* (from which *Navaea* was segregated), *Limonium, Rhamnus* and the current species of *Echium, Sonchus, Euphorbia* and *Convolvulus*, among others.

The routes followed in the colonization of the Canary Islands and the migrations from there have been traced for various taxa. Many genera, *Adenocarpus, Bencomia, Echium, Genista, Hedera, Ixanthus, Sonchus, Teline*, among others, come from the Western Mediterranean. Some genera such as *Argyranthemum* and *Crambe* have moved from Madeira to the Canaries, while others such as *Aeonium, Echium, Sideritis* and *Sonchus* travelled from the Canaries to Madeira, as *Aichryson* and *Pericallis* did to later reach the Azores. *Monanthes* migrated northwards to the Salvages. *Aeonium, Echium* and *Sonchus* were also dispersed from the Canaries to Cape Verde.

Colonization of the African continent from the islands has been confirmed, as in the genus *Aeonium*, with *A. arboreum* and *A. kornelius-lemsii* in NW Africa, *Sonchus* (*Dendrosonchus*) with *S. pinnatifidus*, and *Tolpis*, with the two derived herbaceous species *T. barbata* and *T. virgata*.

Chapter 6
Vegetation of the Canary Islands

6.1 An Overview

The vegetation of the Canary Islands is very diverse and many of this communities are endemic. The widespread and more representative plant formations of the natural landscapes of the islands display a marked zonal distribution, with a clear distribution into altitudinal belts (Fig. 6.1). These belts are unevenly developed on the north and south slopes. The northern slopes are cooler, more humid and rainy, and climatically more diverse than the southern slopes which are warmer and drier. In the western islands, which biogeographically include Gran Canaria, a cloud layer develops in the mid-altitude zone (midlands) of the north slopes because of the influence of the humid northeast trade-wind. It gives rise to a special climatic humid zone on these slopes, with its own altitudinal vegetation pattern, including laurel forests. Potential vegetation maps of all islands are displayed in Figs. 6.164, 6.165, 6.166, 6.167, 6.168, 6.169 and 6.170.

Marine vascular vegetation is poor; it is characterized by three seagrasses: *Halophila decipiens*, a small marine weed with paddle-shaped leaves that grows in deep silty and sandy marine beds; *Cymodocea nodosa*, with well developed stands on submerged sandy beds (these communities belong to the tropical and subtropical class *Halodulo wrightii-Thalassietea testudinum*); and the smaller *Nanozostera noltii* known only from the intertidal zone of Arrecife in Lanzarote (belonging to the temperate and cold sea-water class *Zosteretea marinae*). These latter two plants have a narrow linear band-like leaf morphology.

Brackish water vegetation develops in coastal ponds, also in old artificial irrigation tanks. Its floristic composition is very poor, and limited to some submerged monocots mainly *Ruppia maritima* (class *Ruppietea*), *Zannichellia pedunculata* (class *Potametea*), generally interspersed with filamentous green algae.

Coastal salt marshes, facing temporary marine flooding, are particularly characterized by succulent shrubby *Chenopodiaceae* (*Salicornia perennis, Arthrocnemum spp., Suaeda spp.*), highly specialized and salt-tolerant (belonging to the

M. J. del Arco Aguilar, O. Rodríguez Delgado, *Vegetation of the Canary Islands*,
Plant and Vegetation 16, https://doi.org/10.1007/978-3-319-77255-4_6

Mediterranean, Irano-Turanian and Saharo-Arabian class *Salicornietea fruticosae*). They were relatively common in the eastern islands, but now they are declining owing to urban and tourist developments on the coast.

On **rocky shores** a low size and medium coverage community is characteristic with small cushioned, woody or rosette plants and some perennial herbs. In its floristic composition *Astydamia latifolia*, *Limonium pectinatum*, *Frankenia spp.* and *Crithmum maritimum* stand out as well as *Zygophyllum fontanesii*, a common plant also growing in almost all arid halophile littoral communities. They belong to the littoral Mediterranean and Atlantic class *Crithmo-Limonietea*.

Sandy substrates show various communities mainly on the shores of the eastern islands. The typical shrubby community of the succulent *Traganum moquinii* is found fixing dunes, and the smalll-sized community of *Euphorbia paralias* and *Cyperus capitatus* on mobile and embryo dunes. They belong to the Canary and Western Oceanic Saharan class *Polycarpaeo niveae-Traganetea* moquinii. On nitrophilized sands at the shore line, where sea-borne organic matter accumulates, the *Cakile maritima* and *Salsola kali* community is sparsely established.

Tamarisk thicket is characteristic of backshores and ravine mouths, penetrating inland through their humid and salty beds in the arid territories of some islands such as Gran Canaria and Fuerteventura. It is characterized by *Tamarix canariensis*, which forms dense and intrincate stands with a dark understory with the soil covered by a thick layer of slowly decomposing organic matter (Mediterranean, Irano-Turanian and Saharo-Arabian class *Nerio-Tamaricetea*).

***Euphorbia* scrub and shrublands** are typical succulent communities of the lower zone of the islands. They mainly settle on rocky soils, scarcely developed, under an arid and semiarid climate with little rain (50–300 mm), high temperatures (annual average > 19 °C) and high insolation. They are characterized by several *Euphorbia* species, some with an arborescent pachycaule habit, the dendroid spurges or "tabaibas"; and others with a dominant candelabriform shape provided by cactiform spurges. From the first group *E. balsamifera*, *E. lamarckii, E. berthelotii* (Gomera) and *E. regis-jubae* are widespread; and the second group comprises *E. canariensis or* "cardón" and the rare *E. handiensis*. But in this vegetation also several other shrubs participate, such as *Periploca laevigata*, *Kleinia neriifolia*, *Rubia fruticosa*, *Ceropegia fusca*, *Neochamaelea pulverulenta*, being prominent, together with small bushes and several herbs. As xerophytes, they show several morphological and physiological features linked to drought adaptation: thick stems that are rich in water, leaves transformed into spines, aphylly, summer deciduousness, waxy or hairy leaves, ephemerous herbs, etc. *E. balsamifera* scrub (*e.g.,*

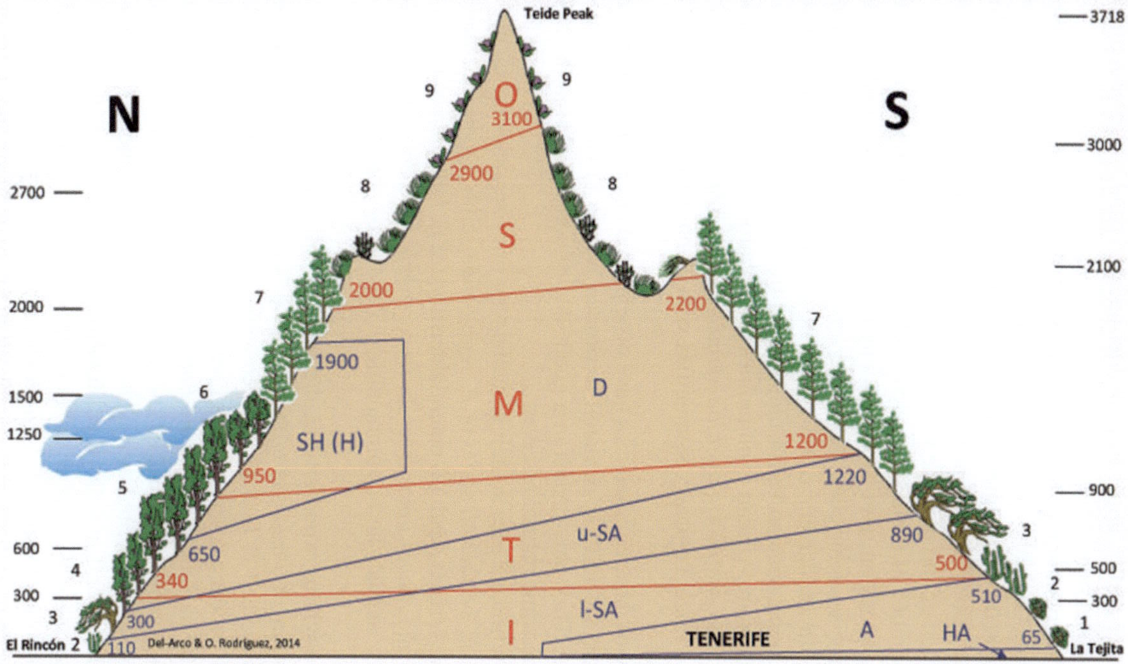

Fig. 6.1 Altitudinal transect or the Canary Islands (Tenerife). Thermotypes (red letters and lines): I: Inframediterranean. T: Thermomediterranean. M: Mesomediterranean. S: Supramediterranean. O: Oromediterranean. Ombrotypes (blue letters and lines): HA: hyperarid. A: arid. l-SA: lower semiarid. U-SA: upper semiarid. D: dry. SH: subhumid. H: humid. Potential Natural Vegetation: 1. *Kleinio neriifoliae-Euphorbion canariensis* (*Euphorbia balsamifera* scrub and *Euphorbia canariensis* shrubland); Coastal cliffs (N), and up to 500 m (S). 2. *Mayteno canariensis-Juniperion canariensis* (thermo-sclerophyllous woodland); Up to 350 m (N) and between 500–900 m (S). 3. *Pruno hixae-Lauretalia novocanariensis* (laurel forest); 350–1500 m (N). 4. *Cisto symphytifolii-Pinion canariensis* (pine forest); 1500–2000 m (N) and 900–2200 m (S). 5. *Spartocytisetum supranubii s.l.* (summit broom scrub); 2000–2700 m (N) and 2200–3250 (S) -only on favourable substrate. 6. *Violetum cheiranthifoliae* (Teide violet community). Height in m a.s.l

Ceropegio fuscae-Euphorbietum balsamiferae on Tenerife), is more drought tolerant than *E. canariensis*, and grows at a lower altitudinal level within the arid and hyperarid Inframediterranean bioclimatic belts. However, *E. canariensis* shrubland (*e.g., Periploco laevigatae-Euphorbietum canariensis* of Tenerife), grows higher up within the upper-arid upper-Inframediterranean and lower-semiarid Inframediterranean bioclimatics belts.

As climatophilous communities they grow on all the islands, with variable altitudinal limits depending on relief, exposure, slope, lithology and climate. In the north of the western islands they are almost restricted to cliffs and steep rocky terrains. They have a wider distribution over the rest of the lower zone of the territory, reaching an average altitude of 500 m in the south of Tenerife.

They are communities of a marked endemic character (geosynvicariant) based on their floristic composition with particular single island endemics, and they are included in the Canary endemic class *Kleinio neriifoliae-Euphorbietea canariensis*. They are related with similar dry communities in marginal areas of the African Continent and the South Arabian Peninsula, derived from the ancient Rand Flora. Their current geovicariant African communities are Saharo-Atlantic communities of the order *Euphorbio echini-Arganietalia spinosae* (class *Quercetea ilicis*), and other Mauritanian Saharo-Tropical communities with *Euphorbia sudanica* and *E. pseudobalsamiferae.*

Nowadays many of these communities have disappeared or are seriously threatened with disappearance in their potential territories, due to urban, industrial or tourist development. They only persist in a good state of conservation in the more rugged places and in some natural reserves.

The main substitutional stages in their potential territories are made up mainly of *E. lamarckii* scrub, *E. berthelotii* scrub or *E. regis-jubae* scrub, *Launaea arborescens* and *Schizogyne sericea* scrub (in potential territory of *E. balsamifera* scrub) or *Artemisia thuscula* and *Rumex lunaria* scrub, *Hyparrhenia hirta* and *Cenchrus ciliaris* meadow, and annual herb communities, depending on the degree of disturbance.

Thermo-sclerophyllous woodland of the insular midlands includes more or less open communities of *Juniperus turbinata* subsp *canariensis*, *Olea cerasiformis*, *Pistacia atlantica* and *P. lentiscus*, which settle in general in well structured soils, e.g. cambisols, under a cooler climate (average annual temperature 15–19 °C) and rainier (250–400 mm) than that of the *Euphorbia* shrubberies. This woodland has a different appearance depending on the dominant tree. Chorological reasons not well known probably lie behind their uneven distribution. In general *P. atlantica* stands tend to be distributed in valleys and foothills, while *O. cerasiformis* and *J. turbinata* subsp. *canariensis* stands withstand drier conditions. It is rich in shrubs, many of them also growing within the *Euphorbia* scrub, among them: *Asparagus umbellatus* and *A. scoparius*, *Euphorbia lamarckii* and *E. regis-jubae* and, *Jasminum odoratissimum*, *Rubia fruticosa*, *Echium giganteum, Erysimum bicolor*, etc. This reafirms the dry conditions the woodland still endures, also shown in some xerophytic characters of the trees such as the coriaceous leaves of *Olea* and the small size of the juniper tree's leaves.

Its potential distribution covers all the islands, although on the easternmost this woodland is only relictual. In the north of the western islands on developed soils it extends as a climatophilous community from the coast to (300) 350 m, while in the south it can reach up to 900 m, mainly within the upper-semiarid Infra- and thermo-mediterranean belts. These limits are largely exceeded as an edaphoxerophilous community on rocky ground. In the eastern islands its potential distribution extends from 400 m upwards on the north slopes and from about 450 m on the south slopes. These woodlands together with associated shrublands and scrublands belong to the Madeira-Canarian endemic class *Rhamno crenulatae-Oleetea cerasiformis*. Because of its structure and the Afroeuropean-Mediterranean and Paleomediterranean origin of a good proportion of its flora it is closely related to the Mediterranean *Quercetea ilicis* class.

Beside this woodland, on sunny and humid slope debris (colluvia) of the lower and middle zone, originating from ancient geomorphological processes, the *Phoenix canariensis* palm community has one of its optimum biotopes, sharing territory with D*racaena draco* and the typical shrubs of the Infra- and Thermomediterranean territories.

Other shrubby communities, such as *Retama rhodorhizoides* broom community on dry colluvial soil, *Euphorbia atropurpurea* scrub on rocky Infra-Thermomediterranean semiarid substrates, and *Rhamno crenulata–Hypericum canariensis* scrub, are the typical mantle communities of this woodland/forest.

The main substitutional stages are dendroid spurge scrub, a *Cistus* community and a nitrophilous shrubby community of *Artemisia thuscula* and *Rumex lunaria*, as well as the perennial hemicryptophytic community of *Cenchrus ciliaris* and *Hyparrhenia sinaica*, together with other herbaceous communities depending on the disturbance degree.

Laurel forest, laurisilva, is an arboreal plant formation, which has its origin in the subtropical Tertiary forest of southern Europe that was forced to migrate southwards as a consequence of the climatic change to a Mediterranean climate at the end of the Pliocene and particularly during the Pleistocene glaciations. This does not rule out an earlier colonization of ancient now submerged Macaronesian islands by subtropical forest. These could have served as stepping stones in the colonization process at an earlier time. This ancient forest, nowadays extinct on the mainland, survives on the islands as a relict. In Macaronesia, it is also present on Madeira and the Azores, although in the latter with significant participation of temperate flora. Madeiran and Canarian laurel forests belong to the class *Pruno hixae-Lauretea novocanariensis* that is endemic to the Canarian-Madeiran subregion. It is characteristic of the northern midlands that are influenced by trade-wind clouds. It is established on deep soils, the most characteristic being an association of humic andosols and cambisols. The forest is subject to an annual rainfall regime between 500 and 1200 mm. However, the presence of clouds is the key factor for its survival, which is more important for helping to maintain its water balance by avoiding evaporation than the water it provides, and above all it mitigates the extreme summer dryness of the Mediterranean climate. The influence of clouds is variable along the altitudinal range, and good indicators of its influence are the epiphytic mosses. The average annual temperature is between 13 and 18 °C and frost only occurs near its highest altitudinal limits. It is a humid forest of predominantly lauriphyllous trees with hypsophyllary buds, a feature in which they differ from other cooler laurisilvas in which trees with scaled buds are dominant. The main tree species are the *Lauraceae Laurus novocanariensis*, *Persea indica*, *Apollonias barbujana* and *Ocotea foetens*, and others of several families such as *Picconia excelsa* (*Oleaceae*), *Visnea mocanera* (*Pentaphylacaceae*), *Ilex perado* subsp. *platyphylla* (*Aquifoliaceae*), *Heberdenia excelsa* (*Myrsinaceae*), etc. Climbers include *Semele androgyna*, *Hedera canariensis*, *Smilax spp.* and *Convolvulus canariensis*, and several ferns including *A. hemionitis*, *Blechnum spicant*, *Culcita macrocarpa*, *Diplazium caudatum*, *Dryopteris oligodonta*, *Polystichum setiferum* and *Woodwardia radicans*. Also abundant are epiphytes, including fungi, bryophytes and lichens, and some vascular plant species like *Davallia canariensis*, *Aichryson laxum* and *Polypodium macaronesicum*. Some species like *Morella faya*, *Erica arborea*, *Ilex canariensis* and *Laurus novocanariensis*, are widely distributed in the laurel forest and mainly constitute the matrix in which other typifying species are interspersed to characterize several tree communities, linked to particular environmental conditions. Laurisilva develops on the windward trade-wind slopes from (300) 350 to 1500 m. In this belt, over basaltic substrate, with the north slope of the central island of Tenerife as a model, the following climatophilous communities of these forests can be distinguished. Their theoretical altitudinal limits vary a lot depending on the local geomorphology and topography, giving rise to a broad fluctuation between climatophilous and edaphophilous behavior:

- from 300 to 600 m we find the dry laurel forest (*Visneo mocanerae-Arbutetum canariensis*), a thermophilic community, more or less dry, that develops under a variable direct influence of the cloud bank. It is characterized by the particular presence of drought-resistant species such as *Apollonias barbujana*, *Picconia excelsa*, and *Visnea mocanera*. It is mainly established in the dry Infra- and Thermomediterrean belts with trade-wind clouds.
- from 600 to 1250 m there is the humid laurel forest (*Lauro novocanariensis-Perseetum indicae*), a mesophilic community under a more steady influence of the clouds and thus rainier, with a dominance of *Laurus novocanariensis*, *Ilex canariensis*, and *Persea indica*. It is mainly established in the subhumid and humid (Infra-) Thermo- and Mesomediterranean belts with trade-wind clouds.
- and from 1250 m to 1500 m the cold (and summer dry) laurel-forest (*Pericallido-Myricetum faya s.l.*) occurs, in territories without clouds in summer because of the trade-wind dynamics, and where plants that are tolerant to cold and dryness dominate, such as the arboreal *Morella faya* (a species that can grow in a range of different habitats), *Erica arborea*, a few *Laurus novocanariensis* individuals at the lowest edge of this zone, and some isolated *Pinus canariensis*. It is mainly established in the subhumid and humid Thermo- and Mesomediterranean belts with trade-wind clouds except in summer.

On salic substrates these communities are replaced by a special pine forest community where plants of the laurel forest and pine forest grow together (e.g. *Morello fayae-Pinetum canariensis* on Tenerife).

Besides these climatophilous communities, there are other potential communities with an edaphophilous behavior:

- At humid ravine bottoms and slopes with heavy fog drip and waterlogged soils the hygrophilous laurel forest (*Diplazio-Ocoteetum foetentis*) has developed. *Ocotea foetens, Ilex perado* subsp. *platyphylla* and *Persea indica* are the characteristic species and the understory is rich in ferns.
- On windswept ridges, depending on the altitude, wind humidity and its constancy throughout the year, other evergreen communities have become established (e.g. *Micromerio lepidae-Ericetum arboreae* – Gomera, or *Ilici canariensis-Erietum platycodonis* –Tenerife and Gomera). They are rich in heathers: *Erica arborea* or *Erica scoparia*, and diverse laurifolious trees.
- The azonal hydrophytic willow community (*Rubo-Salicetum canariensis*), can naturally grow in ravine beds at the lowest altitudinal level of the potential territory of the dry laurel forest where light can penetrate. The expansion of willow into the potential area of laurel forest is also favoured by logging.

The "fayal-brezal" (*Myrico-Ericetum arboreae*) is the typical substitutional shrubland of the laurel forest, mainly extended by anthropic destruction. It is mainly formed by *Erica arborea* or "brezo", *Morella faya*, *Ilex canariensis*, *Daphne gnidium*, *Viburnum rigidum* and *Pteridium aquilinum*. It is at first a small-sized, dense thicket, with straight stems and many sprouts induced by frequent pole cutting. Its size depends on the felling regime; after more or less fifteen years of growing in a well developed soil it can reach about 10 m in height. Today, because of the general abandonment of its use, we find a transition to good secondary laurel forest on old exploitation areas or cultivated lands. Broom scrub of several endemic species of legumes, like *Teline spp., Adenocarpus foliolosus* and *Chamaecytisus proliferus* subsp. *proliferus* s.l. form dry mantle communities of the forest growing on sunny cliffs and crags (*e.g. Telinetum canariensis*). They also spread to disturbed places subject to grazing, ploughing and fires. Bramble patches are typical of disturbed ravine beds and present also as humid mantle community of the laurel forest (*e.g. Rubio periclymeni-Rubetum ulmifolii*).

Besides, the herbaceous vegetation of the natural semi-shaded edges of the forest (*Ranunculo cortusifolii-Geranion canariensis*) show interesting endemics such as *Ranunculus cortusifolius, Geranium reuteri, Luzula canariensis, Carex canariensis* and *Pericallis spp.*

Landscape transformation within the potential territory of the laurel forest has been intense, and consequently the forest has recoiled greatly (Fig. 6.2). Nowadays, the main reserves are Garajonay National Park (World Heritage) on Gomera, the Biosphere Reserve of "El Canal y Los Tilos" on La Palma, the summits of the Rural Park of Anaga, and "Monte del Agua y Pasos" in Teno Rural Park on Tenerife, and the summits of the Frontera Rural Park which includes the Mencáfete Integral Natural Reserve on Hierro.

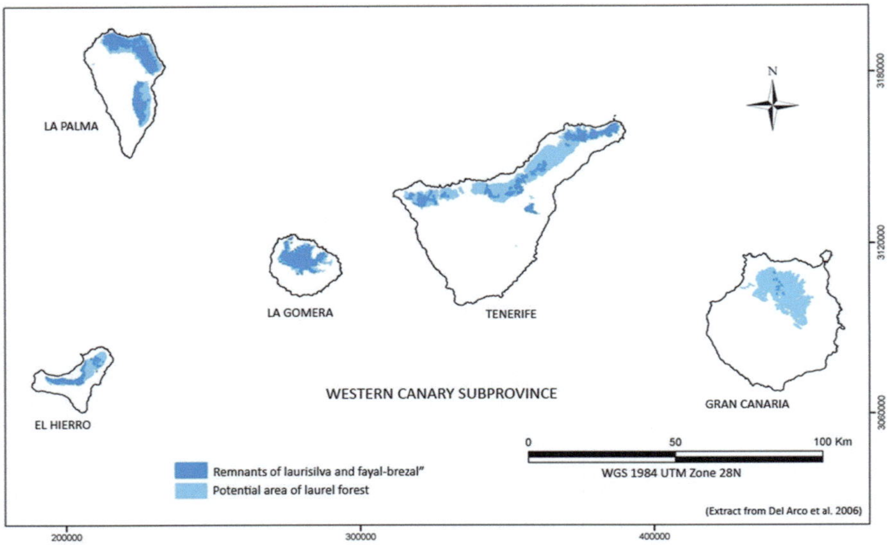

Fig. 6.2 Remnants of laurel forest in the Canary Islands

The **Canary Pine forest**, sometimes also called pine woodland because of its open canopy, is characterized by *Pinus canariensis.* It grows above the cloud zone in the north, and right above the territory of thermo-sclerophyllous woodland in the south, in the dry-humid Thermo- and Mesomediterranean belts without trade-wind clouds. It is subject to an annual rain regime of about 450–600 mm (much higher in the humid pine forest of La Palma), and average annual temperatures of 11–15 °C, with frost and some snow in winter, particularly at higher altitudes. The territory is mainly dry, directly influenced by the warm and dry subsidence layer of the NW trade-wind that generally blows above 1500 m a.s.l. The dryness is further accentuated on the south facing slopes. *Juniperus cedrus* may have been a companion of pine in the past, but selective felling and its sensitivity to fire could lie behind its current absence in the pine forest. The typical pine forest (e.g. *Sideritido solutae-Pinetum canariensis typicum* on Tenerife) has a tree coverage around 60%, and a poor understory of shrubs, introduced herbivores probably being responsible. In general, one can find the same species forming a scrub where the forest is cut back. The main species are *Chamaecytisus proliferus, Cistus grex symphytifolius*, *Lotus spp.*, *Bystropogon origanifolius*, *Argyranthemum spp.* and *Sideritis spp.*, among others. The altitudinal limits of climatophilous pine forest range between 1500–2000 m on the north slope, and 900–2200 (2300) m on the south. These limits are widely exceeded as a consequence of the strong competitiveness of pine on salic substrates and recent lava flows, which is particularly noticeable at lower altitudes, giving rise to special potential edaphophilous pine forests. Wild pine forest is naturaly present on Gran Canaria, Tenerife, La Palma and Hierro. On Gomera, where it probably also grew at a high elevation on the south slope, its current presence is limited to some rocks and salic outcrops. The shrubby broom community of the potential territory of the pine forest, dominated by *Chamaecytisus proliferus,* grows on stony ground and in the dry mantle of the forest. Fires, grazing and wood-cutting probably favored its extension in the past.

This forest has a close Afroeuropean-Mediterranean and Paleomediterranean affinity and together with the legume summit scrub, and summit violet community, is included within the class *Chamaecytiso-Pinetea canariensis* that is endemic to the Canary Islands.

Both in and outside the potential area of the pine forest, there have been forestry plantations with Canary pine and other pines (*Pinus radiata, P. halepensis,* etc.). This has contributed to maintaining this woodland, although sometimes at the expense of other authochtonous plant communities of great biological interest, like juniper woodland, laurel forest or summit broom scrub.

The **legume summit scrub**, *Genista benehoavensis* and *Adenocarpus viscosus* subsp. *spartioides* scrub (*Genisto-Adenocarpetum spartioidis*) on La Palma, and *Spartocytisus supranubius* scrub (*Spartocytisetum supranubii s.l.*) on Tenerife, are today generally recognized as the more characteristic potential vegetation of the Canary summits. Based on testimonial presence of *Juniperus cedrus*, and considering the profound transformation of the landscape through felling, grazing and fires, there is some evidence for an earlier sparse juniper forest (*Juniperetum cedri s.l.*) in its area, which would constitute the climatophilous potential vegetation.

These scrub grow on poorly developed soils, between 2000 and 2900 m on the north and between 2200 and 3100 m on the south slopes of the islands, in an altitudinal belt with an average rainfall between 350–500 mm, average annual temperatures of 6–11 °C, a possible frost period extending from October to May and some snow in winter-time, mainly in dry-humid supra- and oromediterranean belts. At first the scrub has a medium coverage but it becomes very sparse with increasing altitude. These rather harsh environmental conditions cause the appearance of adaptations to cold and drought in frequent cushion-like biotypes. They are related to Mediterranean legume shrubs, and apart from the aforementioned species numerous endemics form part of these communities, like the red flowered *Echium wildpretii*, blue flowered *E. auberianum*, *Argyranthemum teneriffae*, *Descurainia bourgeauana*, *Andryala pinnatifida* subsp. *teydensis*, *Nepeta teydea*, *Pterocephalus lasiospermus, Scrophularia glabrata.*

Violetum cheiranthifoliae grows on slopes and skirts of Mount Teide, above 2700 m on the north and above 3000 m on the south face of Tenerife, in semi-arid and dry oromediterranean belts, although it can descend. Its biotope is screes and lapilli subject to solifluction. Some rubble creepers like *Viola cheiranthifolia, Silene nocteolens* or *Argyranthemum teneriffae* grow in this zone. At higher altitudes, the *Vulpio myuri-Gnaphalietum teydei* constitutes the highest reaching vascular plant community of the archipelago, to be found in the active fumaroles of the summit crater still emitting hot sulfur dioxide and other gases.

The **rupicolous communities** are greatly diversified. The almost naked rock surface or crevices of crags, recent lava flows ("malpaíses"), and secondarily artificial walls and roofs, are suitable environments for their development. They are spread from coast to summit, but their optimum is reached in the humid environment of the north midlands. These communities belong to the Canarian-Madeiran class *Aeonietea*. They are rich in endemic plants mostly from *Crassulaceae* (*Aeonium* -including *Greenovia*-, *Aichryson*, and *Monanthes*) and *Asteraceae* (with several endemic genera as *Allagopappus, Atalanthus, Babcockia, Gonospermum, Sventenia, Vieraea,* besides other non-endemic genera among which *Sonchus* stands out due to its wide diversification. This richness, together with the multiple niches and the micromosaic structure provided by rocky cliffs where many plants have taking refuge, make Canary rocky ecosystems the main reserve of endemic flora on the islands.

Fresh water vegetation is sparse due to the scarcety of this habitat in the archipelago. Apart from the aformentioned Canary willow community, the rush community of *Juncus acutus* and *Scirpus holoschoenus* (*Holoschoeno-Juncetum acuti*), common reed *Pragmites australis* community, and cane-beds of invasive *Arundo donax*, along with diverse herbaceous communities (*Nasturtium officinale*, *Apium nodiflorum*, *Lemna spp.*, etc.), settle the ravine beds and banks, puddles, tanks, reservoirs, irrigation channels and other more or less permanent wet points.

Substitutional vegetation of potential communities is diverse and is today widespread. Many introduced plants (some invasive) occur at disturbed places, and some are so strongly established that sometimes they are accepted as native by the local inhabitants (e.g. *Agave americana, Arundo donax, Nicotiana glauca, Opuntia spp., Pennisetum setaceum, Ricinus communis, Spartium junceum, Tropaeolum majus*, etc.). This will be considered later.

6.2 Vegetation Units

6.2.1 [1]Floating or Rooted Submerged Aquatic Vegetation

6.2.1.1 Fresh Water Vegetation

Charetea fragilis

This class groups oligo- to monospecific submerged hydrophytic communities characterized by *Charophyte* algae growing in fresh or brackish water, oligo- to mesotrophic, generally accompanying the lower layer of other hydrophytic cormophytic communities, at the edges or bottom of the water body but not at great depth. They are widespread in the Holarctic.

[1] **Abreviations used in the text**. L: Lanzarote, F: Fuerteventura, C: Gran Canaria, T: Tenerife, G: La Gomera. H: El Hierro, P: La Palma. (c) with trade-wind clouds, (ces) with trade-wind clouds except in summer, (oc) with overflowing trade-wind clouds, (wc) without trade-wind clouds.

Photos by authors unless otherwise expressed: JRL Juan Luis Rodríguez Luengo, CGR Candelaria Gil Rodríguez, PLP Pedro Luis Pérez de Paz, AFL Ángel Fernández López, JC Juan Curbelo Rivero.

In the Canaries, only poor pioneer settlements have been recorded in usually hard fresh water and in brackish water, in small natural ponds, ravine-bed puddles and artificial irrigation tanks. The *Chara connivens* community is the most common. Other *Chara* species (*C. globularis, C. vulgaris*) and *Nitella hyalina* have been reported for Gran Canaria, although their communities have not been well studied.

Charetalia hispidae
Includes communities in basic and frequently calcic, phosphate poor, hard water.

Charion fragilis
Groups communities in basic oligo-mesotrophic water, calcium carbonate rich.

Chara connivens
This is a community that grows in hard fresh water and in brackish water, in small natural ponds, ravine-bed puddles and artificial irrigation tanks.

It has been reported for La Palma, Tenerife, and Lanzarote, where it is accompanied by *Potamogeton pusillus, Myriophyllum spicatum* or *Ruppia maritima*.

Distribution: L, C, T, P.

Lemnetea minoris
This class groups bryo-cormophytic fresh-water floating or submerged communities, not rooted in the bottom, characterized by species of *Lemna*, *Hydrocharis*, *Azolla*, *Salvinia*, *Wolffia*, *Riccia*, *Pistia, Stratiotes,* among others. Cosmopolitan.

In the Canary Islands they have been poorly studied, and the following units have been recognized.

Lemnetalia minoris
Single order.

Lemnion minoris
Acropleustophytic communities.

Lemnetum gibbae (Fig. 6.3)
This is a community typical of eutrophic to hypertrophic, nitrogen-rich still water, poorly oxygenated, growing in puddles, irrigation tanks, ponds, and channels with sluggish water. In general only the tiny *Lemna gibba* grows as a single species, but sometimes also *Lemna minor*. *Wolfia arriza* can appear, mainly in winter time.

Distribution: C, T, G, P, H.

Lemnetum minoris
This community grows in mesotrophic to eutrophic water bodies generally forming floating monospecific stands of the tiny *Lemna minor*. Sometimes it is mixed with the tiny Tropical American floating fern *Azolla filiculoides*.

Distribution: L, F, C, T, G

***Pistia stratiotes* stands**
We have observed this pantropical plant forming dense stands in irrigation tanks, in the locality of Valle de Aridane on La Palma.

Distribution: P.

***Salvinia molesta* stands**

This invasive species has been recorded growing in irrigation tanks and puddles in Barranco de Azuaje, Gran Canaria, forming dense stands, together with *Lemna minor, L. minuta* and *Azolla filiculoides*.

Distribution: C.

Fig. 6.3 *Lemnetum gibbae*. Barranco del Río. Tenerife

Potametea (Fig. 6.4)

This class groups calm or running fresh or brackish water cosmopolitan communities formed by rooted hydrophytes (*Potametalia pectinate*) or mesopleustophytes (*Utricularietalia vulgaris*), characterized by several species of *Potamogeton, Callitriche, Ranunculus,* etc.

Potametalia pectinati

Groups cosmopolitan communities of rooting hydrophytes in running or standing fresh water.

Potamion pectinati

Groups fresh water communities of elodeids.

***Elodea canadensis* stands**

Dense stands of this invasive species, native to North America, have been observed in irrigation tanks, rooted in the bottom mud or floating in the water body.

Distribution: T, P.

Fig. 6.4 *Potametea*. Barranco de Afur. Tenerife. (Ph: JRL)

***Myriophyllum spicatum* community**
Pure or almost pure stands of this rooting submerged hygrophyte grow in irrigation tanks, and ravine puddles, mixed with floating *Cladophora spp.* and sometimes *Chara connivens* in the lower layer. When flowering, the inflorescences of *Myriophyllum* stand out above the water surface. The species is a native of Europe, Asia and North Africa.

Distribution: C, T, P, H.

***Potamogeton pusillus* community**
This community has been seen in irrigation channels and small flow-regulation tanks connecting them. This small slender hydrophyte roots in the mud and the inflorescences are often held slightly above the water surface. *Chara connivens* sometimes also grows in the lower layer.

Distribution: C, T.

Ranunculion fluitantis
Groups cosmopolitan communities of submerged crowfoots of running fresh water.

***Callitriche stagnallis-Potamogeton nodosus* community**
This is a community that has been observed growing in backwaters in ravine beds, small basins below ravine falls, small reservoirs and irrigation channels. The single presence of *Potamogeton nodosus* is common and when the water depth diminishes it mixes with *Callitriche stagnallis*. In summer these habitats can dry up. In Anaga, Tenerife, it is frequent.

Distribution: C, T, G.

***Ranunculus penicillatus* community**

A community of *R. cf. aquatilis* was recorded for Barranco del Infierno, Tenerife, by Pérez de Paz et al. 1987, but Wolff & Rosinski 1997 think it is probably *R. penicillatus*. The community grows in channels with fast water and is accompanied by *Chara connivens*, and other hydrophytes (*Apium nodiflorum, Lemna minor, Rorippa nasturtium-aquaticum* and *Veronica anagallis-aquatica*).

Distribution: T.

Zannichellion pedicellatae

Groups hard water or oligosaline cosmopolitan communities of filiform or small-size elodeids.

Zannichellietum pedicellatae

A cormophytic community characterized by *Zannichellia pedunculata*, a small-size rhizomatose and stoloniferous hydrophyte with filiform leaves that has been found growing in oligosaline littoral ponds, and small dams and irrigation tanks with hard well-water. It roots in the muddy bottom of the pond, developing a submerged vegetative body partly next to the surface. Floating on the surface, sometimes fixed to the bed, filamentous algal tangles of *Cladophora vadorum* are common on Tenerife.

Distribution: F, C, T, G.

Utricularietalia vulgaris

Groups cosmopolitan communities of temporary rooted or non-rooting meso-pleustophytes that live floating between the bottom and the surface. They are typical of meso- eutrophic calm water.

Ceratophyllion demersi

Groups communities of macropleustophytes with *Ceratophyllum*, growing in calm meso- and eutrophic waters.

Potamo-Ceratophylletum demersi

Fragments of this submerged non-rooting community that is typical of oligo- or eutrophic fresh water have been observed growing in irrigation tanks in North Tenerife (Icod). It is represented by populations of *Ceratophyllum demersum*, a submerged hydrophyte bearing verticillate leaves with thread-like segments and spiny edges, abundantly branched, which form tangled masses in the water body.

Distribution: T.

6.2.1.2 Marine and Salt-Water Vegetation

Halodulo wrightii-Thalassietea testudinum

Groups vascular plant submerged marine meadows, in tropical waters spreading out to the Mediterranean.

Thalassio-Syringodietalia filiformis

Single order.

Syringodio-Thalassion testudinum

Single alliance.

Cymodoceetum nodosae (Fig. 6.5)
Infralittoral benthic thalo-cormophytic community, tropical West-African and Mediterranean, characterized by *Cymodocea nodosa*. This rhizomatous seagrass, with narrow linear leaves up to 600 × 5 mm with 3–5 parallel nerves, forms submerged meadows at 3–20 m depth in clear waters over sandy sediments. Some benthic algae also participate but epiphytic algae are numerous on its leaves, many of them *Corallinaceae*. This community has great importance as a habitat for many marine invertebrates, juvenile fish and fry from rocky substrates which find here refuge from predators. It is considered a European habitat of Community Interest.

Distribution: all islands.

Fig. 6.5 *Cymodoceetum nodosae*. Seabed South of Tenerife. (Ph: CGR)

Halophiletum decipientis (Fig. 6.6)
Infralittoral benthic thalo-cormophytic community that forms meadows in deep waters, about 15–40 m in depth, off the Canary coasts, on sandy or muddy seabeds. The rhizomatous pantropical seagrass *Halophila decipiens*, with typical paired paddle-shaped leaves up to 25×6 mm, characterizes the community, sometimes together with *Cymodocea nodosa*. The seaweeds *Caulerpa racemosa* var. *cylindracea*, at least in subtidal areas of Tenerife, and *Dictyota dichotoma* are also present.

Distribution: C, T.

Fig. 6.6 *Halophiletum decipientis*. Seabed South of Tenerife. (Ph: CGR)

Ruppietea
Bryo-cormophytic halophytic communities of permanent or temporary littoral or continental brackish water.

Ruppietalia maritimae
Single order

Ruppion maritimae
Communities with filiform eloideids of *Ruppia* growing in brackish water bodies that sometimes dry up in summer.

Enteromorpho intestinalis-Ruppietum maritimae (Fig. 6.7)
Thallo-cormophytic community of marine or continental brackish water, characterized by *Ruppia maritima* subsp. *rostellata*. This is a seasonal hydro-geophytic plant, with filiform stems and leaves, and small biomass, which develops in littoral ponds with sapropelic sediments, artificial irrigation tanks, and rain puddles in ravine beds on eastern islands. It frequently grows together with several algal species of *Enteromorpha.*

Distribution: L, F, C, T, C.

Zosteretea marinae
Groups herbaceous benthic communities of *Zostera,* accompanied by some seaweeds, growing in the eu- and infralittoral, up to 5 m deep, on sandy or muddy bottoms of lagoons, estuaries and calm coasts of temperate seas.

Zosteretalia marinae
Single order.

Zosterion marinae
Single alliance.

Fig. 6.7 *Enteromorpho intestinalis-Ruppietum maritimae*. Charco del Cieno. Valle Gran Rey. La Gomera

Zosteretum noltii (Fig. 6.8)
Mesolittoral mono- or oligospecific benthic thallo-cormophytic community of wide Mediterranean distribution, Caspian, Black Sea, Aral Sea and Atlantic Ocean from N Europe to NW Africa and Canaries, characterized by *Nanozostera (Zostera) noltii*. This small rhizomatous unisexual monoecious seagrass, with narrow linear leaves up to 200×1.5 mm with three parallel nerves, forms monospecific meadows in shallow waters that can emerge at low tide, growing over gross sediments poor in organic matter, subject to strong mechanical pressures and danger of desiccation. In the islands, its distribution is limited to the harbor area of Arrecife, Lanzarote, where the seaweeds *Asparagopsis taxiformis* and the exotic invasive *Caulerpa racemosa* var. *cylindracea* participate in the community.

Distribution: L.

Fig. 6.8 *Zosteretum noltii*. Arrecife. Lanzarote. (Ph: CGR)

6.2.2 Amphibious Vegetation of Fresh-Water

Isoeto-Nanojuncetea
Pioneer amphibious vegetation of annual or perennial (mainly geophytic) small herbs growing on periodically flooded, oligotroph or slightly saline, soils. Holarctic, with wide Mediterranean diversification.

Nanocyperetalia
Includes herb communities without Isoetids, with a late summer and fall flowering period, typical of soils with a long-lasting flooded stage.

Nanocyperion
Includes dwarf graminoid ephemeral vegetation, on soils poor in nutrients, flooded with acidic to neutral fresh water.

***Juncus bufonius-Laphangium luteoalbum* community**
Herbaceous community with a late-summer development, formed by small fleeting therophytes and geophytes, which grow in temporarily flooded ground.

Its most representative species are *Laphangium luteoalbum*, *Juncus bufonius* and *Lythrum hyssopifolia*, among others.

Distribution: all islands.

Verbenion supinae
Groups pioneer communities that develop in late summer and fall, on soils periodically flooded with eutrophic, oligosaline water, sometimes slightly nitrified. Mediterraneo-Atlantic and continental regions.

Verbenetum supinae
This is an amphibious, pioneer therophytic community that develops in late summer or fall. It grows in bare soils that are temporarily flooded by eutrophic water as at dam and pond edges that dry up at low water levels.

Verbena supina is frequently accompanied by some transitory therophytes and geophytes (*Laphangium luteoalbum*, *Cyperus longus*, and *Lythrum hyssopifolia*) among other herbs.

Distribution: Endemic to C, T.

Magnocarici elatae-Phragmitetea australis (Fig. 6.9)
It groups perennial aquatic vegetation that colonizes the edge of fresh and brackish water bodies. It is characterized by helophytes rooting in temporary or always submerged soil. The communities are dominated by perennial graminoids, sedges, forbs and herbs. Cosmopolitan.

Magnocaricetalia elatae
Communities of tall fresh water sedges. Eurosiberian and Mediterranean.

Magnocaricion elatae
Communities of sedges typical of mesotrophic to dystrophic, frequently peaty, soils.

Fig. 6.9 *Magnocarici elatae-Phragmitetea australis*. Charca de Maspalomas. Gran Canaria

Mentho-Caricetum calderae (Fig. 6.10)
This is a community that grows at stream and pond edges in the Supramediterranean belt of Tenerife, between 2100 and 2400 m a.s.l.

It is characterized by *Mentha longifolia* and *Carex paniculata* subsp. *calderae*, the latter forming typical dense tufts in the landscape. They are marginally accompanied by plants of the typical Teide broom shrubland, other specific companions being rare.

Distribution: Endemic to T.

Fig. 6.10 *Mentho-Caricetum calderae*. Las Cañadas. Tenerife

Nasturtio officinalis-Glyceretalia fluitantis (Fig. 6.11)
Includes pioneer communities of low rhizomatous helophytes, growing at the edge of calm or running fresh water, or semi-flooded soils.

Rorippion nasturtii-aquatici
Groups prostrate communities at the edge of nutrient-rich fresh water.

Helosciadietum nodiflori
This is a small-size helophytic community growing in wetland habitats flooded by calm or sluggish shallow eutrophic waters that are subject to a very low water level. They are common in ravine beds, ponds and channels.

It is characterized by *Apium nodiflorum, Rorippa nasturtium-aquatícum,* and *Veronica anagallis-aquatica*.

Distribution: P, G, T, C.

Phragmitetalia australis
Groups communities of tall helophytic graminoids (*Cyperaceae, Graminae, Juncaceae, Typhaceae*), from the edge of fresh water bodies. They make up the strip of helophytes nearest the water. Cosmopolitan.

Fig. 6.11 *Nasturtio officinalis-Glyceretalia fluitantis*. Barranco del Infierno. Tenerife

Phragmition australis
Alliance with a wide Eurosiberian distribution.

Bolboschoenenion maritimi
It includes reed communities of shallow fresh water, tolerant to desiccation.

Cypero alopecuroidis-Bolboschoenetum maritimi
This is a helophytic community characterized by *Cyperus alopecuroides* and *Scirpus maritimus* subsp. *maritimus*, the latter being more frequent. It has been observed on clayey pond-edges on Tenerife, accompanied by *Paspalum distichum* and *Panicum repens*.

Distribution: T.

Phragmitenion australis
It groups communities of common reed and bulrush of deep waters, sensitive to desiccation.

***Arundo donax* community** (Fig. 6.12)
Physiognomic community characterized by the large-size helophyte *Arundo donax*, which grows in ravine beds with some running water or with a high water table most of the year, forming dense thickets. *A. donax* was introduced in the islands for cultivation and it expanded very quickly, occupying mainly the biotope of the mainly felled Canary willow community (*Rubo-Salicetum canariensis*).

Ageratina adenophora and *Rubus ulmifolius* also grow in the community, over all in zones in contact with laurel forest. In places where water still runs, like at edges of irrigation channels and ravines, cane coexists with spontaneous stands of *Colocasia esculenta*.

Distribution: F, C, T, G, P, H.

Fig. 6.12 *Arundo donax* community. Vallehermoso. La Gomera

***Colocasia esculenta* community** (Fig. 6.13)
This is a physiognomic community characterized by the introduced eastern tropical herbaceous perennial *C. esculenta* ("taro", although known in the Canary Is. as "ñame"), a hydrophyte with characteristic big heart-shaped leaves and edible tubers. It forms clumps along anthropized streams, pond and puddle edges. Its expansion has been favored by cultivation.

Distribution: C, T, G, P.

***Cyperus eragrostis* community**
This herbaceous perennial, introduced from America, forms characteristic clumps in wet ravine beds and pond edges.

Distribution: C, T, G, P.

***Cyperus involucratus* community**
This tropical-African, introduced species can be seen growing in anthropized wet ravine bottoms, where it stands out because of its tall slim stems and radial leaves, resembling an open umbrella. It was previously identified as *C. alternifolius* community, but now *C. involucratus* includes *C. alternifolius* Baker, not *C. alternifolius* (Willd.) Endl.

Distribution: C, T, G, P.

***Typha domingensis* community**

This worldwide spread helophyte in temperate and tropical regions forms clumps in fresh water flooded ravine beds, puddles, pond edges and channels. Sometimes it is accompanied by *Phragmites australis*.

Distribution: F, C, T, G.

Fig. 6.13 *Colocasia esculenta* community. Barranco de Afur. Tenerife

6.2.3 Coastal Halophilous and Sand Dune Vegetation

6.2.3.1 Vegetation of Coastal Sand Dunes

Cakiletea maritimae

Includes pioneer halo-nitrophilous annual forb communities of shore strand lines, beaches, and coastal sand dunes. Holarctic.

Cakiletalia integrifoliae

Groups the European-Atlantic, North-Atlantic, and Baltic communities.

Cakilion maritimae

Groups the Mediterranean halo-nitrophilous annual forb communities.

Salsolo kali-Cakiletum maritimae (Fig. 6.14)

Halo-psammophilic and nitrophilous association mainly formed by warm medium-size therophytes sometimes tending towards perennial. On sandy beaches and sandy or pebbled seashores it usually forms a more or less continuous fringe coinciding with deposits of organic matter thrown up by the sea, especially during storms caused by spring tides. This is an association poor in species representing the natural vegetation of these environments. The intense human disturbance of the vast majority of beaches and sandy shores favors this community in secondary sandy locations away from the high-tide line. In the Canaries, although it is at its distribution limit, the association can be found on several beaches.

Characteristic species: *Cakile maritima, Chamaesyce peplis, Glaucium flavum, Polygonum balansae, P. maritimum, Salsola kali* and *Suaeda maritima*.

Distribution: L, F, C, T.

Fig. 6.14 *Salsolo kali-Cakiletum maritimae*. Fuerteventura. (Ph: PLP)

Polycarpaeo niveae-Traganetea moquinii
Phytosociological class that groups open psammophilous communities of perennial grasses, dwarf scrub or shrubs in the Canary Islands (deviant in Madeira) and the Oceanic Sahara, growing on littoral or inland deep regosols, mobile and fixed dunes subject to the hyperarid and arid Inframediterranean bioclimate.

Zygophyllo fontanesii-Polycarpaetalia niveae
Single order.

Polycarpaeo niveae-Euphorbion paraliae
Open communities of perennial grasses and small chamaephytes, colonizing mobile unstable littoral sands of the hyperarid and arid Inframediterranean belts. Canarian and Oceanic-Saharan distribution.

Euphorbio paraliae-Cyperetum capitati (Fig. 6.15)
This is an association settling on mobile embryo dunes and other more or less unstable flat sandy substrates, influenced by the marine saline atmosphere.

It is mainly characterized by *Cyperus capitatus* and *Euphorbia paralias*, but *Medicago marina*, *Polygonum balansae* var. *tectifolium* and *Polygonum maritimum* also participate.

On La Gomera it is rare, with a local distribution in Punta Llana; on Tenerife it is also limited in a fragmentary way to the littoral sands of the southern coast; but it is well distributed on Gran Canaria, Fuerteventura and Lanzarote.

The subass. *typicum* grows in the most unstable sand. The subass. *zygophylletosum fontanesii* has been proposed for more fixed sand, and is differentiated by *Zygophyllum* (*Tetraena*) *fontanesii* and *Polycarpaea nivea*. The subass. *atripliceto-sum glaucae* shows halo-nitrophilous behavior, being differentiated by *Atriplex glauca* and *Senecio grex crassifolius*.

Distribution: L, F, C, T, G.

Fig. 6.15 *Euphorbio paraliae-Cyperetum capitati*. Coastal sands in Fuerteventura. (Ph: PLP)

Polycarpaeo niveae-Lotetum lancerottensis (Fig. 6.16)
This is a nitro-psammophilous community developed on sandy-rocky or compact sandy soils, of variable depths, characterized by chamaephytes in open or disturbed areas of the Canary sabulicolous communities (*Polycarpaeo niveae-Traganetea moquinii*).

Polycarpaea nivea, and several endemic *Lotus* (Sect. *Pedrosia*) species (*L. arinagensis*, *L. kunkelii*, *L. lancerottensis* and *L. sessilifolius*) are characteristic, being insular variants. The subass. *ononidetosum hesperiae,* endemic to Lanzarote and Fuerteventura, is characterized by the chamaephyte *Ononis hesperia* which grows on sandy soils ("jables") in the process of fixation (torripsamments) formed from calcareous sand of organic marine origin.

Additionally, the *Cyperus capitatus* facies has been described growing in humid soils, along with the halophilous *Zygophyllum fontanesii* facies, and in Fuerteventura the nitro-halophilous variant with *Salsola divaricata.*

Distribution: L, F, C, T, G.

Fig. 6.16 *Polycarpaeo niveae-Lotetum lancerottensis*. El Médano. Tenerife

Traganion moquini
Chamaephytic and nanophanerophytic psammophilous communities typical of littoral and inland sand dunes of the hyperarid and arid Inframediterranean belts. Canarian and Oceanic-Saharan distribution.

Frankenio capitatae-Zygophylletum gaetuli (Fig. 6.17)
This is a chamaephytic scrub that settles on microdunes of a sea-wind swept narrow platform at Punta de Jandía (Fuerteventura).

It corresponds to the terminal community of the hyperarid Inframediterranean psammophilic edaphoxerophilous vegetation series of the *Zygophyllum* (*Tetraena*) *gaetulum* scrub on microdunes of Fuerteventura: *Frankenio capitatae-Zygophyllo gaetuli sigmetum.*

Its characteristic species are: *Frankenia capitata*, *Polycarpaea nivea*, *Zygophylllum fontanesii* and *Z. gaetulum*. Besides these, some species that occur widespread on the island, like *Launaea arborescens* and *Salsola vermiculata,* also frequently grow there.

Distribution: Endemic to F.

Fig. 6.17 *Frankenio capitatae-Zygophylletum gaetuli*. Punta de Jandía. Fuerteventura

Polycarpaeo niveae-Pulicarietum burchardii

This is a small chamaephytic halo-psammophilic scrub growing on mobile sands subject to marine spray at some isolated littoral sites (La Pared, Islote de Cofete, Punta de Jandía) on Fuerteventura, within the hyperarid and arid Inframediterranean bioclimate.

Its characteristic taxa are *Pulicaria burchardii* subsp. *burchardii, Ononis hesperia, Polycarpaea nivea, Frankenia laevis* and *Lotus lancerottensis* among others, which are commonly accompanied by some species of the *Kleinio-Euphorbietea* like *Lycium intricatum, Kickxia heterophylla, Helianthemum canariensis* and others belonging to the *Pegano-Salsoletea* like *Launaea arborescens, Atriplex glauca* subsp. *ifniensis* and *Chenoleoides tomentosa.*

Distribution: Endemic to F.

Traganetum moquinii (Fig. 6.18)
This is a chamaephytic-nanophanerophytic psammophilous association, more or less nitrophilous, oligospecific, typical of littoral or high inland dunes, within the hyperarid and arid Inframediterranean belts.

It is characterized by a dense growth of *Traganum moquinii* usually accompanied at its edges by *Atriplex glauca* subsp. *ifniensis*, *Launaea arborescens, Polycarpaea nivea* or *Zygophyllum fontanesii*.

It is the terminal community of the hyperarid and arid Inframediterranean psammophilic edaphoxerophilous vegetation series of *Traganum moquinii* and *Atriplex glauca* subsp. *ifniensis* shrublands on dunes: *Tragano moquinii sigmetum*.

The subass. *zygophylletosum fontanesii* has been found in sheltered SE areas of the municipality of Arrecife, the capital of Lanzarote, and also the more nitrophilous variant with *Salsola vermiculata* occurs on this island.

Distribution: Endemic to L, F, C, T.

Fig. 6.18 *Traganetum moquinii*. Playa de Famara. Lanzarote

6.2.3.2 Coastal Halophilous Vegetation

Crithmo-Limonietea
Vegetation class that includes chasmo-comophytic, halophilic, and anemogenic permanent communities of littoral rocks and cliffs. They are formed by cushion-like as well as rosette chamaephytes, hemicryptophytes or graminoids. Atlantic-European and Mediterranean.

Crithmo-Limonietalia
Mainly includes endemic-rich communities, especially of the genus *Limonium*. Mediterranean.

Frankenio-Astydamion latifoliae
Includes halophilic and anemogenic littoral rupicolous chamaephytic communities. Endemic to Canarian-Madeiran biogeographic subregion. All Canary Islands.

Frankenio ericifoliae-Astydamietum latifoliae (Fig. 6.19)
Rupicolous association which colonizes lithosols of littoral rocks and cliffs subjected to steady saline sea-wind on windward shores of western islands, under the semiarid-dry ombrotype. In the arid littoral of south Tenerife and Gran Canaria it is replaced by *Frankenio ericifoliae-Zygophylletum fontanesii*.

This is a permanent community of small size and medium coverage with frequent cushion-like chamaephytes as well as rosette chamaephytes and hemicryptophytes.

In the association are frequent: *Astydamia latifolia, Crithmum maritimum, Frankenia ericifolia* and *Limonium pectinatum*.

The subass. *limonietosum imbricati* was described from the North of La Palma, growing on saline lithosoils. *Limonium imbricatum* is its differential species, sometimes acompanied by *Reichardia ligulata*.

Distribution: Endemic to C, T, G, P, H.

Fig. 6.19 *Frankenio ericifoliae-Astydamietum latifoliae*. Coast of Bajamar. Tenerife

Frankenio ericifoliae-Zygophylletum fontanesii (Fig. 6.20)
Rupicolous association of lithosols of littoral rocks and cliffs on the hyperarid and arid coast of south Gran Canaria and Tenerife, and all around the coasts of Lanzarote and Fuerteventura. It is found in contact with the climatophilous territory of

Inframediterranean *Euphorbia balsamifera* scrublands. On these extremely arid coasts the existing salinity supplied by sea-wind is increased because of the scarcity of washing episodes through rainfall. It allows the appearance within the community of some halophytes like *Zygophyllum fontanesii*, a species which also grows in a wide variety of saline habitats on the islands, accompanying the typical rocky aero-halophytes.

Representative plants are: *Atractylis preauxiana, Frankenia capitata, F. ericifolia, Limonium papillatum* var. *papillatum, L. pectinatum, Reichardia crystallina* besides *Z. fontanesii.*

Apart from the typical subassociation (subass. *zygophylletosum fontanesii*), the subass. *suaedetosum verae* has been described for Fuerteventura, and is well distributed in the north littoral. *Suaeda vera*, a common plant on saline soils, acts as differential species, and sometimes becomes dominant at heavily aero-saline sites.

Distribution: endemic to L, F, C, T.

Fig. 6.20 *Frankenio ericifoliae-Zygophylletum fontanesii*. Montaña Roja. El Médano. Tenerife

Sarcocornietea fruticosae

This is a class that groups coastal and inland chamaephytic or nanophanerophytic communities of salt marshes and salt-pans. They are formed by succulent chamaephyte communities able to tolerate long salt or brackish water flooding, or communities with esparto grass together with many endemics of *Limonium* only able to tolerate light temporary flooding. Mediterranean and Atlantic-European.

Sarcocornietalia fruticosae

Succulent chamaephytic and nanophanerophytic halophilic communities of shore and inland salt-marshes subject to prolonged flooding. Mediterranean and Cantabro-Atlantic territories.

Arthrocnemion macrostachyi
Coastal and inland halo-xerophilic communities not or briefly reached by tides. Mediterranean.

Arthrocnemenion macrostachyi
Includes communities without direct tidal control, structured by *Arthrocnemum macrostachyum.*

Zygophyllo fontanesii-Arthrocnemetum macrostachyi (Fig. 6.21)
This is an oligospecific permanent community, dominated by the succulent subfrutescent chamaephyte *Arthrocnemum macrostachyum*, settled at the upper level of the Canary salt marshes. It is established in coastal depressions, with strongly saline soils subject to sea-water circulation. At its upper distributional limit it can withstand long periods without direct tidal action, while at the lower limit, it can very exceptionally be immersed during spring tides.

Characteristic species: *A. macrostachyum* and *Zygophyllum fontanesii*. The latter is a species with wide tolerance to salinity, which grows in several littoral associations.

Distribution: Endemic to L, F.

Fig. 6.21 *Zygophyllo fontanesii-Arthrocnemetum macrostachyi*. Costa de Orzola. Lanzarote

Sarcocornion fruticosae
Coastal halophilic communities, temporarily flooded. Mediterranean and Cantabro-Atlantic territories.

Sarcocornienion perennis
Includes communities with slight tidal control, structured by *Salicornia fruticosa.*

Halimiono portulacoidis-Salicornietum perennis (Fig. 6.22)
This is a community mostly characterized by *Salicornia perennis*, a decumbent chamaephyte whose upright stem tops form a more or less dense lawn on frequently flooded coastal depressions. These depressions are subject to permanent sea-water circulation as a consequence of tidal oscillation. On high tide the community can resist short immersion periods. The soils always remain wet at shallow depths. The community occurs in the low-lying areas of the Canary salt marshes. Apart from *Salicornia perennis*, other species less frequent are: *Arthrocnemum macrostachyum, Halimione portulacoides*, and *Suaeda fruticosa*.

In Lanzarote the community grows on the coast east of Órzola, although it shows a greater development in the salt marsh near the resort La Santa Sport, within the municipality of Tinajo. On Fuerteventura it is found in the North, at Charco de Bristol. Besides the typical subassociation (*typicum*) the subass. *limonietosum canariensis* has been described, exclusive to the islet of Lobos, characterized by *Limonium ovalifolium* subsp. *canariense*. This subassociation grows at the lower level of the salt marsh and remains flooded throughout the year.

Distribution: Endemic to L, Lobos, F.

Fig. 6.22 *Halimiono portulacoidis-Salicornietum perennis*. Costa de Órzola. Lanzarote

Suaedion verae
Groups littoral and inland halo-nitrophilic associations, able to tolerate slight flooding, with a dominance of the succulent nanophanorophyte *Suaeda vera*. Mediterranean and Cantabro-Atlantic territories.

Frankenio capitatae-Suaedetum verae (Fig. 6.23)
This is a halophilic and scarcely nitrophilous association, distributed over the central and eastern islands. It develops in coastal depressions on incipient saline soils originating from rising brackish water or sea water filtration.

It is physiognomically characterized by *Suaeda vera*, a chamaephytic species with thick fleshy leaves, with a wide Mediterranean distribution. This species forms thickets in which the Canary or Canarian-Saharan species *Frankenia capitata* and *Zygophyllum fontanesii* appear sparsely.

Distribution: Endemic to L, F, C.

Fig. 6.23 *Frankenio capitatae-Suaedetum verae*. El Cotillo. Fuerteventura

***Zygophyllum fontanesii and Suaeda mollis* community**
A chamaephytic community mainly characterized by these two widespread species in halophilic environments. It develops in highly saline disturbed littoral places, and so it is common in abandoned saltpans.

It is well represented in Salinas del Janubio, on Lanzarote, and it is scattered along the E and SE coast of Gran Canaria.

Distribution: Endemic to L, C.

Thero-Salicornietea
This is a class that includes pioneer annual coastal and continental halophytic or halo-nitrophilous communities that develop on tidal mud-flats and temporary flooded salt marshes and salt-pans. They are characterized by saltwort (*Salicornia*) and other therophytic succulent plants. Holarctic.

Thero-Suaedetalia
Littoral and inland halophytic communities.

Thero-Suaedion
Littoral and inland halo-nitrophilous communities.

Cressetum villosae
This is an annual pioneer halo-nitrophilic association that develops in summer and fall on saline soil. It grows at the edges of the depressions or littoral drainage channels flooded in winter and spring and dry in summer, where *Frankenio boissieri-Suaedetum spicatae* is established. Along these edges there is accumulation of organic matter from decomposition of seagrasses, and also saline upwelling. Coastal human activity and grazing also contribute to its nitrification.

It is a practically monospecific community characterized by *Cressa cretica* var. *villosa.* Up to now it has been found only in "Juncalillo del Sur", Gran Canaria.

Distribution: Endemic to C.

Frankenio boissieri-Suaedetum spicatae
This is a pioneer halo-nitrophilic association that develops in summer and fall on saline soil. It dwells at the bottom of depressions or littoral drainage channels flooded in winter and spring that dry up in summer, and in inland salt-marshes and edges or tails of reservoirs containing brackish water. It is established in arid Inframediterranean territories.

The community shows a poor floristic composition. *Suaeda spicata* is dominant, but generally accompanied by the chamaephyte *Frankenia boissieri* that acts as a differential species of similar communities in the Mediterranean area. Some common therophytes of saline environments like *Spergularia marina, Messembryanthemum nodiflorum* and *Patellifolia patellaris* are to be found growing within the association.

Distribution: Endemic to C, F.

6.2.3.3 Littoral and Riverine Thickets

Nerio oleandri-Tamaricetea
Phytosociological class that groups microforest, shrublands and tall grasslands in and along water courses, ramblas (seasonal stream beds), or places temporary flooded with fresh or brackish water, typical of the Mediterranean, Saharo-North-Arabian and Irano-Turanian regions.

Tamaricetalia
Single order in the Mediterranean region.

Tamaricion africanae
Fresh more or less hard water Tamarisk communities.

Atriplici ifniensis-Tamaricetum canariensis (Fig. 6.24)
This is a species-poor community, characterized by *Tamarix canariensis*, typical of backshores, ravine mouths, salty moist ravine beds, and endorheic flatlands generally near the coastline, subject to variable salinity, within the arid and semiarid Inframediterranean belts.

It generally forms a dense and intricate monospecific thicket, with a dark understory, and soil covered by a thick layer of slowly decomposing organic matter. A few species are able to growth at the edges, outstanding among these are *Atriplex glauca subsp. ifniensis, Lycium intricatum, Salsola divaricata* and *Schizogyne sericea.*

It is the terminal community of the arid and semiarid brackish and hard water edaphohygrophilous Western Canary vegetation series of *Tamarix canariensis* microforest with *Atriplex glauca* subsp. *ifniensis* in ramblas and on shores: *Atriplici ifniensis-Tamarici canariensis sigmetum.*

Distribution: C, T, G, P.

Fig. 6.24 *Atriplici ifniensis-Tamaricetum canariensis*. Barranquera de Valle de Guerra. Tenerife

Tamaricion boveano-canariensis
Halophillic Tamarisk communities.

Suaedo verae-Tamaricetum canariensis (Fig. 6.25)
This is a species-poor community, characterized by *Tamarix canariensis*, typical of backshores, ravine mouths, salty moist ravine beds, and endorheic flatlands near the coastline, within the hyperarid and arid Inframediterranean belts. It is subject to variable hydromorphy and salinity, the latter higher than those of the Atriplici-*Tamaricetum canariensis.*

It is the terminal community of the hyperarid and arid halophilic edaphohygrophilous Central and Eastern Canary vegetation series of *Tamarix canariensis* with *Suaeda vera* on littoral and ramblas*: Suaedo verae-Tamarici canariensis sigmetum*.

The community, favored by the hyperarid and arid ombrotype of Fuerteventura and the south of Gran Canaria, penetrates inland up the water course beds, provided they have brackish water filtering through at the surface or at shallow depths. This moisture favors *Suaeda vera*, a species that also participates in the salt marshes. In less saline soils, *T. canariensis* is interspersed with the palm tree community (*Periploco-Phoenicetum canariensis*).

Distribution: C, F.

Fig. 6.25 *Suaedo verae-Tamaricetum canariensis*, growing in an arid ravine in Fuerteventura. (Ph: PLP)

6.2.4 Chasmophytic, Epiphytic and Scree Vegetation

6.2.4.1 Rock Crevice Chasmophytic Vegetation

Adiantetea

Includes comophyte and chasmophyte bryo-pteridophytic communities, with few spermatophytes, subject to a constant flow, drip or splashing of water rich in calcium carbonate and bicarbonate, which can form calcareous tuff and travertines. Wide Mediterranean, submediterranean Temperate and arid Tropical distribution.

Adiantetalia capilli-veneris
Single order in our territory.

Adiantion capilli-veneris
Communities of very moist rocky crevices and calcareous tuffs with abundant water flow.

Adianto-Hypericetum coadunati
This is a community related to the *Eucladio-Adiantetum capilli-veneris*. It grows within dry and subhumid Thermo- and Mesomediterranean belts on the north slope of Gran Canaria, occupying also damp but sunnier places.

The chamaephyte *Hypericum coadunatum* is the most characteristic species, together with *Adiantum capillus-veneris, Epilobium parviflorum* and *Cystopteris fragilis* among the vascular plants. Apart from these vascular plants, also several moss species grow there.

Distribution: Endemic to C.

Camptolometum canariensis (Fig. 6.26)
Sciophilous community with a wide distribution in Gran Canaria, growing on overhanging porous cliffs and dripping water crevices, with a preferably northern orientation. It is a species-poor community with an appearance of hanging shreds in such biotopes.

It is almost exclusively characterized by *Camptoloma canariensis*. *Adiantum capillus-veneris* and *Samolus valerandi* are occasionally involved.

Distribution: Endemic to C.

Fig. 6.26 *Camptolometum canariensis*. Cliffs of the West of Gran Canaria

Eucladio-Adiantetum capilli-veneris (Fig. 6.27)
This is a sciophilous community characterized by hydrophytic vascular plants and bryophytes, which grows in wet or oozing caves, water channels, crevices or cavities in vertical or overhanging walls, with percolation of water over the rock surface in very moist surroundings.

They are frequently associated to deposits of calcium carbonate, and the lower moss carpet is often encrusted in chalk tuff.

Adiantum capillum-veneris and *Samolus valerandi* are the most characteristic vascular species. Among the mosses: *Eucladium verticillatum* and *Gymnostomum aeruginosum*.

Distribution: L, C, T, G, P.

Fig. 6.27 *Eucladio-Adiantetum capilli-veneris*. Caldera de Taburiente. La Palma

Asplenietea trichomanis
Chasmophytic communities typical of rock crevices receiving water from rain. They are formed by non-nitrophilous pteridophytes, bryophytes, and spermatophytes (small vivacious hemicryptophytes, and chamaephytes). Widely diversified. Holarctic.

Notholaeno marantae-Cheilanthetalia maderensis
Mainly pteridophytic ultra-mafic rupicolous communities. Mediterranean.

Cheilanthion pulchellae (Fig. 6.28)
Includes Canary chasmophytic ultra-mafic rock (igneous substrates rich in iron and magnesium) communities, mainly consisting of small-sized ferns. Arid and dry Infra- and Mesomediterranean territories. All islands.

Adianto pusilli-Cheilanthetum pulchellae

This is an association dominated by small ferns, which grows in rock crevices and small cavities typical of Infra- Thermo- and Mesomediterranean lava flows and other rocky places, typically with a local northern aspect.

Characteristic species: *Adiantum reniforme* subsp. *pusillum, Cheilanthes pulchella, Notholaena marantae* subsp. *subcordata, Cosentinea vellea.* On Tenerife *Monanthes brachycaulos* is a companion species.

Distribution: Endemic to (L), C, T, G, P, H.

Fig. 6.28 *Cheilanthion pulchellae*. Cumbres de Arico. Tenerife

***Cheilanthes guanchica* community**

This is a community described for lava rocks on El Hierro, preferentially distributed throughout southeast and southwest slopes within the semiarid to dry Infra- and Thermomediterranean belts.

It is characterized by *Cheilanthes guanchica, Cosentinia vellea, Notholaena marantae, Asplenium trichomanes,* accompanied by *Umbilicus gaditanus* (in nutrient-rich environments), and some plants of the *Aeonietea* like *Aeonium hierrense, Sonchus hierrensis* and *Aichryson punctatum.*

An *Aeonium spathulatum* community on the island was described by Stierstorfer 2005, but considering its other characteristic species (*Cheilanthes guanchica* and *Cosentinia vellea*), and the distribution pattern, it is difficult to differentiate it from the *Cheilanthes* community, if not for the presence of this *Aeonium.*

These communities mentioned show similarity with the other communities of the *Cheilanthion pulchella* described for the archipelago and an overall study of them is required.

Distribution: H

***Notholaena marantae* community** (Fig. 6.29)

This is a rupicolous community originally described from lava flows in the South of La Palma.

It is characterized by *Notholaena marantae* subs. *subcordata* and other rupicolous plants of the *Asplenietea,* like *Cosentinia vellea,* or the *Aeonietea*, like *Aeonium spathulatum, Sonchus hierrensis, Todaroa aurea* and *Tolpis laciniata*, apart from other plants of diverse phytosociological ascriptions. It seems to correspond to a community of the *Cheilanthion pulchellae* masked within a typical rupicolous vegetation complex on a lava flow.

The community is present in the western islands, but a detailed study is required.

Distribution: Endemic to C, T, G, P, H.

Fig. 6.29 Notholaena marantae community. Lava flows of Santiago del Teide. Tenerife

Parietarietea judaicae

Perennial nitrophilous and subnitrophilous chasmophytic communities of rural, urban or pastoral environments. They are mainly formed by scapous hemicryptophytes, suffruticose chamaephytic and a few micro- and nanophanerophytic spermatophytes, and some small ferns, that grow on walls or at their bases, and in caves. Cosmopolitan.

Parietarietalia judaicae

Single order in our territory.

Cymbalario muralis-Asplenion quadrivalentis

Chasmophytic nitrophilous communities growing on shady or moist walls and caves characterized by *Cymbalaria muralis* and several small ferns. Extratropical cosmopolitan.

Adianto capilli-veneris-Parietarietum judaicae

This is a community that grows on regularly humid or oozing walls and slopes like those next to irrigation channels, bridges or roof drain spouts.

Cymbalaria muralis, Adiantum capillus-veneris, Parietaria judaica and *Erigeron karvinskianus* are characteristic species.

Distribution: Endemic to C, T.

Parietario judaicae-Centhrantion rubri
Chasmophytic nitrophilous communities growing on drier walls and in caves. Extratropical cosmopolitan.

Umbilico gaditani-Parietarietum judaicae (Fig. 6.30)
This is an herbaceous nitrophilous community, frequently on moist and shady places, that grows on terrace walls, urban plots, walls and cornices of old buildings, and compacted clayey-stony road and track cuttings.

Characteristic species: *Centranthus ruber, Cymbalaria muralis, Erigeron karvinskianus, Parietaria judaica* and *Umbilicus gaditanus*, among others.

Distribution: Endemic to L, C, T.

Fig. 6.30 *Umbilico gaditani-Parietarietum judaicae*. Acentejo. Tenerife

Anomodonto viticulosi-Polypodietea cambrici
Ombrophilous and sciophilous comophytic and epiphytic communities made up of small bryophytes, ferns and a few spermatophytes. They mainly grow in shaded forest environments on trunks, branches, moist earthy rocks and slopes. Holarctic.

Anomodonto viticulosi-Polypodietalia cambrici
Single order.

Polypodion cambrici
Mediterranean and Europe-Atlantic associations characterized by the *Polypodium cambricum* complex.

Bartramio strictae-Polypodienion cambrici
Communities that grow in environments poor in basic nutrients.

Davallio canariensis-Polypodietum macaronesici (Fig. 6.31)
This is a bryophytic-pteridophytic, mainly epiphytic community, which grows on branches, forks and coarse bark of trees in the laurel forest, within the dry to humid Infra- to Mesomediterranean belts (c).

The most characteristic species are the ferns *Davallia canariensis* and *Polypodium macaronesicum*, always accompanied by several bryophytes.

Distribution: C, T, G, P, H.

Fig. 6.31 *Davallio canariensis-Polypodietum macaronesici*. Anaga laurel forests. Tenerife

Hymenophyllion tunbrigensis
Mainly epiphytic associacions characterized by *Hymenophyllum*. Mediterranean and Atlantic Europe.

Hymenophylletum tunbrigensis (Fig. 6.32)
This is a local and rare bryophytic-pteridophytic community that grows on stem-base bark and lignotubers of *Erica platycodon* within the community *Ilici-Ericetum platycodonis* in the subhumid and humid Thermo- and Mesomediterranean belts (c) (Tenerife: Chinobre, Pijaral; Gomera: Carbonero-Inchereda and Zarza-Epina-Teselinde). The undergrowth habitat of the *Ilici-Ericetum* is one of the most humid environments of the laurel forest and it allows the development of highly moisture-demanding species.

The characteristic fern is *Hymenophyllum tunbrigensis* together with several bryophytes, *Frullania tamarisci, Plagiochila exigua, P. punctata* and *Microleujena molleri* being the most frequent.

Distribution: T, G.

Selaginello denticulatae-Anogrammion leptophyllae
Edaphically indifferent communities growing on shady earthy slopes. Wide Mediterranean distribution.

Fig. 6.32 *Hymenophylletum tunbrigensis*. Anaga laurel forests. Tenerife

Selaginello denticulatae-Anogrammetum leptophyllae (Fig. 6.33)
This is a small-sized bryophytic-cormophytic community that colonizes shady earthy substrates forming carpets, mainly in territories of the subhumid Infra- to Mesomediterranean belts (c). It also grows in humid cavities of lava flows in other belts.

The most characteristic species are the ferns *Selaginella denticulata* and *Anogramma leptophylla* which are accompanied by several bryophytes. Some spermatophytes like *Parietaria debilis*, *Romulea columnae* subsp. *grandiscapa* or *Aichryson tortuosum* (Lanzarote) may also be present.

Distribution: Temperate and Mediterranean. All Canary Islands.

Fig. 6.33 *Selaginello denticulatae-Anogrammetum leptophyllae*.Vueltas de Taganana. Tenerife

Aeonietea (Figs. 6.34 and 6.35)
Permanent comophytic and chasmo-comophytic communities, made up of succulent and rossete-succulent chamaephytes of the genus *Aeonium, Aichryson* and *Monanthes* (*Crassulaceae*), some *Asteraceae* and several endemic species from other families. They mainly grow on rocks, walls and wall crevices. Canarian-Madeiran biogeographic subregion.

Soncho acaulis-Aeonietalia
Single Order. Canaries and Madeira.
Distribution: All islands.

Soncho acaulis-Aeonion
This alliance groups comophytic and chasmo-comophytic communities, with a dominance of several kinds of succulent, rosular, and marcescent chamaephytes. They colonize cliffs, recent lava flows and other rocky surfaces of varying inclination and cohesiveness, but can also grow in anthropogenic locations, such as on walls and roofs. They occur from hyper-arid Inframediterranean to subhumid Thermomediterranean.

Aeonium species usually preponderate, but also other succulent chamaephytic *Crassulaceae* stand out (*Aichryson* and *Monanthes*), as well as a number of rosular chamaephytic *Compositae* (*Babcockia*, *Hypochoeris*, *Chrysoprenanthes*, *Sonchus, Sventenia*, *Tolpis*, *Vieraea*, etc.). Endemic of the Canary Province.

Distribution: All islands.

Fig. 6.34 *Aeonietea*. Cliffs of Interián. Tenerife

Fig. 6.35 *Aeonietea*. Casas de la Cumbre. Anaga. Tenerife

Aeonietum canariensis (Fig. 6.36)
This community occurs from Teno along the north slopes of Tenerife and reaches the south slopes of Anaga. It mainly grows in dry Inframediterranean to subhumid Thermomediterranean belts particularly with northern exposure, on cliffs and ledges with small earthy surface accumulations, in the potential territory of dry and humid laurel forest. In the Anaga Mountains it yields to the *Aeonietum lindleyi* towards the lower belts, and to the *Aeonietum cuneati* towards the upper, more rainy belts.

The most characteristic plants of this association, not too rich in species, are *Aeonium canariense* subsp. *canariensis* and *Monanthes anagensis*. More or less common are *Aeonium ciliatum*, *Aichryson laxum*, *Habenaria tridactylites*, *Monanthes polyphylla*, *Monanthes brachycaulos*, *Sonchus acaulis* and *Sonchus congestus*. In some nitrogen-rich places *Sonchus congestus* is abundant.

Distribution: Endemic to T.

Aeonietum christii (Fig. 6.37)
A rupicolous community, endemic to El Hierro, well developed on cliffs and earthy rock surfaces, with a preferably direct exposure to NE trade-winds. It grows in upper-semiarid Inframediterranean till dry-subhumid Thermomediterranean belts, within the climatophilous territories of juniper woodland and laurel forest.

Characteristic taxa are *Aeonium canariense* subsp. *christii* (*Aeonium longithyrsum*), *Habenaria tridactylites*, *Sonchus hierrensis* and *Tolpis proustii*.

Distribution: Endemic to H.

Fig. 6.36 *Aeonietum canariensis*. Chamorga. Anaga. Tenerife

Fig. 6.37 *Aeonietum christii*. San Andrés. El Hierro

Aeonietum christii-davidbramwellii

This is a chasmo-comophytic permanent community characterized by the abundance of *Aeonium canariense* subsp. *christii* (=*Aeonium palmense*), widely distributed in La Palma. It is a species-poor community, which grows in walls, cliffs and cliff ledges within the semiarid to humid Infra- to Mesomediterranean belts with or without clouds, climatophilous territory of *Euphorbia canariensis* shrubland, juniper woodland and laurel forest.

Characteristic species are *Aeonium canariense* subsp. *christii* (=*A. palmensis*), *A. davidbramwellii* (=*A. ciliatum* subsp. *praegeri*), *Carlina falcata, Lobularia palmensis, Reichardia ligulata* and *Sonchus hierrensis* var. *benehoavensis.*

Apart from the typical subassociation, the subass. *aeonietosum goochiae* has been described, characterized by *A. goochiae*, especially for humid and shaded environments in the Northern half of the island. These appear within the subhumid and humid Thermo- and Mesomediterranean belts (laurel forest and pine forest territory). The subass. *aeonietosum nobile*, characterized by the red-flowered *A. nobile*, shows thermophilic behavior and occurs on the western and eastern slopes of the island within the semiarid and dry Infra- and Thermomediterranean belts.

Distribution: Endemic to P.

Aeonietum cuneati (Fig. 6.38)

This community grows on earthy cliffs and ledges in windward Tenerife within the subhumid (and humid) Thermo- and Mesomediterranean bioclimatic belts of the most humid ridges of Anaga and Teno, although it is scarce in the latter. It succeeds the *Aeonietum canariense* in altitude, and is typical of the highest altitude of humid laurel forest (Ixantho-Laurion), particularly in potential territory of the *Ilici-Ericetum platycodonis*.

The most frequent plants of this association are *Aeonium ciliatum, Phyllis nobla, Aichryson laxum*, and *Monanthes laxiflora*, generally growing together with some common comophytes like *Davallia canariensis* and *Polypodium macaronesicum*, and other plants of the forest mantle and fringe like *Hypericum glandulosum, Isoplexis canariensis, Canarina canariensis, Scrophularia smithii,* etc.

Distribution: Endemic to T.

Fig. 6.38 *Aeonietum cuneati*. Chamorga. Anaga. Tenerife

Aeonietum lancerottensis (Fig. 6.39)
A chamaephytic-nanophanerophytic community, endemic to Lanzarote, that grows within the upper-arid to semiarid Infra- and Thermomediterranean belts, characterized by the pink flowered *Aeonium lancerottense.* It grows on cliffs, steep rocks and recent malpaises. In the latter, in the central zone of the island between La Vegueta and Mozaga belonging to Timanfaya lava flows, it colonizes crevices, and earthy surfaces of rock, and forms a oligospecific vascular plant community, accompanied by a rich lichenic vegetation mainly characterized by *Ramalina bourgeauana* and *Tornabea scutellifera.* In the North of the island the community contains more species, like *Carlina salicifolia* subsp. *lancerottensis* and *Helichrysum gossypinum*, among others.

The community becomes more species-rich through the participation of nitrophilous plants, and occasionally forms a dense scrub, in some disturbed malpaises near to tracks and roads, old fig cultivation areas and in proximity to urban areas.

Distribution: Endemic to L.

Fig. 6.39 *Aeonietum lancerottensis*. Haría. Lanzarote

Aeonietum latifolii

Chasmo-comophytic community that grows preferably in the much washed clayey scarps of the humid north zone of La Gomera, within the climatophilous territory of juniper woodland and laurel forest.

It is characterized by *Aeonium canariense* subsp. *latifolium* and its most constant companion species is the island's endemic *Sonchus ortunoi*.

Distribution: Endemic to G.

Aeonietum lindleyi (Fig. 6.40)

A chamaephytic-nanophanerophytic community, endemic to Tenerife, which occurs in semiarid Infra- and Thermomediterranean belts of Anaga, and extends along the north coast of the island until Puerto de la Cruz. It thrives on ancient basaltic rocks, more or less steep, small scarps, and sunny xeric cliffs. The community often occurs adjacent to the remnants of the typical climatophilous vegetation of these territories, *Euphorbia canariensis* shrubland (*Periploco-Euphorbietum canariensis*) or *Juniperus turbinata* subsp. *canariensis* woodland (*Junipero-Oleetum cerasiformis*), and it gives rise to mixed patches. *Opuntia* spp. is sometimes abundant within the community at disturbed grazed areas.

The association is well characterized by the Tenerife endemic *Aeonium lindleyi*, accompanied by other local endemics like *Aeonium haworthii*, *Sideritis dendrochahorra* and *Sonchus tectifolius*.

Apart from the typical subass. *aeonietosum lindleyi*, the subass. *aeonietosum canariensis* has been described. It is specific of shady places and small earthy landings or ledges locally facing north. *Aeonium canariense* subsp. *canariensis* that has its optimum on upwards slopes in more humid belts under the t-w.c. influence, acts as differential.

Distribution: Endemic to T.

Fig. 6.40 *Aeonietum lindleyi*. Taganana. Tenerife

Aeonietum undulato-percarnei

This is a community initially described as anthropogenic, growing on old house roofs (*e.g*., in Teror and San Mateo) or other rupicolous urban or rural biotopes in the semiarid-subhumid Infra-Mesomediterranean belts on the north slope of Gran Canaria. But it also occurs on natural rocks and ledges in the North of Gran Canaria, where it is widespread. It has its optimum between 300 and 850 m, within the climatophilous territories of *Olea* woodland and laurel forest.

Besides *Umbilicus gaditanus*, the association is characterized by *Aeonium percarneum* and *A. undulatum*, endemic to this island, widely distributed from coast to summit. In some localities of the NE *A. arboretum* (=*A. manriqueorum*) is common, giving rise to small urban populations. The community is species-poor at its lowest limit but with a regular presence of *A. percarneum*. As the community penetrates into the potential territory of laurel forest the more water demanding *A. undulatum* becomes more widespread. The presence of *Umbilicus gaditanus* indicates nitrophilization.

Distribution: Endemic to C.

Aeonietum valverdensis

This is a chamaephytic and small-sized nanophanerophytic permanent community of chasmophytes that grow on pahoehoe lava flows at the Southern side of El Hierro, within the arid and semiarid Inframediterranean belts.

Characteristic plants of the association are *Aeonium valverdense, Cheilanthes maderensis* and *Phagnalon purpurascens.*

On these lightly sloping lava flows the community is immersed in a rupicolous vegetation complex with many anemochore pioneer plants of the *Kleinia-Euphorbietea* like *Kleinia neriifolia, Periploca laevigata,* and of the *Pegano-Salsoletea* like *Schizogyne sericea* and *Rumex lunaria*. Among the companion species, apart from several therophytes, the chamaephyte *Micromeria hyssopifolia* and the hemicryptophytes *Hyparrhenia sinaica* and *Aristida adscensionis* subsp. *coerulescens* also stand out.

Distribution: Endemic to H.

Aeonietum virginii

This is a chasmo-comophytic community characterized by *Aeonium canariense* subsp. *virgineum*. It grows on cliffs, natural and artificial walls, cliff ledges and shallow earthy accumulations, preferably with a direct exposure to humid NE trade-winds. It occurs from 300 to 800 m, from upper-semiarid Inframediterranean until subhumid Thermomediterranean (c), within the climatophilous territories of thermo-sclerophyllous woodland and laurel forest.

In the lower zones, the community is almost mono-specific, and it becomes enriched with other rupicolous taxa as it gains in altitude: *Aeonium undulatum, Carlina salicifolia, Davallia canariensis, Polypodium macaronesicum, Sonchus acaulis, Sonchus congestus*, etc. In western Gran Canaria it contains some local endemics like *Lotus callis-viridis* or *Sonchus brachylobus*. It has a wide distribution in the valleys of the northern half of the island, from Teror ravine in the E to the inland cliffs of Tasartico and Horgazales in the W, passing the Tamadaba massif and Andén Verde. It is especially abundant in the ravines of Silva, Azuaje, La Virgen and Moya.

Distribution: Endemic to C.

Aeonietum viscati

A rupicolous chasmo-comophytic association, endemic to the semiarid Inframediterranean and semiarid-dry Thermomediterranean belts of the NE half of Gomera, within the climatophilous territories of *Euphorbia canariensis* shrubland and juniper woodland.

It is characterized by *Aeonium lindleyi* subsp. *viscatum*, a vicariant subspecies to *A. lindleyi* subsp. *lindleyi* of Tenerife (some authors make these two taxa synonymous). Like the latter, it grows along with mature fragments of the climatophilous and edaphophilous communities of those territories (cardón shrubland and juniper woodland).

Distribution: Endemic to G.

Aeonio decoris-Atalanthetum pinnati

This is a rupicolous chasmo-comophytic community growing in crevices on the southern scarps of La Gomera, between 50 and 750 m. It grows in the Infra- and Thermomediterranean territories of potential *Euphorbia canariensis* shrubland and juniper woodland. The variant with *Aeonium saundersii* occurs on slightly eroded lava flows without fissures, and it often forms a seamless covering over the rock.

Characteristic species are *Aeonium decorum, Ceropegia krainzii, Sideritis gomerae* and *Atalanthus pinnatus* (= *Taeckholmia pinnata;* = *Sonchus leptocephalus)*, among others.

Distribution: Endemic to G.

Chrysoprenantho pendulae-Atalanthetum pinnatae

This is an association with an optimum in the semiarid Infra- and Thermomediterranean belts of Gran Canaria, preferably in north-facing places that are not too dry. It occurs on cliffs and rocky environments, mainly between 200 and 400 m, within the climatophilous potential of *Euphorbia canariensis* shrubland and thermo-sclerophyllous woodland, from which it receives some elements.

Among its characteristic species are the widely distributed *Allagopappus viscosissimus* (preferably with a southern distribution), *Chrysoprenanthes pendula* (*Prenanthes pendula*), *Ruta oreojasme* and *Atalanthus pinnatus* (this species is almost absent on the northern slopes), as well as local species like *Dendriopoterium menendezii* and *Descurainia artemisioides* limited to the Tamadaba Massif and Agaete valley.

Distribution: Endemic to C.

Parietarietum filamentosae (Fig. 6.41)

This is a species-poor association that grows in shaded collapsed debris and cavity mouths with an accumulation of crumbly soft rocky material, preferably in semiarid Thermomediterranean territory.

The main characteristic species is *Parietaria filamentosa*, that occurs on the SE slope of Tenerife with *Micromeria teneriffae* and *Pericallis lanata* (El Tablero, Valle de Güímar, Agache). The community is also found in Teno, and probably on La Palma and Gomera it constitutes insular geographic races.

Distribution: P, G, T.

Fig. 6.41 *Parietarietum filamentosae*. Ladera de Güímar. Tenerife

Pericallido lanatae-Sonchetum gummiferi (Fig. 6.42)
This is a dry-semiarid Thermomediterranean association, endemic to the eastern ravines of Tenerife, extending from the Agache district to the slopes of El Tablero. It preferably grows under a northern exposure, mostly in potential juniper woodland territory. On ascending in the Güímar Valley to the upper Thermomediterranean and particularly the Mesomediterranean belt (with a dominance of pine forest), the community yields to the *Aeonietum aizoi*.

Characteristic species are *Pericallis lanata, Sonchus gummifer* and other local endemics like *Monanthes minima* and *Crambe arborea.*

Distribution: Endemic to T.

Fig. 6.42 *Pericallido lanatae-Sonchetum gummiferi*. Barranco de Badajoz. Güímar. Tenerife

Phyllido viscosae-Aeonietum sedifolii
This is a rupicolous community typical of the northern and western slopes in the Teno region of Tenerife. It is particularly developed leeward of NE trade-winds, within the semiarid-dry Infra- Thermomediterranean belts, in a mosaic with the permanent community *Euphorbietum atropurpureae*.

It is characterized by *Phyllis viscosa*, *Aeonium sedifolium*, *A.* x *burchardii* (*A. pseudurbicum x sedifolium*) and *Tolpis crassiuscula*.

Distribution: Endemic to T.

***Polycarpaea smithii* community**
This is an oligospecific community that grows on steep porous strata of lapilli and alluvial material, which retain a certain amount of moisture, mainly in semiarid to subhumid Infra and Thermomediterranean belts. It is more or less common in the center and North of La Palma and rare on El Hierro.

It is characterized by the endemic *Polycarpaea smithii*. According to the companion species, such as *Adiantum capillus-veneris, Aeonium hierrense, Sonchus hierrensis* or *Tolpis laciniata,* it seems to occupy an intermediate position between the *Aeonietea* and *Adiantetea*, but further studies are required.

Distribution: Endemic to P, H.

Reichardio famarae-Helichrysetum gossypini

This is a chasmophytic association endemic to the vertical basaltic northern walls of the Famara cliffs on Lanzarote. It grows in rock crevices, and is characterized by *Reichardia famarae* and *Helichrysum gossypinum.*

This community may be in contact in Famara with the *Aichrysetum tortuosi*, but the latter has a comophytic root system exploiting superficial soil layers.

Distribution: Endemic to L.

Soncho congesti-Aeonietum holochrysi (Fig. 6.43)

This is a chasmo-comophytic association that grows in cracks, on small basaltic platforms, walls and on rocky slopes of roadsides, in the North coast of Tenerife, between 25–500 m, within the lower-semiarid Inframediterranean and dry Thermomediterranean belts. It usually contains nitrophilous species.

Sonchus congestus and *Aeonium arboreum* subsp. *holochrysum* are the main characteristics. Some other species of the *Aeonietea,* like *Davallia canariensis, Hypericum reflexum, Lobularia canariensis, Aichryson laxum, Aeonium ciliatum, Monanthes brachycaulos,* and *M. laxiflora,* can also occur, although in smaller numbers. The community shows nitrophilous affinities, as is clear from the composition of companion species in many stands in which several species of diverse nitrophilous communities also participate: *Rumex lunaria, Artemisia thuscula, Argyranthemum frutescens, Gonospermum fruticosum, Bituminaria bituminosa, etc.*

Distribution: Endemic to T.

Fig. 6.43 *Soncho congesti-Aeonietum holochrysi.* Acentejo. Tenerife

Soncho radicati-Aeonietum tabulaeformis (Fig. 6.44)
This is a chasmo-comophytic, Infra- and Thermomediterranean association that grows in the North of Tenerife, extending from Anaga to Teno Bajo. It occurs on windward cliffs and steep rocks below 800 m.

Characteristic taxa are *Aeonium tabulaeforme*, *Sonchus radicatus*, *Aeonium arboreum* subsp. *holochrysum* var. *holochrysum*, *Monanthes laxiflora*, *Monanthes polyphylla* and *Sonchus congestus*.

Besides the typical subass. *aeonietosum tabulaeformis*, the subass. *hypochoeridetosum oligocephalae* is distinguished on the moist cliffs of north Teno. It is differentiated by the local endemic *Hypochoeris oligocephala* and the fern *Adiantum capillus-veneris*.

Distribution: Endemic to T.

Fig. 6.44 *Soncho radicati-Aeonietum tabulaeformis*. Tierra del Trigo. Los Silos. Tenerife

Umbilico gaditani-Aeonietum urbici (Fig. 6.45)
This is a rocky community that grows in cliffs and scarps of the north slope of Tenerife, also found in old house roofs and other rupicolous urban or rural biotopes, from where it was firstly described. The old roofs of the city of La Laguna are a good example of that anthropogenic location. It preferably grows in semiarid and dry Infra- and Thermomediterranean areas.

Aeonium urbicum, Umbilicus gaditanus, and *Sonchus acaulis* are the most characteristic species which grow together with several other common rupicolus species of the North.

Distribution: Endemic to T.

Fig. 6.45 *Umbilico gaditani-Aeonietum urbici*. Roofs in La Laguna. Tenerife

Vieraeo laevigatae-Polycarpaeetum carnosae (Fig. 6.46)
This is an association spread over the N facing sea cliffs of El Fraile, and the SW facing cliff of Los Gigantes (Teno). Although it colonizes ombrophobe environments, seasonal waterfalls and the frequent trade and marine winds moisten the crevices in which it thrives in contact with *Euphorbia aphylla* scrub (*Ceropegio dichotomae-Euphorbietum aphyllae*). *Adiantum capillus-veneris* also grows in this association at the foot of the Los Gigantes cliff, because of the proximity to hydrophilic communities favored by water flowing or filtering through the veins of red-ochres "almagres" and basaltic dikes.

Characteristic species are *Polycarpaea carnosa* and *Vieraea laevigata*.

Distribution: Endemic to T.

Fig. 6.46 *Vieraeo laevigatae-Polycarpaeetum carnosae*. Interián. Tenerife

Aeonion aureae

Comophytic and chasmo-comophytic associations, colonizing the rock surface and crevices, with its optimum representation within Meso- and Supramediterranean belts.

It is characterized by several *Crassulaceae* of the *Greenovia* section of *Aeonium* (*A. aizoon, A. aureum, A. diplocyclum*) together with some other *Aeonium* of cold environments like *A. simsii* and *A. spathulatum* as well as hybrids between them, and some other rupicolous species like *Festuca agustini, Tolpis calderae* and *Silene pogonocalyx*.

At higher elevations, *Aeonium aureum* succeeds the more thermophilic and better developed communities of *Soncho-Aeonion* in the middle and lowest belts. Western Canary Biogeographic subprovince.

Distribution: Endemic to C, T, G, P, H.

Cheilantho guanchicae-Aeonietum smithii (Fig. 6.47)

This is an association endemic to the territory neighboring the cirque of Las Cañadas del Teide (1200–2300 m), in the dry-subhumid Mesomediterranean and dry Supramediterranean belts. It grows in crevices on vertical basaltic and phonolitic rocks, especially facing south, in potential territory of pine forest and summit broom shrubland.

Characteristic species are *Aeonium smithii*, *Aeonium spathulatum*, *Cheilanthes guanchica*, *Notholaena marantae*, *Descurainia gonzalesii*, *Erysimum scoparium*,

Hypericum reflexum, *Pericallis lanata*, *Rhamnus integrifolia* and *Silene bertheloti-ana*, among others.

Distribution: Endemic to T.

Fig. 6.47 *Cheilantho guanchicae-Aeonietum smithii*. Las Cañadas. Tenerife

Aeonietum aizoi
This is a local community that grows in rocks between 1000 and 1800 m on the summits and SE slope of Tenerife, preferably in the Güímar Valley. It has its optimum within subhumid Thermo- and Mesomediterranean belts without summer clouds and in the dry Mesomediterranean, climatophilous territories of summer dry laurel forest and pine forest.

Characteristic species are *Aeonium aizoon, Hypericum reflexum, Festuca agustini, Pericallis lanata, Crambe scaberrima, Pimpinella dendrotragium* and frequently *Carlina salicifolia* and *Tolpis lagopoda*.

Distribution: Endemic to T.

Aeonietum aureae (Fig. 6.48)
This is an association growing on the north and east facing highlands of Tenerife, above the sea of clouds. It usually occurs on rock surfaces and steep, often fissured basaltic cliffs, in the potential territory of summer-dry laurel forest and humid pine forest.

Characteristic species are *Aeonium spathulatum*, *Festuca agustini*, *Aeonium x lambii, Aeonium aureum*, *Hypericum reflexum* and *Monanthes brachycaulos*.

It has been also reported for Gran Canaria by Quintana et al. 2006 in the laurel forest area, but the relevés they present seem to be better placed within the *Aeonietum undulato-percarnei* as a variation of the latter.

Distribution: Endemic to Tenerife.

Fig. 6.48 *Aeonietum aureae*. Ayosa. Tenerife

Aeonietum diplocyclae

This is a rupicolous community extending over the northern slope of La Palma at altitudes between 1000 and 1900 m. It occurs in the subhumid-humid lower-Mesomediterranean belts (c), in potential laurel forest territory, and the subhumid-humid upper Mesomediterranean belts, in potential pine forest territory.

Characteristic species are *Aeonium spathulatum* var. *cruentum*, *Aeonium diplocyclum, Pimpinella dendrotragium*, *Silene pogonocalyx* and *Teline stenopetala*, among others. The presence of plants belonging to *Aeonietum christii-davidbramwellii*, like *Aeonium canariense* subsp. *christii* and *Lobularia palmensis* is frequent at its lowest altitudes.

Distribution: Endemic to P.

Aeonietum aureo-caespitosi (Fig. 6.49)
This is a chasmophytic community that grows on cliffs and in natural or artificial walls of the midlands and summits of Gran Canaria, particularly on the geological edifice of Roque Nublo. It occurs in the subhumid Thermomediterranean (c) towards dry Mesomediterranean zone, from 800 to 1800 m, although its optimum is above 1000 m, in the climatophilous territories of laurel forest and pine forest. It succeeds the *Aeonietum virginii* at higher elevations, also under the t-w.c. influence.

Characteristic plants are *Aeonium simsii* and *Aeonium aureum*, which show a typical cushion-like habit, formed by a dense accumulation of their sessile rosettes. A rich cover of bryophytes and lichens usually accompanies them. Other common species are *Aeonium spathulatum*, *Babcockia platylepis*, *Festuca agustini*, *Monanthes brachycalulos*, *Silene tamaranae*, *Tolpis lagopoda*, etc.

Distribution: Endemic to Gran Canaria.

Fig. 6.49 *Aeonietum aureo-caespitosi*. Summits of Gran Canaria

Soncho hierrensis-Aeonietum diplocyclae (Fig. 6.50)
This is a rupicolous community of El Hierro and La Gomera. In the North of El Hierro it replaces the *Aeonietum christii* at higher elevation, and the *Aeonietum subplanii* on La Gomera. It descends over the southern slopes, preferable with a local north aspect, till about 700 m, because of its greater resistance to drought.

Its most frequent species are *Davallia canariensis*, *Aeonium diplocyclum*, *Polypodium macaronesicum*, *Sonchus hierrensis* and *Tolpis proustii*.

The subass. *phyllidetosum noblae*, described from El Hierro, preferably grows in the upper region of the central and E part of El Golfo in a more humid environment. It is rich in endemic plants. Differential species are *Phyllis nobla, Silene sabinosae, Ceterach aureum, Aichryson punctatum, Festuca agustini* and *Cerastium sventenii*. Interspersed with this community grow numerous species typical of the laurel forest fringe and mantle in crevices.

Distribution: Endemic to G, H.

Fig. 6.50 *Soncho hierrensis-Aeonietum diplocyclae*. Garajonay. La Gomera. (Ph: AFL)

Tolpidetum calderae

This is a rupicolous community, floristically very rich, restricted to the rocky scarps of the northern summits of La Palma. It is found at altitudes between 1700 and 2400 m, in the upper Mesomediterranean and subhumid-humid Supramediterranean belts.

Among its characteristic species are *Argyranthemum haouarytheum*, *Erysimum scoparium* var. *lindleyi*, *Micromeria lasiophylla* subsp. *palmensis*, *Teline stenopetala* subsp. *sericea* and *Tolpis calderae*. Besides these, *Festuca agustini*, *Aeonium diplocyclum*, *Pimpinella dendrotragium*, *Bethencourtia palmensis*, *Silene pogonocalyx* are common.

Distribution: Endemic to P.

Aichryso laxi-Monanthion laxiflorae (Fig. 6.51)

This alliance includes oligospecific (chasmo-)comophytic communities of earthy rock surfaces. They are characterized by species of *Aichryson, Monanthes* and *Sedum*. The rarity of *Aeonium, Sonchus* s.l. and other species is notheworthy and more characteristic of chasmo-comophytic communities of the *Soncho acaulis-Aeonion.* In moist and shaded places they incorporate transgressive ferns of the *Anomodonto-Polypodietea.* From arid Inframediterranean to subhumid Mesomediterranean. Canary Biogeographic Province.

Fig. 6.51 *Aichryso laxi-Monanthion laxiflorae*. Several small species of *Monanthes* are characteristic of the communities of this Alliance. *M. icterica*. Tierra del Trigo. Tenerife

Aichrysetum immaculati

This is an oligospecific comophytic community that grows on usually moist steep earthy slopes and cliffs of the subhumid-humid Thermo-Mesomediterranean belts (c) in Anaga and Agua García-Santa Úrsula on Tenerife (700–1200 m).

It is characterized by *Aichryson pachycaulon* subsp. *immaculatum*, a dominant plant in the community, and it often grows alone. *Aichryson laxum, Monanthes anagensis* and *M. laxiflora* are present, especially when the community is in contact with other rupicolous communities. Some shrubby species of laurel forest mantle communities are sometimes interspersed. Mosses, liverworts and hornworts are usually abundant companion species.

Distribution: Endemic to T.

Aichrysetum pachycauli

This is a comophytic community typical of rocks oozing water of the highest and most humid North-facing crests of Jandía, within the upper-semiarid (and dry) Thermomediterranean belts (c).

It is characterized by *Aichryson pachycaulon* subsp. *pachycaulon, Monanthes laxiflora* var. *microbotrys* and *Polypodium macaronesicum* that are accompanied by abundant bryophytes and lichens.

Distribution: Endemic to F.

Aichrysetum punctati

This is a comophytic mono- or oligospecific community that grows on steep earthy slopes and cliffs with seasonal dripping in the dry-subhumid Infra- Mesomediterranean belts (c) in the western Canary Islands (200–1400 m).

It is characterized by *Aichryson punctatum*, and sometimes accompanied by *Polypodium macaronesicum, Davallia canariensis* or *Asplenium (Ceterach) aureum,* but these three fern species are rarely abundant.

Distribution: Endemic to C, T, G, P, H.

Aichrysetum tortuosii

This is a dense chamaephytic, chasmophytic, carpet-like community growing in crevices subject to humid NE trade-winds on the highest parts of Lanzarote, within the semiarid Thermomediterranean belt.

Characteristic species are *Aichryson tortuosum, Monanthes laxiflora* var. *microbotrys* and *Sedum nudum* subsp. *lancerottense*.

There is a floristic impoverishment of the community on southern slopes giving rise to monospecific stands of *Aichryson tortuosum.*

Distribution: Endemic to L.

Davallio canariensis-Aichrysetum laxi (Fig. 6.52)

This is a partially ruderal community that grows on shaded steep earthy slopes and walls of agricultural terraces mainly in the dry-subhumid Thermo- and Mesomediterranean belts (c), within the potential territory of laurel forest. The abundance of humid crevices filled with fertile soil favor the establishment of a number of saxicolous species. Fertilization of the fields contributes to their nitrophilisation.

Davallia canariensis, Aichryson laxum and *Polypodium marcaronesicum* are the characteristic species to which other rupicolous endemics are added (*e.g., Aeonium arboreum* subsp. *holochrysum, A. percarneum, Sonchus congestus, Sonchus acaulis, Monanthes brachycaulon, etc.*

It is a community floristically close to the *Davallio canariensis-Polypodietum macaronesicum* but with a different habitat.

Distribution: Endemic to C, T, G, P, H.

Fig. 6.52 *Davallio canariensis-Aichrysetum laxi*. Chamorga. Anaga. Tenerife

Lolio saxatile-Aichrysetum bethencourtiani
This is a chasmo-comophytic community that grows in rocky ledges, crevices and rocky cavities on vertical North-facing cliffs, preferentially between 650 and 800 m, within the upper-semiarid (and dry) Thermomediterranean belts with frequent t-w.c. on the Jandía Massif. It can be considered as vicariant of the *Aichrysetum tortuosi* from Lanzarote.

It is characterized by *Lolium saxatile, Aichryson tortuosum* var. *bethencurtianum, Minuartia platyphylla, Monanthes laxiflora* var. *microbotrys* and *Polypodium macaronesicum.*

A poor xeric variant has been described for drier places with a high presence of *Roccella spp.*

Distribution: Endemic to F

Monanthetum anagensis
This is an oligospecific comophytic community that grows on open exposed stony slopes, cliffs and walls, preferably within dry-subhumid Thermomediterranean belts (c) in the North of Anaga.

It is characterized by *Monanthes anagensis*, an endemic species of Anaga, Tenerife, which dominates the community. *Monanthes laxiflora* var. *laxiflora, Monanthes brachycaulos,* and *Aichryson parlatorei* are noteworthy as companion species.

Distribution: Endemic to T.

Monanthetum pallentis (Fig. 6.53)
This is an especially comophytic mono- or oligospecific community that grows in somewhat moist and semi-shaded steep earthy rock slopes or small crevices filled with earth, generally with a northern aspect, within the arid Inframediterranean and lower-Thermomediterranean belts on the South and West slopes of Tenerife.

It is characterized by *Monanthes pallens*. In typical comophytic habitats, it is sometimes accompanied by species of the *Anomodonto-Polypodietea* like *Polypodium macaronesicum* or *Davallia canariensis,* whereas in filled crevices species of the *Asplenietea rupestris* like *Cheilanthes catanensis* or *C. pulchella* tend to be commoner.

Distribution: Endemic to T.

Fig. 6.53 *Monanthetum pallentis*. Los Derriscaderos. Granadilla. Tenerife

Soncho-Monanthetum laxiflorae
This is a chasmo-comophytic community typical of steep alveolar or jointed basalts with scarce soil that preferably grows within the dry – subhumid Infra-Thermomediterranean belts (c).

Characteristic species are *Sonchus gomerensis* and *Monanthes laxiflora*, and frequent are *Aeonium aureum, Dicheranthus plocamoides, Silene bourgeaui, Aeonium castello-paivae, Monanthes polyphylla* subsp. *amydros, Tolpis laciniata* and *Crambe gomera.*

Distribution: Endemic to G.

6.2.5 Synanthropic, Fringe and Megaforbic Vegetation

6.2.5.1 Synanthropic Vegetation

Artemisietea vulgaris
Biennal or perennial ruderal nitrophilous communities growing on nitrogen- and/or phosphorus-rich soils because of farming activity. They are common in rural environments, livestock folds, disturbed soils, abandoned cultivation fields, etc. Holarctic. As neophyte vegetation, nowadays it shows a cosmopolitan distribution.

Onopordenea acanthii
Pioneer ruderal-nitrophilous vegetation formed by tall herbs and biennial thistles (several spiny species of *Asteraceae*) typical of dry-subhumid Mediterranean territories.

Carthametalia lanati
Ruderal-nitrophilous communities of biennial herbs, perennial grasses, and thistles growing on very disturbed soils, mainly in semiarid Inframediterranean to humid Mesomediterranean territories. Wide Mediterranean distribution.

Bromo madritensis-Piptatherion miliacei
Mediterranean subnitrophilous communities of tall biennial herbs and perennial grasses like those belonging to *Piptatherus* that thrive on wasteland and abandoned agricultural lands, on deep soils.

They are typical of the Mediterranean region, and occur from the semiarid Inframediterranean to dry-subhumid Thermo- and Mesomediterranean belts.

Distribution: All islands.

***Dittrichia viscosa* community**
This corresponds to dense stands of *Dittrichia viscosa* on deep wet soils. It is common in depressions that are temporarily flooded and in the tail of artificial dams and reservoirs.

This community stands out in the landscape during summer and fall due to the profuse yellow bloom of the dominant plant.

Distribution: F, C, T, G, H.

Pericallido murrayi-Bituminarietum bituminosae
This is a vivacious herbaceous and nanophanerophytic community that grows in abandoned agricultural fields on El Hierro, preferentially in the dry-subhumid Thermo- and Mesomediterranean belts. It is related to the *Piptathero miliacei-Foeniculetum vulgaris,* differing by the absence of *Piptatherum miliaceum* on the island (*s.* Stierstorfer 2005).

Pericallis murrayii is characteristic, being a typical plant of the laurel forest fringe that expands into abandoned fields, and grows together with *Bituminaria bituminosa, Foeniculum vulgare* and *Sonchus hierrensis*.

Their plants are sometimes used as fodder and pure community stands are difficult to distinguish because they grow in an intricate mixture with annual herbaceous communities of the *Stellarietea* and *Tuberarietea*.

Distribution: H.

Piptathero miliacei-Foeniculetum vulgaris (Fig. 6.54)
This is a vivacious herbaceous and subnitrophilous community in which the hemicryptophytic grass *Piptatherum miliaceum* and the scapaceous hemicryptophyte *Foeniculum vulgare* stand out. Besides, there are some suffrutescent chamaephytes like *Bituminaria bituminosa* and *Dittrichia viscosa* as well as a number of rather tall subnitrophilous therophytes. It occurs in the semiarid Inframediterranean to dry-subhumid Thermo- and Mesomediterranean belts (c).

The community forms a vivacious pasture on abandoned agricultural land and wasteland in potential laurel forest areas. It is still abundant in the rural landscape of Thermomediterranean areas of the thermo-sclerophyllous woodland, but almost absent in arid Inframediterranean areas where it mainly appears on moist soils.

Distribution: L, F, C, T, G, P.

Fig. 6.54 *Piptathero miliacei-Foeniculetum vulgaris*. La Laguna. Tenerife

Urtico piluliferae-Silybion mariani

These are spring-blooming herbaceous perennial communities, generally with a dominance of large thistles like *Carduus tenuiflorus, Cynara cardunculus* var. *ferocissima*, *Scolymus hispanicus*, *S. maculatus* and *Silybum marianum*. They usually form dense communities with high biomass, indifferent to substrate, although they require adequate moisture. Typical of path edges, roadside slopes and abandoned cultivation fields. Mediterranean and Eurosiberian.

Distribution: All islands.

Conio maculati-Silybetum mariani

This is a megaforbic, nitrophilous, spring-blooming herbaceous community, with a dominance of annual or biennal tall thorny thistles (*Compositae*), as well as some other tall hemicryptophytes. It generally grows on fresh deep soils that can become flooded, and grows especially in the dry to subhumid Thermo- and Mesomediterranean belts (c), in potential territory of laurel forest.

Characteristic species are *Carduus tenuiflorus*, *Conium maculatum* and *Silybum marianum*, among others.

Distribution: C, T, G, P, H.

Scolymo maculati-Cynaretum ferocissimae (Fig. 6.55)

This is an herbaceous association formed by large-flowered spring-blooming thistles, with a rather quick development depending on the thermotype, but in general briefly present. It is more xeric than the *Conio-Silybetum mariani* and grows particularly in the upper-semiarid Thermomediterranean belts (within the potential dominance of thermo-sclerophyllous woodland) to dry-subhumid Thermo- and Mesomediterranean belts (c) (within the potential dominance of laurel forest).

Cynara cardunculus, *Scolymus hispanicus* and *S. maculatus* are characteristic species. Other thorny *Compositae* like *Carduus tenuiflorus* and *Galactites tomentosa* are involved.

Distribution: All islands.

Fig. 6.55 *Scolymo maculati-Cynaretum ferocissimae*. Tamadaya. Tenerife

Pegano harmalae-Salsoletea vermiculatae
Nitrophilous and halo-nitrophilous scrublands and shrublands from the Mediterranean, Sahara-North-Arabian, and Iran-Turanian biogeographic regions. It also includes some shrubby nitrophilous anthropogenic communities, mainly formed by tropical neophytes.

Chenoleoidetalia tomentosae
Nitrophilous and halo-nitrophilous scrublands from hyperarid-arid Infra- and Thermomediterranean territories. Canary and Sahara-Atlantic distribution.

Chenoleoidion tomentosae
Single alliance.

Chenoleoideo tomentosae-Salsoletum vermiculatae (Fig. 6.56)
This is a nitrophilous, chamaephytic-nanophanerophytic scrub. It substitutes the vegetation of the arid Inframediterranean belt of the low parts of Lanzarote, on stony superficial soils that may contain some organogenic sand. It is a scrub traditionally subject to goat and dromedary grazing.

The nanophanerophyte *Salsola vermiculata* dominates together with the chamaephyte *Chenoleoides tomentosa*. In addition, *Lycium intricatum* is common.

Several variants are recognized:

The aero-halophilous *Suaeda ifniensis* variant, also characterized by *Salsola divaricata,* is linked to marine spray. The variant with *Launaea arborescens*, typical of disturbed soils or soils with a loose structure, grows at road edges, in small ravines, on uneven slopes, abandoned farmland, etc., where *Salvia aegyptiaca* also grows. This variant is enriched in the N of the island with *Atriplex halimus*, as well as *Lavandula pinnata*, typical of the slope debris of the Famara massif.

Several facies are also recognized: the *Salsola vermiculata* facies, widespread in the SE, which occurs on disperse old farmlands, where *Helianthemum canariense*, *Heliotropium ramosissimum* and *Lycium intricatum* are common. The *Chenoleoides tomentosa* facies has mainly developed in deep semi-compact sandy soils with a steady aero-saline influence. The *Salsola tetrandra* facies grows in the lower-arid and hyperarid ombrotypes. The *Lavandula pinnata* facies is abundant on the NW slope of the Famara massif. Other facies are those of *Opuntia dillenii*, *Kleinia neriifolia*, and *Periploca laevigata.*

Distribution: Endemic to L.

Fig. 6.56 *Chenoleoideo tomentosae-Salsoletum vermiculatae*. Caleta de Famara. Lanzarote

Chenoleoideo tomentosae-Suaedetum mollis

This is a low halo-nitrophilous chamaephytic scrub, typical of the arid Inframediterranean belt, which has spread through grazing. It develops on haplo-calcic and petro-calcic soils, sometimes with a shallow sand cover. It is frequent in coastal places swept by NE trade-winds, which contributes to soil salinity.

Characteristic species are *Chenoleoides tomentosa*, *Frankenia capitata* and *Suaeda mollis*.

One can recognize the variant with *Salsola vermiculata*, in contact with the *Chenoleoideo-Salsoletum vermiculatae*, which grows in less degraded soils and with a less developed crust. The anemogenous variant with *Gymnocarpos decandrus* is well represented in Punta de Jandía (Fuerteventura) and Punta de Gando (Gran Canaria). In Lanzarote, the aero-halophilous variant with *Suaeda ifniensis* is well represented in Punta de Tierra Negra.

The subass. *atractyletosum preauxianae* occupies very windy places on the East and South coast of Gran Canaria. *Atractylis preauxiana, Convolvulus caput-medusae* and *Herniaria fontanesii* are differentials. The subass. *suaedetosum verae* was described from Fuerteventura, differentiated by the presence of *Suaeda vera*. It may constitute an ecotone with the *Frankenio-Suaedetum verae*.

Distribution: L, F, C.

Forsskaoleo angustifoliae-Rumicetalia lunariae
Nitrophilous and sub-nitrophilous chamaephytic and nanophanerophytic communities growing on disturbed soils of the arid Inframediterranean to subhumid Mesomediterranean belts. Canarian-Madeiran biogeographic subregion.

Launaeo arborescentis-Schizogynion sericeae
Nitrophilous and sub-nitrophilous, chamaephytic and nanophanerophytic, arid Inframediterranean Canary associations growing on disturbed soils, lapilli, and volcanic ash in the territory of potential climatophilous *Euphorbia balsamifera* scrub.

Distribution: All Canary Islands.

Artemisietum ramosae (Fig. 6.57)
This is an oligospecific, halo-nitrophilous, ashen hued, chamaephytic scrub. It grows in endorreic micro-basins with eutric cambisols with a compact silty texture, retraction crusts, and the characteristic superficial black efflorescence of sodium humates. Here, it constitutes a halo-nitrophilous serial stage of *Euphorbia balsamifera* scrub.

It is characterized by the Canary endemic *Artemisia ramosa* and *Launaea arborescens*.

Distribution: Endemic to C, T.

Fig. 6.57 *Artemisietum ramosae*. Malpaís de Rasca. Tenerife

Cenchro ciliaris-Launaeetum arborescentis (Fig. 6.58)
This is a low scrub settled on soils covered by a sand layer, both volcanic and organogenic. It largely characterizes the southern and central landscape of Lanzarote, and its expansion is favored by human action.

Characteristic species are *Launaea arborescens*, the hemicryptophytic grass *Cenchrus ciliaris*, *Heliotropium ramosissimum*, *Lotus lancerottensis*, *Polycarpaea nivea* and *Salvia aegyptiaca*. The outstanding presence of *Cyperus capitatus* is related to the accumulation of organogenic sand over the substrate.

Distribution: Endemic to L.

Fig. 6.58 *Cenchro ciliaris-Launaeetum arborescentis*. Caleta de Famara. Lanzarote

Herniarietum canariensis
This is a small-sized chamaephytic community, which develops on somewhat disturbed anthropic pumice lithosols.

It is characterized by *Herniaria canariensis*, which is accompanied by widespread species belonging to the *Pegano-Salsoletea,* like *Argyranthemum frutescens* subsp. *gracilescens*, *Launaea arborescens* and *Schizogyne sericea*, as well as *Helianthemum canariense*, *Lotus glaucus*, etc.

Distribution: Endemic to T.

Launaeo arborescentis-Schizogynetum sericeae (Fig. 6.59)
This is a widespread Canary association, linked to the arid Inframediterranean, mainly in areas of potential *Euphorbia balsamifera* scrub. It occurs on more or less nitrogen-rich places, especially on loose soil with a sandy texture, abandoned agricultural land, slopes, ditches, etc. It can resist high saline concentrations in the soil. Fruticose chamephytes are dominant, although also some nanophanerophytes coming from *Euphorbia* scrub are involved.

The most frequent species are *Launaea arborescens, Schizogyne sericea, S. glaberrima, Lotus sessilifolius* var. *pentaphyllus, L. sessilifolius* var. *sessilifolius, Lycium intricatum, Salsola divaricata, Gymnocarpos decandrus*, among others.

Besides the typical subass. *schizogynetosum sericeae*, several subassociations have been described: subass. *reichardietosum crystallinae*, typical of coastal aerosaline sites; subass. *zygophylletosum fontanesii* with halophile characteristics, typical of both littoral sands and cliffs; subass. *artemisietosum ramosae* typical of small clayey basins in contact with the *Artemisietum ramosae; and* subass. *salsoletosum divaricatae*, typical of coastal areas with a strong halonitrophilous influence.

In addition, we recognize some facies due to the particular abundance of some species: those of *Argyranthemum gracile, Plocama pendula* in porous permeable soils – very common on Gran Canaria, and the widespread *Opuntia spp.,* mainly that of *O. dillenii.*

Distribution: Endemic to C, T, G.

Fig. 6.59 *Launaeo arborescentis-Schizogynetum sericeae*. Montaña Grande. Güímar. Tenerife

***Pelargonium capitatum-Launaea arborescens* community**

Pioneer community on recent lapilli, characterized by *Launaea arborescens* and the xenophyte *Pelargonium capitatum,* within the arid Inframediterranean belt. First recorded for the historical Timanfaya volcanic area (1730–1736), now it is also found around Masdache. *P. capitatum* also grows in humid habitats with the rush *Juncus acutus* subsp. *leopoldii.*

Distribution: L.

Artemisio thusculae-Rumicion lunariae

Nitrophilous and sub-nitrophilous chamaephytic and nanophanerophytic communities growing on disturbed soils on semiarid Inframediterranean to subhumid Mesomediterranean territories, within the climatophilous territories of potential *Euphorbia canariensis* shrubland, potential thermo-sclerophyllous woodland, and reaching the lower territory of potential laurel forest or even pine forest on the southern slope. The development of nitrophilous therophytic communities is common at their margins, and these form a mixed mosaic with them given suitable open areas. Canary biogeographic province.

Distribution: All islands.

Artemisio thusculae-Plantaginetum arborescentis (Fig. 6.60)

This is a chamaephytic scrub, very similar to the *Artemisio-Rumicetum lunariae*, typical of ancient basaltic outcrops, that develops on lithosols and debris in the potential territory of laurel forest (dry to humid Infra- to Mesomediterranean with t-w.c belts).

Characteristic species are *Artemisia thuscula*, *Plantago arborescens* and *Argyranthemum broussonetii.*

Distribution: C, T, G.

Fig. 6.60 *Artemisio thusculae-Plantaginetum arborescentis*. Jardina. La Laguna. Tenerife

Artemisio thusculae-Rumicetum lunariae (Fig. 6.61)
This is a nitrophilous shrubby community, with a dominance of nanophanerophytes and chamaephytes, widespread in upper-arid and semiarid Inframediterranean belts (potential *Euphorbia canariensis* shrubland) and upper semiarid Infra- and Thermomediterranean belts (potential thermo-sclerophyllous woodland), and finicolous in dry-humid Thermomediterranean (c) belts (potential laurel forest). It occurs on disturbed, partly stabilized soils, typically of agricultural lands and wastelands.

Common plants in this association are *Argyranthemum frutescens* subsp. *frutescens*, *Artemisia thuscula*, *Lavandula canariensis*, *Rumex lunaria* and *Salvia canariensis*.

The subass. *rumicetosum lunariae* (typical) occurs on loose organic soils and almagres. The subass. *salsoletosum divaricatae* is typical of disturbed coastal aero-saline sites.

It shows a wide physiognomic variety, or facies, depending on the relative abundance of some of its species. The *Rumex lunaria* facies settles along humid road and track cuttings and disturbed land. The *Artemisia thuscula* facies tends to be dominant in dryer territories and abandoned agricultural land. On Gran Canaria, the *Salvia canariensis* facies is very abundant, over all the uplands. On La Palma the *Echium brevirame* facies is common in fractured loose soils. On El Hierro the *Erysimum bicolor* facies is typical of the wind-swept summits of the Risco de

Bascos (NW of El Hierro) subject to overflowing the NE trade-wind. Other common facies are: *Argyranthemum frutescens*, *A. gracile*, *Bosea yervamora*, *Lavandula canariensis* and *Pennisetum setaceum*.

Distribution: C, T, G, P, H.

Fig. 6.61 *Artemisio thusculae-Rumicetum lunariae*. Acantilados de Acentejo. Tenerife

Bidenti pilosae-Ageratinetum adenophorae (Fig. 6.62)
This is a community of perennial herbaceous plants and small chamaephytes, typical of slopes, tracksides and ditches, in humid Infra- to Mesomediterranean territories of potential laurel forest.

Bidens pilosa and *Ageratina adenophora* are dominant.

Distribution: C, T, G, P.

Fig. 6.62 *Bidenti pilosae-Ageratinetum adenophorae*. Tierra del Trigo. Tenerife

Gonospermo fruticosi-Lavateretum acerifoliae
This is a western Canary association that thrives on rocks, ledges and at the foot of cliffs, on some disturbed nitrogen-rich soils in the semiarid Infra- and Thermomediterranean territory of thermo-sclerophylous woodland.

The most characteristic species are *Gonospermum fruticosum* and *Lavatera acerifolia*.

Distribution: Endemic to T.

Ceballosio fruticosae-Withanietum aristatae
This is an association formed by tall nanophanerophytes that grow in anthropogenic environments of upper-arid and semiarid Inframediterranean territories, particularly in the potential *Euphorbia canariensis* shrubland.

The most characteristic species are *Ceballosia fruticosa* and *Withania aristata.*

Distribution: Endemic to T.

Salvio canariensis-Pterocephaletum dumetori (Fig. 6.63)
This is a subnitrophilous substitutional scrub, typical of the central summits of Gran Canaria (Tejeda and Tirajana calderas, and Artenara) preferably with a S or SW aspect between 1400–1800 m. It occurs within the subhumid upper-Thermomediterranean and Mesomediterranean belt out of the t-w.c. area. Expansion of the community is favored by grazing.

Characteristic species are *Pterocephalus dumetorus, Salvia canariensis* and *Artemisia thuscula*, in general accompanied by a species cohort of the *Andryalo-Ericetalia* like *Teline microphylla, Argyranthemum adauctum* subsp. *canariensis, Andryala pinnatifida, Erysimum albescens, Micromeria benthami, Adenocarpus foliolosus* and other heliophilous species like *Echium onosmifolium, Carlina texedae* and *Aeonium percarneum.*

A facies of *Salvia canariensis* with a particular N exposure has also a wide distribution over the summits of the association's northern area.

Distribution: Endemic to C.

Fig. 6.63 *Salvio canariensis-Pterocephaletum dumetori*. Tirajana. Gran Canaria

Rumex lunaria communities

A pioneer community of *Rumex lunaria* is distinguishable on El Hierro, growing in recent malpaíses and lapilli fields within the semiarid Inframediterranean belt. Although a pioneer, it behaves as a permanent community in thick recent volcanic substrates, as it keeps on growing for centuries.

Another permanent community of *Rumex lunaria* grows in holocenic alluvial deposits in the Caldera de Taburiente, La Palma, within the subhumid Thermomediterranean belt.

Distribution: P, H.

Tricholaeno teneriffae-Rumicetum lunariae

This is a shrubby community that develops on recent lapilli cones of the NE sector of Gran Canaria, in the Infra- and Thermomediterranean belts. The lapilli substrate favors its development, due to the isolation effect of the surface layer in retaining moisture within shallow soil cover.

The two characteristic plants are *Rumex lunaria*, which shows preference for deeper soils, and *Tricholaena teneriffae* with a clear preference for sandy recent volcanic soils retaining moisture.

Distribution: Endemic to C.

Asteriscion sericei

Anemophilous and nitrophilous scrublands spread by human disturbance and grazing, characterized by *Asteriscus* species, typical of the summits of the islands of the Eastern Canary subprovince, extending also to Gran Canaria.

Andryalo variae-Asteriscetum sericei

This is an endemic association of Fuerteventura dominated by the island endemic *Asteriscus sericeus*. It occupies the highest elevations of the island, in the semiarid Thermomediterranean belt of the climatophilous potential of *Olea cerasiformis* woodland.

This scrub could constitute a permanent anemophilous community on ridges, and is favored in its spread by anthropic activity e.g., intense grazing, and so it easily colonizes clayey-stony disturbed soils.

Andryala glandulosa subsp. *varia* and *Asteriscus sericeus* are characteristic species, together with *Launaea arborescens* and *Lycium intricatum*. Some widespread plants of the *Kleinio-Euphorobietea* like *Kleinia neriifolia, Euphorbia regis-jubae* or *Micromeria varia* subsp. *rupestris* are common.

Distribution: Endemic to F.

Lavandulo pinnatae-Asteriscetum intermedii

This is a chamaephytic substitutional community, typical of the highest potential zones of *Euphorbia balsamifera* scrub and thermophilous *Olea* woodland which originally capped the island of Lanzarote.

It has its optimum in the upper-arid to semiarid Inframediterranean and semiarid Thermomediterranean belts. It is established on old agricultural land with deep soils (*Asteriscus intermedius* facies), generally on eroded slopes (*Lavandula pinnata* facies), and on stony slopes, ravines, ledges, etc.

It becomes floristically impoverished towards the South, the absence of *Lavandula pinnata* being noticeable, since it is endemic to the Famara Massif. Frequently it mixes with communities of the *Artemisietea vulgaris*, both the *Silybion mariani* and *Bromo-Piptatherion miliacei*.

The facies of *Rumex lunaria*, characterized by this Canary endemic species introduced in Lanzarote, grows in young volcanic soils with abundant lapilli.

Distribution: Endemic to L.

Asterisco stenophylli-Ononidetum ulicinae

This is a nitrophilous scrub occurring on disturbed somewhat stabilized soils, between 400 and 1000 m, over an extensive Infra- and Thermomediterranean area of the W, S and SW of Gran Canaria. It is a substitutional community for *Euphorbia* scrub and thermo-sclerophylous woodland.

The more frequent species are *Asteriscus graveolens* subsp. *stenophyllus*, *Euphorbia regis-jubae*, *Kleinia neriifolia* and *Ononis angustissima* subsp. *angustissima* (*O. ulicina*).

A variant with *Artemisia ramosa* can be distinguished in hollows and gullies with a great accumulation of clay-silty fine materials. It is well represented between the ravines of Mogán and Veneguera. Previously it had been called *Odontospermum stenophyllum* and *Artemisia ramosa* community (Del Arco et al. 2006).

Distribution: Endemic to C.

Nicotiano glaucae-Ricinetalia communis
Anthropogenic shrubby nitrophilous communities mainly formed by tropical fast-growing and tall soft-wooded neophytes. They thrive from the arid Inframediterranean to subhumid Mesomediterranean belts, and are frequent in disturbed ravine beds, volcanic ash quarries, abandoned arable fields, and roadside slopes. The American *Nicotiana glauca* and the tropical-African *Ricinus communis* are their most representative plants. Mediterranean distribution.

Distribution: All Canary Islands.

Nicotiano glaucae-Ricinion communis
Single alliance.

Polycarpo tetraphylli-Nicotianetum glaucae
This is an arid association widespread in the Canaries, which develops in very anthropized places, as well as on disturbed soils and in rubble and quarries. It is very frequent in quarries for the extraction of lapilli. It preferably grows in territories of potential *Euphorbia* communities, within the arid-semiarid Infra- and Thermomediterranean belts, reaching deep into the territory of the thermo-sclerophyllous woodland.

The tall phanerophyte *Nicotiana glauca* is dominant. It is accompanied by other shrubs of the *Launaeo-Schizogynion* like *Forsskaolea angustifolia* and *Launaea arborescens*.

In soils with more water availability the *Ricinus communis* variant is common.

Distribution: Endemic to all islands.

Tropaeolo majoris-Ricinetum communis (Fig. 6.64)
This nitrophilous association develops on humus rich deep soils, within dry-subhumid Thermomediterranean belts, and wet ravine beds and watercourses down to the sea in the Inframediterranean. It generally replaces the Canary willow community (*Rubo-Salicetum canariensis*).

This is a species-poor community, characterized by *Ricinus communis*, a thermophilous phanerophyte of African origin today widespread in the Tropics and warm Mediterranean countries, and usually the Neotropic xenophyte *Tropaeolum majus*.

Distribution: Endemic to F, C, T, G, P.

Fig. 6.64 *Tropaeolo majoris-Ricinetum communis*. Tierra del Trigo. Los Silos. Tenerife

Polygono-Poetea annuae (Fig. 6.65)
Annual pioneer ephemeral, rarely small creeping perennial communities, nitrophiles of anthropic-zoogenic origin, typical of well-trodden urban and rural paths, in climates with one rainy season. Cosmopolitan.

Polygono arenastri-Poetalia annuae
Single order in this territory. Holarctic, extratropical territories.

Euphorbion prostratae
Summer annual therophytic communities, of tropical and subtropical origin that develop as neophytes in warm and cool Holarctic regions.

Polycarpo-Alternantheretum
This is an anthropic community that grows among cobblestones and asphalt fissures of urban streets mainly in sunny places of the Inframediterranean belt up to 300 (max. 600) m. It mainly starts to grow in midwinter and reaches its optimum in summer and the beginning of the fall.

It is characterized by the original Central and South American neophyte *Alternanthera caracasana*, a prostrate species that grows together with *Polycarpon tetraphyllum, Euphorbia prostrata,* to a lesser extent *E. chamaecyse,* together with *Coronopus didymus, Oxalis corniculata, Polygonum aviculare, Poa annua* and *Amaranthus deflexus. Conyza bonariensis, Eleusine indica, E. tristachya* and *Amaranthus lividus* are common companion species.

Distribution: P, G, T, C.

Fig. 6.65 *Polygono-Poetea annuae*. Taganana. Tenerife

Polycarpion tetraphylli
Pre-summer therophytic communities growing on light soil. Western Mediterranean and Thermo-Atlantic distribution.

Polycarpo tetraphylli-Cotuletum australis
This is a nitrophilous pioneer therophytic association that grows in well-trodden soils among cobblestones, asphalt fissures, paths and dirt roads in Thermo- and Mesomediterranean territories of the islands.

It is characterized by *Polycarpon tetraphyllum*, the neophyte *Cotula australis*, native of Australia and New Zealand but today distributed worldwide, and *Sagina apetala* together with *Poa annua, Crassula tillaea* and *Polygonum aviculare. Coronopus didymus, Trifolium glomeratum* and *Oxalis corniculata* are common companion species.

The subass. *cotuletosum australis* (*typicum*) shows preference for urban environments.

The subass. *solivetosum stoloniferae*, differentiated by *Soliva stolonifera* and *Trifolium suffocatum*, grows on very well-trodden paths and dirt roads, generally in rural or more natural environments.

Distribution: Endemic to P, T, C.

Saginion procumbentis
Pioneer, vivacious, semi-sciophilous, mesohygrophilous communities rich in bryophytes. Eurosiberian, extending to Mediterranean regions.

Poo infirmae-Saginetum stoloniferi
This is an oligo-specific community that grows on permanently warm and humid soils, in the shade of the top (between 3650 and 3700 m a.s.l.) of a lava flow that stretches down from the summit of Mount Teide towards the La Rambleta cable-car station. Sometimes it is subject to trampling by visitors.

It is characterized by *Sagina procumbes* subsp. *stolonifera*, *Poa infirma* and unidentified species of *Bryum* and *Amphidium*.

Distribution: Endemic to T.

Sclerochloo durae-Coronopodion squamati
Annual communities of the temperate Eurosiberian and Mediterranean regions, growing on clayey calcareous soils.

Polycarpo tetraphylli-Coronopodetum squamati
This is a nitrophilous pioneer reptant therophytic association that grows in well-trodden clayey compact soils, in dry to subhumid Thermomediterranean territories of northern Tenerife.

It is characterized by *Polycarpon tetraphyllum, Coronopus squamatus, Polygonum aviculare* and *Poa annua; Lolium rigidum* is frequent.

Distribution: Endemic to T.

Stellarietea mediae
A wide class which includes ruderal, nitrophilous or subnitrophilous, and arvensis communities of annual fleeting herbs. Distributed all over the world except tropical territories. Cosmopolitan.

Chenopodio-Stellarienea
Includes ruderal, nitrophilous and seminitrophilous communities.

Chenopodietalia muralis
Includes strongly nitrophilous annual herbaceous communities rich in anthropophytes, cosmopolitan but mostly in Mediterranean territories.

Chenopodion muralis
Includes spring-summer ruderal communities, highly nitrophilous, Infra- to Supramediterranean. They are common in abandoned cultures, among rubbish or debris or in waste places.

Malvenion parviflorae
Includes arid to subhumid Thermo- to Mesomediterranean communities blooming after fall and winter rains.

Chenopodio muralis-Malvetum parviflorae (Fig. 6.66)
This is a strongly nitrophilous therophytic community with winter and spring development. It is widespread in rather nitrophilous urban and rural environments, sometimes with loose soil, frequently visited by livestock, in arid to subhumid Infra- and Thermomediterranean belts.

The following characteristic species are particularly notable: *Chenopodium murale*, *Malva parviflora*, *Sisymbrium irio* and *Urtica urens*.

Distribution: All islands.

Fig. 6.66 *Chenopodio muralis-Malvetum parviflorae*. El Médano. Granadilla. Tenerife

Mesembryanthemion crystallini

Includes halonitrophilous annual herbaceous communities with a dominance of succulent prostrate *Aizoaceae*, growing on rocky or loamy barren soils, or disturbed soils, mostly in littoral but also in inland hyperarid to semiarid Thermomediterranean territories.

Mesembryanthemetum crystallini (Fig. 6.67)

This is a therophytic, desertic and xeric Inframediterranean association, especially in the arid and hyperarid ombrotype. It usually has a halonitrophilous character, dense cover, and winter growth. It mainly appears in anthropic environments like abandoned cultivated land, path edges, rubbles, etc. It is very common on old tomato fields, giving rise to dense polychrome or reddish carpets after the winter rains.

Mesembryanthemum crystallinum or *M. nodiflorum* dominate in its floristic composition, and usually *Patellifolia patellaris* and *Aizoon canariense* are frequent, with a dominance of the latter at well-trodden places.

Distribution: All islands.

Fig. 6.67 *Mesembryanthemetum crystalline*. El Médano. Tenerife

Sisymbrietalia officinalis

Includes Mediterranean and Temperate viatical and fleeting ruderal nitrophilous therophyte communities, with a spring or summer growth.

Hordeion leporini

Groups the Mediterranean communities of the previous order.

Bromo-Hirschfeldietum incanae

This is a ruderal and roadside therophytic community, a little nitrophilous, typical of dry to subhumid Infra- and Thermomediterranean belts, even semiarid on eastern islands. It is frequent on wastelands and abandoned cultivation lands. It starts developing with fall-winter rains, and its optimum is in spring.

Among its floristic composition: *Avena barbata*, *Bromus rigidus*, *Galactites tomentosa*, *Hirschfeldia incana* and *Hordeum murinum* subsp. *leporinum*.

Distribution: All islands.

Euphorbio terracinae-Anacycletum coronatii

This a nitrophilous ruderal and roadside association growing in a wide variety of soils (sandy, clayey, etc.). It is especially common in the arid Inframediterranean belt of Lanzarote. It colonizes road edges in the proximity of villages and cultivated land, even entering the latter after recent abandonment.

Its main characteristic is *Anacyclus radiatus* subsp. *coronatus*.

Distribution: Endemic to L.

Thero-Brometalia

Includes subnitrophilous Mediterranean annual communities, spring blooming.

Echio plantaginei-Galactition tomentosae (Fig. 6.68)
A western-Mediterranean Alliance of oceanic character mostly developed in subhumid-humid ombrotypes and eutrophic or mesotrophic soils.

Galactito tomentosae-Trachynietum distachyae
This is a subnitrophilous therophytic association, rich in Gramineae and Fabaceae, together with other herbaceous plants, with spring development, which grows in wastelands, path edges and abandoned cultivation land. It occurs in upper-semiarid to dry Inframediterranean, and dry-subhumid Thermomediterranean belts, in the Western Canary subprovince. It is usually found in contact or in mosaic with other annual or vivacious communities of a more or less nitrophilous character like the *Bromo-Hirschfeldietum* (*Hordeion leporini*) or *Piptathero-Foeniculetum vulgaris* (*Bromo-Piptatherion miliacei*).

Its more common species are *Avena barbata*, *Trachynia distachya*, *Bromus rigidus*, *Echium plantagineum*, *Galactites tomentosa, Medicago polymorpha*, *Plantago lagopus* and *Trifolium glomeratum*, among others.

Distribution: C, T, G, P.

Fig. 6.68 *Echio plantaginei-Galactition tomentosae*. Los Rodeos. Tenerife

***Raphanus raphanistrum-Bromus diandrus* community**

This is a subnitrophilous therophytic community described from high elevations on El Hierro, in subhumid and humid Thermo- and Mesomediterraneean territories, growing in mesotrophic abandoned fields and crofts, with spring development.

Raphanus raphanistrum, Bromus diandrus, Echium plantagineum and *Ornithopus compressus* are their characteristic species.

The community grows in a mosaic of diverse herbaceous communities and it is closely related to the *Galactito tomentosae-Trachynietum distachiae*. Further studies for a better differentiation are required.

Distribution: H.

Resedo lanceolatae-Moricandion

Subnitrophilous or nitrophilous short-lived half-size therophyte communities, with spring development which grow in disturbed fields, paths, and abandoned fields, in a desertic or xeric Mediterranean bioclimate.

Bupleuro semicompositi-Mairetetum microspermae

This is a transitory therophytic grassland, which develops on disturbed organic sands. The ruderalization of wind-blown (aeolian) organic dune systems ('jable' sandfields) due to cultivation, grazing or tourism, favor the appearance of *Stellarietea* species in the sand ecosystem.

The more frequent species are *Bupleurum semicompositum, Cutandia memphitica, Erodium grex laciniatum, Ifloga spicata, Mairetis microsperma, Medicago littoralis, Plantago aschersonii, Rumex bipinnatum* and *Triplachne nitens*, among other companion species like *Reichardia tingitana, Launaea nudicaulis* or *Reseda lancerottae*.

Near coastal zones, affected by the saline atmosphere, a *Plantago coronopus* form with fleshy leaves grows, while inland, with less saline influence, other species like *Cutandia memphitica* or *Mairetis microsperma* are more characteristic.

Distribution: Endemic to L, F?.

Iflogo spicatae-Stipetum capensis (Fig. 6.69)

This is a (sub)nitrophilous therophytic community with a grassland physiognomy, dominated by *Stipa capensis*. It has a fleeting winter-spring development. It settles in abandoned cultivated land and also among the shrubs frequented by livestock, especially in the hyperarid and arid Inframediterranean belts.

Several therophytes participate in its floristic composition, particularly *Astragalus solandri, Bromus madritensis, Carrichtera annua, Erodium neuradifolium, Ifloga spicata, Lamarckia aurea, Lotus glinoides, Malva parviflora, Medicago laciniata, M. polymorpha, Stipa capensis* and *Volutaria canariensis*, among others.

Distribution: L, F, C, T.

Fig. 6.69 *Iflogo spicatae-Stipetum capensis*. Montaña Grande. Güímar. Tenerife

Launaeo nudicaulis-Resedetum lancerotae
This is a small-sized therophyte community with a fleeting winter-spring development that mostly grows in volcanic sand layers, but also in clayey soil within the arid Inframediterranean belt.

Its main characteristic species are *Reseda lancerotae, Cuscuta planiflora, Launaea nudicaulis, Ifloga spicata* and *Lolium parabolicae*. Other common companion species are *Calendula aegyptiaca* and *Echium bonnetii*.

Distribution: L, F.

Senecionetum incrassati
This is a short-lived (sub)nitrophilous therophytic community developing in winter-spring, growing along dirt road edges and in disturbed volcanic lands, within the arid Inframediterranean belt, described from Faro de la Orchilla on El Hierro.

Its main characteristic species are *Senecio incrassatus, Spergularia fallax, Ifloga spicata* and *Chenopodium coronopus*. Other nitrophilous species of diverse phytosociological adscription are present, like *Chenopodium murale, Malva parviflora, Forsskaolea angustifolia* or *Polycarpaea divaricata*.

Distribution: H.

Senecioni coronopifolii-Echietum bonnetii (Fig. 6.70)
This is a subnitrophilous therophytic community with a fleeting winter-spring development that preferentially grows on lithosols of volcanic loose material and abandoned cultivation fields within the arid Inframediterranean belt.

Its main characteristic species are *Senecio glaucus* subsp. *coronopifolius, Echium bonnetii, Volutaria canariensis, Lobularia lybica, Anagallis arvensis, Plantago ovata, Ifloga spicata* and *Notoceras bicorne*.

Distribution: T.

Fig. 6.70 *Senecioni coronopifolii-Echietum bonnetii*. Montaña Roja. El Médano. Tenerife

***Volutaria tubuliflora* community**

This is a therophytic ruderal community growing on roadsides and in abandoned cultivation fields within the arid Inframediterranean belt. Unlike the *Iflogo-Stipetum capensis* it has a spring flowering period.

The main characteristic species are *Volutaria tubuliflora, Erodium neuradifolium, Lotus glinoides, Launaea nudicaulis, Reichardia tingitana, Stipa capensis, Avena barbata, Eragrostis barrelieri,* among others.

Distribution: L

Stellarienea mediae

Includes herbaceous communities of cultivated fields (arvensis communities), among these cereal fields.

Aperetalia spicae-venti

Weed communities of cereal fields on sandy or sandy-clayey soils.

Scleranthion annui

Weed communities of cereal fields on sandy or sandy-clayey soils.

***Raphanus raphanistrum-Scandix pecten-veneris* community**
Community of weeds that grows among cereal (oat, barley, rye, wheat) and other winter-spring cereal crops. Common in subhumid Thermo- and Mesomediterranean northern territories.

Raphanus raphanistrum and *Scandix pecten-veneris* are characteristic among other weeds like *Calendula arvensis, Fumaria muralis, Oxalis pes-caprae, Galium verrucosum, Sinapis alba, S. arvensis, Spergula arvensis* and *Galactites tomentosa.*

Distribution: C, T.

Solano nigri-Polygonetalia convolvuli
Eurosiberian and Mediterranean weed communities of non-cereal summer crops.

Fumarion wirtgenii-agrariae
Thermophilic Mediterranean weed communities with a winter-spring flowering period typical of both rain-fed and irrigated crops.

***Convolvulus arvensis* community**
This is a weed community of uncertain phytosociological ascription, growing on abandoned old cereal fields on Lanzarote, within the arid and semiarid Inframediterranean belts.

Several species are characteristic, most of them of the *Stellarietea*: *Convolvulus arvensis, Melilotus sulcatus, Calendula arvensis, Lamium amplexicaule, Reichardia tingitana, Euphorbia terracina, Avena sterilis* subsp. *ludoviciana, Sinapis arvensis, Bromus rubens, etc.*

Distribution: L

***Sinapis arvensis – Fumaria muralis* community**
A weed community of cultivated fields and edges of abandoned crops typical of northern Thermomediterranean subhumid-humid areas.

Apart from *Sinapis arvensis* and *Fumaria muralis*, other plants like *Calendula arvensis, Oxalis pes-caprae* and *Bromus rigidus* are common.

Distribution: C,T.

Polygono-Chenopodion polyspermi
Weed vegetation of non-cereal crops with a summer-fall development. In the Mediterranean region they thrive on irrigated soils.

Digitario ischaemi-Setarienion viridis
Communities growing on poor sandy soils.

Setario verticillatae-Echinochloetum cruris-galli
This is a therophytic graminoid community that colonizes irrigated fruit and vegetable fields with a sandy soil texture, mainly in Thermo- and Mesomediterranean territories. It is frequent in tomato fields.

Frequent species are *Setaria adhaerens, Echinochloa crus-galli, Lamarckia aurea, Lophochloa pumila, Schismus barbatus, Digitaria sanguinalis, Chenopodium album, Amaranthus hybridus* and *Galinsoga parviflora*, among others.

Distribution: L, F, C, T, G, P, H.

6.2.5.2 Fringe and Megaforbic Vegetation

Cardamino hirsutae-Geranietea purpurei
Class that includes fleeting annual communities which grow in spring-summer at the edge of forests and shrublands, in lightly nitrified semi-shaded environments rich in organic nutrients, in Mediterrranean and Temperate territories. Holarctic.

Cardamino hirsutae-Geranietalia purpurei
Single order.

Geranio purpurei-Torilidion neglectae
Includes Infra- Thermomediterranean communities in the Canary territory.

Carduo clavulati-Urticetum stachyoidis
This is a subnitrophilous therophytic community, which grows at path edges and at the foot of rocks mainly within the dry-subhumid Infra- and Thermomediterranean belts (c) potential territory of laurel forest.

It is characterized by the Canary endemics *Urtica stachyoides* and *Carduus clavulatus* together with *Drusa glandulosa* and *Erucastrum canariense,* apart from other common herbaceous species of the *Tuberarietea* (*Rumex bucephalophorus* subsp. *canariensis*) or nitrophilous species of the *Stellarietea* (*Erodium chium, Sisymbrium irio* or *Geranium molle*).

Distribution: Endemic to (H), (P), T, (C).

Drusa glandulosa-Parietaria debilis **community**
This is a subnitrophilous therophytic semi-sciophilous community that grows in environments that are rich in organic matter, within the upper-semiarid Inframediterranean (thermo-sclerophyllous woodland territory) and dry Infra-Thermomediterranean with t-w.c (dry laurel forest) belts.

It is characterized by *Drusa glandulosa*, *Parietaria debilis* and *Erucastrum canariense* which are usually accompanied by several nitrophilous species of the *Stellarietea*.

Distribution: (All islands), H, T.

Galio aparines-Torilidetum neglectae
This is a scio-nitrophilous therophytic community, which grows in the anthropic environment of the laurel forest potential territory.

Its characteristic species are *Galium aparine, Torilis arvensis* subsp. *neglecta, Geranium purpureum, G. rotundifolium, Theligonum cynocrambe*, generally accompanied by several species of the *Stellarietea* (*Geranium molle, Stellaria media, Bromus rigidus, Anagallis arvensis, Fumaria muralis, Sonchus oleraceus, Oxalis pes-caprae, Achyranthes aspera, Mercurialis annua).*

Distribution: C, T, G, P.

Geranio pusilli-Anthriscion caucalidis
Includes Meso- and Supramediterranean Western-Mediterranean and Canarian communities, and Infra- to Mesotemperate Cantabrian-Atlantic communities.

***Arabidosis thaliana-Myosotis gracillima* community**
Observed in some northern localities of Tenerife within the subhumid Mesomediterranean belt in the territory of humid Canary pine forest (*Sideritido solutae-Pinetum canariensis*).

The characteristic species are *Arabidopsis thaliana, Myosotis ramosissima* subsp. *gracillima, Geranium purpureum, G. rotundifolium, Theligonum cynocrambe, Galium spurium* subsp. *aparinella, G. muralis,* and *Cardamine hirsuta.*

Distribution: T.

Myosotidetum canariensis
This is a subnitrophilous therophytic community that grows in semi-sciophilous laurel forest edges and path edges, within subhumid-humid Thermo-Mesomediterranean belts.

Characteristic plants are *Myosotis discolour* subsp. *canariensis, Drusa glandulosa, Luzula elegans* and *Geranium purpureum.* Apart from these, *Parietaria cf. judaica* and *Galium parisiense* have high presence.

Distribution: Endemic to H

Trifolio medii-Geranietea sanguinei
Includes humicolous and non-nitrophilous perennial herb communities, growing in spring-summer in semi-shady places at the edges of forests and shrublands. Holarctic.

Origanetalia vulgaris
Single order.

Ranunculo cortusifolii-Geranion canariensis
Groups Canary and Madeiran prerennial herb fringe communities of laurel and pine forests.

Ranunculo cortusifolii-Geranietum canariensis (Fig. 6.71)
This is a mesophytic, semi-sciophilous and humicolous forest fringe community. It is mainly made up of hemicryptophytic vivacious herbs growing in potential laurel forest (of the *Ixanto-Laurion novocanariensis* and *Visneo-Apollonion* alliances), within the subhumid-humid Thermo- and Mesomediterranean belts (c).

Among its characteristic species the following are frequent: *Calamintha sylvatica* subsp. *ascendens*, *Carex divulsa*, *Geranium reuteri* (=*G. canariense*), *Myosotis latifolia*, *Origanum vulgare*, *Pericallis appendiculata*, *P. cruenta*, *P. webbii*, *P. tussilaginis*, *Ranunculus cortusifolius* and *Rumex maderensis.*

Distribution: Endemic to C, T, G, P.

Fig. 6.71 *Ranunculo cortusifolii-Geranietum canariensis*. La Esperanza. Tenerife

Rumici maderensis-Pimpinelletum dendrotragii (Fig. 6.72)
This is a semi-sciophilous community that grows in moist environments like ravine beds, and rock and cliff bases, within dry-subhumid Mesomediterranean belts, as a perennial herbaceous fringe of the pine forest.

The characteristic species are *Rumex maderensis*, *Pimpinella dendrotragium* and *Ranunculus cortusifolius*.

Distribution: Endemic to T.

Fig. 6.72 *Rumici maderensis-Pimpinelletum dendrotragii*. Pinares sobre Anocheza. Güímar. Tenerife

6.2.6 *Grassland and Meadow Vegetation*

6.2.6.1 Therophytic Grasslands

Tuberarietea guttatae
Class that includes non-nitrophilous, therophytic pioneer communities of spring or early summer development. Mediterranean and Eurosiberian (Submediterranean).

Tuberarietalia guttatae
Includes acidophilous or calcifugal non-psamophilous communities.

Tuberarion guttatae
Includes Infra- to Supramediterranean communities on sandy-clayey soils.

Hypochaerido glabrae-Tuberarietum guttatae (Fig. 6.73)
This is a short-lived therophytic community, growing in winter-spring in sandy-clayey soils, sometimes shallow or eroded, poor in organic matter. It inhabits (upper-semiarid) dry-subhumid Infra- to Mesomediterranean belts, outside the trade-wind cloud zone.

A species-rich community; its characteristic species are *Tuberaria guttata, Hypochoeris glabra, Vulpia myuros, Trifolium arvense, T. campestre, Vicia disperma, Galium parisiense, Aira caryophyllea, A. cupaniana, Leontodon taraxacoides* subsp. *longirostris, Asterolinon linum-stellatum, Arenaria leptoclados,* among others.

Distribution: Endemic to L, C. T, G, P, H.

Fig. 6.73 *Hypochaerido glabrae-Tuberarietum guttatae*. Sobre Las Rosas. La Esperanza. Tenerife

Tuberario guttatae-Ophioglossetum lusitanici

This is a therophytic community growing in winter and early spring in shallow silty or even clayey soils, near rock outcrops, and among pasture lands, in the upper parts of El Hierro in the dry to subhumid Thermo- and Mesomediterranean belts.

Character species of the associacion on El Hierro are *Ophioglossum lusitanicum, Crassula tillaea, Asterolinon linum-stellatum, Luzula elegans, Stipa capensis* and *Vicia disperma.* Apart from these other common species of the *Tuberarietea* occur, such as *Tuberaria guttata, Hypochoeris glabra, Trifolium campestre* and *Leontodon taraxacoides* subsp. *longirostris* among several therophytes of diverse phytosociological ascription.

Distribution: H. Mediterranean.

Vulpio myuri-Laphangietum teydei (Fig. 6.74)

This is a therophytic community characterized by *Vulpia myuros* and the Tenerife endemic *Laphangium teydeum* (*Gnaphalium teydeum*), which are accompanied by xero-resistent bryophytes growing in the humid edges of fumaroles and hot soils of La Rambleta, on the peak of Mt Teide. It is found on oligotrophic gleys, semi-flooded by water condensed from vapor, and becoming dry in summer due to strong insolation, high temperature and nil rainfall.

Distribution: Endemic to T.

Fig. 6.74 *Vulpio myuri-Laphangietum teydei.* Teide peak. Tenerife

Malcolmietalia
Non-nitrophilous, therophyte pioneer communities growing on deep sandy soils and coastal or continental paleodunes in the western Mediterranean.

Ononidion tournefortii
Canaries and SW Morocco.

Ononido tournefortii-Cyperetum capitati
This is an annual non-nitrophilous grassland which develops on mobile marine sands during the rainy season.

Ononis tournefortii and *Cyperus capitatus* are the characteristic species. This association was originally described from the imported Saharan sand of Las Teresitas beach, Tenerife, and later recorded in Western Africa where it is abundant.

Distribution: T.

6.2.6.2 Perennial Xerophytic and Mesophytic Grassland

Lygeo spartii-Stipetea tenacissimae
Class that groups vivacious thermophilic grasslands, dominated by hemicryptophytic or chamaephytic grasses, growing on deep, non-saline nor gleyed, permeable, base-rich soils. They are typical seral stages of the Mediterranean *Quercetea ilicis* forests and of the scrublands and shrublands of the *Kleinio-Euphorbietea* and woodlands of the *Rhamno-Oleetea*. Wide distribution in (arid) semiarid to subhumid Infra- to Supramediterranean bioclimatic belts. Mediterranean distribution.

Hyparrhenietalia hirtae
Includes savannoid grasslands, rich in species of *Hyparrhenia, Andropogon* and *Heteropogon*, growing on base-rich soils somehow disturbed and sometimes subnitrophilous. They have their optimum from the semiarid to subhumid Infra- to Mesomediterranean belts. Mediterranean.

Hyparrhenion sinaicae
Single alliance in this territory.

Cenchro ciliaris-Hyparrhenietum sinaicae (Fig. 6.75)
This is a community characterized by a high presence of three hemicryptophytic grasses: *Cenchrus ciliaris, Hyparrhenia sinaica* and *Aristida adscensionis.* It can occur widespread and may reach a high density in the arid-semiarid Infra- and Thermomediterranean belts. It is frequently found mixed with other communities and in its open spaces, in wet periods, a high number of therophytes can develop. As a substitutional stage it settles on dryer and cleared places in the potential territory of the *Kleinio neriifoliae-Euphorbietalia canariensis* (*Euphorbia* scrub and shrublands) and the *Oleo-Rhamnetalia crenulatae* (thermo-sclerophyllous woodland),

giving rise to an evolved successional grassland frequently interspersed with communities of the *Micromerio-Cistion canariensis*.

This grassland grows on all Canary Islands, although it is very rare on Lanzarote and Fuerteventura, where it is relegated to clayey-rocky terrain of the potential territory of the *Kleinio-Euphorbietea canariensis*. It is frequent in areas cleared for livestock grazing, especially on slopes with well developed soil, which contributes to increasing the desertic appearance of the more xeric locations in the lower belt. It also occurs on wind-blown (aeolian) organic coastal dunes and inland sand fields. Finally, its preference is clear for path edges and ditches of asphalted roads with scarce nitrification, as well as abandoned cultivated land edges.

Apart from the typical subass. *hyparrhenietosum sinaicae* two more subassociations are recognized:

Subass. *tricholaenetosum teneriffae*: a hemicryptophytic grassland, of medium or high cover, dominated by *Tricholaena teneriffae*, which preferably grows on lapilli of recent volcano cones in the arid Inframediterranean belt. It is locally abundant in the south of Tenerife and Gran Canaria, where it can reach 600 m a.s.l. It is rarer on the southern slopes of Lanzarote and Fuerteventura (between Pájara and Puerto de la Peña), La Gomera and El Hierro. Its presence on La Palma seems doutfull. The community also grows on aeolian basaltic sands in Tenerife, being particularly well represented between Playa del Socorro and Montaña Grande, in Güímar. There it grows together with *Launaeo-Schizogynetum sericeae* vegetation, in the potential climatophilous territory of *Euphorbia balsamifera scrub.*

Subass. *eremopogonetosum foveolati*: a hemicryptophytic grassland, with dense cover, occurring on sandy substrates of the lower-arid to hyperarid Inframediterranean belts. It is particularly present in El Médano and La Tejita, Tenerife. *Eremopogon foveolatus* and *Tetrapogon villosus* are its differential species. In these environments *Hyparrhenia sinaica* is rarer and its presence is reduced to the more humid soils of small gully beds.

The presence of invasive *Pennisetum setaceum* in the *Cenchro-Hyparrhenietum* habitat is more and more frequent. Its dense populations characterize the facies of *Pennisetum setaceum*.

Distribution: All islands.

Fig. 6.75 *Cenchro ciliaris-Hyparrhenietum sinaicae*. Malpaís de Güímar. Tenerife

Poetea bulbosae
Oceanic western Mediterranean and Temperate Submediterranean pasturelands, developed under livestock grazing, formed by dwarf perennial grasses, and diverse prostrate chamaephytes and hemicryptophytes, like *Poa bulbosa*, clovers and plantain. They are very productive especially in winter, but they dry out in summer, and quickly grow green again with the first rain in the fall. A large number of adapted therophytes grow during spring in the gaps in the dense carpet.

Poetalia bulbosae
Single order.

Molineriello minutae-Trifolion subterranei
Includes communities developed in siliceous substrate, typical of dry to subhumid Thermo- to Oromediterranean and Temperate Submediterranean.

Trifolio subterranei-Poetum bulbosae
This is a dense small-sized, perennial pastureland, growing on manured and sheep-grazed subhumid Mesomediterranean territories, where sheep activity has persisted for a long time.

Characteristic species are the endemic *Poa pittardiana*, dominant in the community, *Trifolium subterraneum* and *Romulea columnae* subsp. *grandiscapa,* accompanied by some annual species of the *Tuberarietea guttatae* like *Hypochoeris oligocephala* and *Vulpia myurus,* and some other nitrophilics belonging to the *Stellarietea mediae* like

Sherardia arvensis, Erodium cicutarium, Cerastium glomeratum and *Muscari comosum*, in a lesser or higher proportion depending on the grazing intensity.

Distribution: C T, G, H.

6.2.6.3 Wet Soil Meadows

Molinio caeruleae-Arrhenatheretea elatioris
A large class that includes mesophile to hygrophilous manured meadows and pasture communities occurring on deep soils. It is widely spread by grazing and human activities in Temperate, Boreal and Mediterranean territories all over the world, although it has a Euroasiatic optimum and origin.

Arrhenatheretalia elatioris
Mesophilic eutrophic grazed and mowed meadows created and maintained through agro-pastoral techniques. Optimum Temperate distribution, but also Mediterranean on soils with additional summer moisture.

Arrhenatherion elatioris
Mesophilic eutrophic grazed, manured, and mowed meadows maintained for a long period, on soils with a light temporary water saturation. Temperate and northern Mediterranean distribution.

***Arrhenatherum elatius* subsp. *bulbosum* community**
This is a perennial grassland, linked to bygone scything and livestock grazing, which sparsely grows in humid and subhumid territories, especially of potential laurel and pine forest.

It is dominated by *Arrhenatherum elatius* subsp. *bulbosum, Agrostis castellana* and *Brachypodium sylvaticum. Calamintha sylvatica* subsp. *ascendens*, *Prunella vulgaris* and *Origanum virens* are frequently also found.

Although present in several parts of La Palma (Puntallana mountains, Los Sauces and Garafía) it is well represented on the high land above Las Breñas and El Paso.

Distribution: P.

Holoschoenetalia vulgaris
Mediterranean and Submediterranean Temperate hygrophilic meadows and reeds.

Molinio arundinaceae-Holoschoenion vulgaris
Mediterranean reed meadows.

Scirpo globiferi-Juncetum acuti (Fig. 6.76)
This is a helophyte community that grows on damp or flooded soils in winter and spring, but subject to moderate summer drying in its upper horizons. It has its optimum in ravine beds of the Western Canary subprovince. On Lanzarote it is very rare.

Juncus acutus and *Scirpus holoschoenus* subsp. *globiferus* are its more characteristic taxa.

Distribution: L, C, T, G, P.

Fig. 6.76 *Scirpo globiferi-Juncetum acuti*. Caldera de Taburiente. La Palma

Plantaginetalia majoris

Eutrophic hygrophilic pastures rich in nitrogen- and phosphorus-rich nutrients.

Paspalo-Polypogonion viridis

Mediterranean decumbent graminoid meadows, frequently flooded.

***Aster squamatus* community**

This is a therophytic community that grows in small clayey temporary flooded depressions after rain episodes.

It is characterized by *Aster squamatus, Phalaris minor* and *Polypogon monspeliensis*.

Distribution: L.

Cyperetum laevigati

This is a community that grows both in humid fresh-water and salt-water depressions on Gran Canaria. It was described for saline sand dune depressions of Maspalomas, giving rise to dense monospecific populations; also recorded inland.

Apart from *Cyperus laevigatus* subsp. *laevigatus* only scarce individuals of *Suaeda vermiculata* and *Launaea arborescens* are to be found.

Community of uncertain phytosociological ascription.

Distribution: Endemic C.

***Cyperus laevigatus-Apium graveolens* community**

This is a dense cespitose community that mainly grows on oozing red-ochre strata. Sometimes it has the appearance of a small reed-bed.

It is characterized by *Cyperus laevigatus, Apium graveolens* and *Polypogon monspeliensis*.

Distribution: L, (F).

***Cyperus rotundus-Cynodon dactylon* community**
This is an herbaceous community observed growing on nitrophilized clay soils, temporary flooded by a sewage outlet near Playa Honda.

Distribution: L.

Paspalo distichi-Polypogonetum viridis
This is a grass community growing on nitrogen-rich wet or flooded soils, especially present in damp ravine or watercourse edges subject to intense grazing.

Its characteristic species are *Cynodon dactylon, Cyperus eragrostis, Polypogon viridis* and *Rumex crispus*, among others.

Distribution: F, C, T, G, P, H.

Potentillion anserinae
Meso-eutrophic reedy hygrophilous alluvial meadows.

***Juncus effusus-Epilobium parviflorum* community**
This is a community typical of moist soils that has been described from subhumid Thermomediterranean areas of North Tenerife (Agua García).

It is characterized by *Juncus effusus* and *Epilobium parviflorum* accompanied by *Agrostis castellana, Galium elongatum, Holcus lanatus* and some other companions like *Mentha pulegium, Myosotis latifollia* and *Galium aparine.*

Distribution: Endemic to T.

6.2.7 Classes of the Climatophilous Vegetation

6.2.7.1 Euphorbia, Plocama, and Other Desertic and Xeric Scrub and Shrublands

Kleinio neriifoliae-Euphorbietea canariensis
Succulent scrub and shrubby potential or seral communities characterized by different *Euphorbia* species (spurges), some dendroids ("tabaibas"), some cactiforms ("cardones") or succulent aphyllae with slender pencil-stems ("tolda"), or *Plocama pendula* ("balo"), which mainly grow in the hyperarid, arid and semiarid Infra- and Thermomediterranean belts. Endemic class of the Canary biogeographic province (Table 6.1).

Kleinio neriifoliae-Euphorbietalia canariensis
Single order.

Kleinio neriifoliae-Euphorbion canariensis
Includes succulent potential scrublands and shrubby communities (climatophilous, and edaphophilous) characterized by different *Euphorbia* species, some dendroids

Table 6.1 Characteristic species of the vegetation of *Kleinio neriifoliae-Euphorbietea canariensis*

Kleinio neriifoliaea-Euphorbietea canariensis				
Scientific name	**Family**	**Common name**	**Biotype**	**Distribution**
Allagopappus canariensis	Asteraceae	"madama de risco"	Ch/N	H?,G,T,C
Argyranthemum coronopifolium	Asteraceae	"magarza azulada"	Ch	P,T
Argyranthemum gracile	Asteraceae	"magarza grácil"	Ch	T
Argyranthemum sundingii	Asteraceae	"magarza de Sunding"	Ch	T
Asparagus arborescens	Asparagaceae	"esparragón"	N	H,P,G,T,C,F,L
Asparagus nesiotes subsp. *purpuriensis*	Asparagaceae	"esparraguera majorera"	N/Li	F,L
Asparagus pastorianus	Asparagaceae	"esparraguera espinablanca"	N	G?,T,C,F,L
Asparagus umbellatus	Asparagaceae	"esparraguera común"	N/Li	H,P,G,T,C,F
Asteriscus intermedius	Asteraceae	"tojia"	Ch/N	F,L
Atalanthus capillaris	Asteraceae	"balillo fino"	N	G,T,C
Atalanthus microcarpus	Asteraceae	"balillo tinerfeño del sur"	N	T
Atalanthus pinnatus	Asteraceae	"balillo alpispillo"	N	G,T,C,F,L?
Bystropogon odoratissimus	Lamiaceae	"poleo de cardonal"	N	T
Campylanthus salsoloides	Plantaginaceae	"romero marino"	N	P?,G,T,C,F,L
Caralluma burchardii	Apocynaceae	"cuernúa"	Ch	C?,F,L
Ceropegia dichotoma	Apocynaceae	"cardoncillo verde"	Ch	H,P,T
Ceropegia fusca	Apocynaceae	"cardoncillo gris"	Ch	P?,T,C
Convolvulus floridus	Convolvulaceae	"guaidil"	N/Mi	H,P,G,T,C,F,L
Echium brevirame	Boraginaceae	"arrebol"	N	P
Euphorbia aphylla	Euphorbiaceae	"tolda"	Ch/N	G,T,C,F?
Euphorbia balsamifera subsp. *balsamifera*	Euphorbiaceae	"tabaiba dulce"	N	H,P,G,T,C,F,L
Euphorbia berthelotii	Euphorbiaceae	"tabaiba picuda"	N	G
Euphorbia canariensis	Euphorbiaceae	"cardón"	N	H,P,G,T,C,F,L?
Euphorbia handiensis	Euphorbiaceae	"cardón de Jandía"	N	F
Euphorbia lamarckii subsp. *lamarckii*	Euphorbiaceae	"tabaiba amarga"	N	T
Euphorbia lamarckii subsp. *wildpretii*	Euphorbiaceae	"tabaiba amarga"	N	H,P,G,T
Euphorbia regis-jubae subsp. *regis-jubae*	Euphorbiaceae	"tabaiba salvaje"	N	C,F,L
Helianthemum canariense	Cistaceae	"jarilla turmera"	Ch	H,P,G,T,C,F,L
Kickxia sagittata	Plantaginaceae	"picopajarito costero"	Ch	T,C,F,L
Kickxia scoparia	Plantaginaceae	"picopajarito de escoba"	Ch	P,G,T,C
Kleinia neriifolia	Asteraceae	"verode"	N	H,P,G,T,C,F,L

(continued)

Table 6.1 (continued)

Kleinio neriifoliaea-Euphorbietea canariensis				
Scientific name	**Family**	**Common name**	**Biotype**	**Distribution**
Micromeria hyssopifolia var. *kuegleri*	Lamiaceae	"tomillo burro"	Ch	T
Micromeria teneriffae var. *cordifolia*	Lamiaceae	"tomillo de costa"	Ch	T
Neochamaelea pulverulenta	Rutaceae	"orijama"	N	H,G,T,C
Pancratium canariense	Amaryllidaceae	"lágrimas de la virgen"	G	H,P,G,T,C,F,L
Parolinia intermedia	Brassicaceae	"dama de Tenerife"	N	T
Parolinia ornata	Brassicaceae	"dama de Gran Canaria"	N	C
Periploca laevigata	Apocynaceae	"cornical"	N/Li	H,P,G,T,C,F,L
Plocama pendula	Rubiaceae	"balo"	N	H,P,G,T,C,F
Reichardia famarae	Asteraceae	"cerraja de Famara"	Ch	F,L
Reichardia ligulata	Asteraceae	"cerraja de costa"	Ch	H,P,G,T,C,F
Reseda scoparia	Resedaceae	"gualdón canario"	Ch	P,G,T,C
Rubia fruticosa subsp. *melanocarpa*	Rubiaceae	"tasaigo negro"	N/Li	G,T,C
Rutheopsis herbanica	Apiaceae	"tajame"	G	F,L
Scilla haemorrhoidalis	Liliaceae	"cebolla almorrana menor"	G	H,P,G,T,C,F,L
Seseli webii	Apiaceae	"apio marino"	G	H,P,G,T,C
Sideritis dendrochaorra	Lamiaceae	"chajorra de Anaga"	Ch/N	T
Transgressive species with significant presence in some communities of the class				
Aeonium lancerotense	Crassulaceae	"bejeque de malpaís"	Ch/N	L
Aeonium percarneum	Crassulaceae	"bejeque rosado"	Ch/N	C
Aeonium valverdensis	Crassulaceae	"bejeque herreño"	Ch/N	H
Astydamia latifolia	Apiaceae	"servilleta"	H	H,P,G,T,C,F,L
Ceballosia fruticosa	Boraginaceae	"duraznillo"	N	H,P,G,T,C,F,L
Gymnocarpos decandrus	Caryophyllaceae	"matocosta milengrana"	C/N	G,T,C,F,L
Jasminum odoratissimum	Oleaceae	"jazmín silvestre"	N	H,P,G,T,C,F
Lycium intricatum	Solanaceae	"espino de mar"	N	P,G,T,C,F,L
Salsola divaricata	Chenopodiaceae	"matabrusca negra"	N	P,G,T,C,F,L
Schizogyne sericea	Asteraceae	"salado blanco"	N	H,P,G,T,C,F?,L

Common names according to Machado and Morera (2005)

("tabaibas"), some cactiforms ("cardones") or succulent aphyllae with slender pencil-stems ("toldas") of the hyperarid to semiarid Infra- and Thermomediterranean belts. Canarian distribution.

Helianthemo canariensis-Euphorbienion balsamiferae
Nanophanerophytic potential crassicaule shrublands characterized by pachycaule dendroid *Euphorbia balsamifera*, typical of the hyperarid to lower-semiarid Inframediterranean belts, growing throughout the Canary Islands. The alliance also includes the cactiform *E. handiensis* scrub growing on Fuerteventura in colluvia of the hyperarid and arid Inframediterranean belts.

Asterisco intermedii-Euphorbietum balsamiferae (Fig. 6.77)
This is a low nanophanerophytic crassicaule scrub and shrubland, summer deciduous, endemic to Lanzarote, which occurs from the hyperarid to lower-semiarid Inframediterranean belts.

It is characterized by the pachycaule *Euphorbia balsamifera* together with *Asparagus nesiotes* subsp. *purpuriensis*, *Caralluma burchardii* subsp. *burchardii*, *Euphorbia regis-jubae*, *Helianthemum canariense*, *Kleinia neriifolia*, *Rubia fruticosa* and *Rutheopsis herbanica*.

Although with a large potential island distribution today it is dispersed in several locations: Llanos del Rubicón in the S, several malpaíses and volcanic cones near El Golfo in the SW, on lava flow gaps of the National Park unaffected by the Timanfaya eruptions in the W, and from coast to summit in several localities of the Famara Massif. In the N of the island its best stand covers much of the Malpaís de la Corona.

It corresponds to the terminal community of the hyperarid-semiarid desertic-xeric Inframediterranean climatophilous series of *E. balsamifera* in Lanzarote (*Asterisco intermedii-Euphorbio balsamiferae sigmetum*).

The subass. *euphorbietosum balsamiferae* (*typicum*) is the mature stage of the hyperarid and arid Inframediterranean vegetation. It is species-poor and shows a floristic cohort of aero-halophile species which decreases as elevation goes up. The following facies: *Campylanthus salsoloides*, *Euphorbia regis-jubae*, and *Kleinia neriifolia* can be recognized. Its substitutional scrublands belong to the *Launaeo-Schizogynion sericeae*, and are mainly dominated by *Chenoleoides tomentosa, Lycium intricatum* and *Salsola vermiculata*.

The subass. *aeonietosum lancerottensis* is the mature stage of the semiarid Inframediterranean vegetation, and occupies the theoretical potential territory of the absent *Euphorbia canariensis* shrubland on Lanzarote. It is more species-rich than the previous subassociation. Among its differential species the following stand out: *Aeonium lancerottense, Asparagus horridus, Asteriscus intermedius, Ferula lancerottensis* and *Periploca laevigata*. When disturbed it is substituted by scrub belonging to the *Artemisio-Rumicion lunariae*, with predominance of several endemic taxa like *Andryala grex pinnatifida*, *Argyranthemum maderense*, *Asteriscus intermedius*, *Lavandula pinnata*.

Distribution: Endemic to L.

Fig. 6.77 *Asterisco intermedii-Euphorbietum balsamiferae*. Malpaís de La Corona. Lanzarote

Ceropegio fuscae-Euphorbietum balsamiferae (Fig. 6.78)
This is a nanophanerophytic crassicaule scrub or shrubland, summer deciduous, which has a preference for the desertic bioclimate of the SW, S and SE slopes of Tenerife, from the NW tip of Punta de Teno around the southern coast as far as Santa Cruz, reaching its maximum elevation in the S (about 500 m). However, this area is not uniform and is sometimes interrupted by the edaphophilous descent of the cardón shrubland *Periploco laevigatae–Euphorbio canariensis sigmetum* over recent lava flows (malpaíses) (Arafo, Güímar, Las Galletas, Guía de Isora), and on the cliffs of Teno. Also, the *Ceropegio Euphorbietum balsamiferae* climbs up as an edaphophile over the abundant salic pyroclasts (mainly pumice) and salic lava flows within the semiarid xeric Inframediterranean belt in this south sector of the island.

It is dominated by the pachycaule species *Euphorbia balsamifera*, but also characterized by *Ceropegia fusca*, *Euphorbia lamarckii*, *Kleinia neriifolia*, *Lycium intricatum*, *Neochamaelea pulverulenta*, *Plocama pendula, Schizogyne sericea, Helianthemum canariense*, *Lotus sessilifolius* and *Reseda scoparia*. Sometimes *Euphorbia canariensis* penetrates into arid territories and grows within *E. balsamifera* scrub, generally on basaltic outcrops.

It corresponds to the terminal community of the hyperarid and arid desertic Inframediterranean climatophilous series, as well as the semiarid xeric Inframediterraean edaphophilous series over pumice, of *E. balsamiferae* on Tenerife (*Ceropegio fuscae-Euphorbio balsamiferae sigmetum*).

The subass. *euphorbietosum balsamiferae* (typical) is the most common within the territory. A *Plocama pendula* facies can be identify on porous soils, lapilli, and fractured lava flows.

The subass. *salsoletosum divaricatae* grows near the coast, in places subject to saline air. *Salsola divaricate* and *Lycium intricatum* are its differential species.

Nowadays the community has greatly retreated as a consequence of agricultural, urban, and industrial expansion over its area – with particular impact from tourism, but some good plots still remain, mainly on the SE coast. The main substitutional scrub is the nitrohalophilous *Launaeo arborescentis–Schizogynetum sericeae*; also the herbaceous nitrohalophilous community *Mesembryanthemetum crystallini* is widely spread, mainly over abandoned tomato-growing areas, as well as subnitrophilous communities of the *Resedo lanceolatae–Moricandion*. Also the *Euphorbietum lamarckii* grows in less disturbed places.

Distribution: Endemic to T.

Fig. 6.78 *Ceropegio fuscae-Euphorbietum balsamiferae*. Malpaís de Güímar. Tenerife

Echio breviramis-Euphorbietum balsamiferae
This is a nanophanerophytic oligospecific crassicaule scrub or shrubland, summer deciduous, settled in the hyperarid and arid Inframediterranean belts of La Palma. It potentially covers a narrow coastal strip in the W, S tip and extreme SE of the island. Today it is well represented to the north of Tazacorte (in the W), in a very rough

territory. In the remainder of this potential area its presence is limited due to proliferation of cultivated land and recent volcanic eruptions. However, dispersed locally on windy promontories and capes ("prois"), cliffs and warm slopes with a western exposure, there are some (usually limited) locations outside this area in which the association is present. The floristic composition and the lichen communities allow this association to be distinguished from the *E. balsamifera* facies of *E. canariensis* shrubland.

It is a species-poor association characterized by the pachycaule *Euphorbia balsamifera*, together with *Echium brevirame*, *Euphorbia lamarckii*, *Kleinia neriifolia* and *Rubia fruticosa*.

It corresponds to the terminal community of the hyperarid and arid desertic Inframediterranean series of *Euphorbia balsamifera* on La Palma (*Echio breviramis-Euphorbio balsamiferae sigmetum*).

Within its potential territory, the *Schizogyno sericeae-Euphorbietum wildpretii* subas. *echietosum breviramis* preferentially occupies places with abundant lapilli or sand in canyons and watercourses as well as colluvial deposits on sea cliffs.

The subas. *schizogynetosum sericeae* constitutes an alluvial permanent community of desertic water courses well represented at the mouth of the main ravine Barranco de las Angustias.

Distribution: Endemic to P.

Euphorbietum balsamiferae (Fig. 6.79)
This is a nanophanerophytic oligospecific crassicaule scrub or shrubland, up to 2 m high, summer deciduous, endemic to Gran Canaria, which preferably grows in the desertic bioclimate. Its potential area corresponds to the hyperarid and arid desertic Inframediterranean belts.

It is characterized by the pachycaule species *Euphorbia balsamifera*, together with *Asparagus pastorianus*, *Ceropegia fusca*, *Euphorbia regis-jubae*, *Helianthemum canariense*, *Kleinia neriifolia*, *Neochamaelea pulverulenta*, *Rubia fruticosa* and *Scilla haemorrhoidalis*, among others. Sometimes *Euphorbia canariensis* penetrates into arid territories and grows within *E. balsamifera* scrub, generally on rocky outcrops.

It corresponds to the terminal community of the hyperarid and arid desertic Inframediterranean climatophilous series of *E. balsamifera* in Gran Canaria, which surrounds the island. It potentially covers a narrow band on the N and NW cliffs where it can reach up to 100–200 m, broadening in the E, and in the S even exceeding 500 m. It also shows a great penetration inland in the Aldea basin in the W. On N and NW slopes, a good part of its potential area is occupied by the edaphoxerophilous series of *Euphorbia aphylla* (*Astydamio–Euphorbio aphyllae sigmetum*). On E and S slopes, the edaphophilous sabulicolous Inframediterranean western Canary *geosigmetum* is found on sandy ground. In addition, the cardón shrubland (*Aeonio percarnei–Euphorbio canariensis sigmetum*) is found as an edaphophilous series on the lava flows of the Amurga massif on the S slope, within the Paleocanarian geological zone. The remaining territory of those belts corresponds to the potential area of the *Euphorbietum balsamiferae*.

The subass. *euphorbietosum balsamiferae* is the typical one. The variant with *Euphorbia canariensis* is to be found in contact with *E. canariensis* shrubland or is determined by an increase in rock content in the substrate. The *Pinus canariensis* facies characterizes populations of pines which reach down to very low elevations, within the *E. balsamifera* scrub, favored by salic substrates, as happens from Lomo Jaboneros to Lomo Cortadores (Arguineguín ravine).

The subass. *salsoletosum divaricatae* grows near the coast, in places subject to saline air. *Salsola divaricata* and *Lycium intricatum* are its differential species.

The subass. *plocametosum pendulae* develops at ravine edges in contact with the desertic rivular series *Plocametum pendulae*.

Nowadays the community has greatly retreated, as a consequence of agricultural, urban and industrial expansion around the island's lowland periphery. The best remnants are located on the SW slope between Puerto Rico and Punta de la Aldea. The main substitutional scrublands belong to the *Euphorbietum regis-jubae,* as an evolved successional stage, and to the *Launaeo–Schizogynion*, in more disturbed places, made up of *Launaea arborescens*, *Schizogyne glaberrima*, *S. sericea*, *Artemisia ramosa* etc., and to the nitrohalophilous *Chenoleo tomentosae–Suaedetum vermiculatae*.

Distribution: Endemic to C.

Fig. 6.79 *Euphorbietum balsamiferae*. Galdar. Gran Canaria

Euphorbietum handiensis (Fig. 6.80)
This is a xeric crassicaule endemic scrub of Fuerteventura, with the physiognomy of cactiform spurge (*Euphorbia handiensis*) scrub, typical of hyperarid and arid Inframediterranean belts.

Apart from *Euphorbia handiensis*, *Gymnocarpos decandrus* and *Lycium intricatum*, several species belonging to the arid nitrophilous scrub (*Chenoleoideo-Suaedetum mollis*) are companions: *Suaeda mollis, Launaea arborescens* and *Salsola vermiculata*.

It corresponds to the terminal community of the hyperarid and arid Inframediterranean edaphophilous colluvial series of *E. handiensis* on Fuerteventura (*Euphorbio handiensis sigmetum*).

Its local area is restricted to colluvial soils within the climatophilous area of *Euphorbia balsamifera* scrub, contacting at higher elevations with that of *Euphorbia canariensis* shrubland. It only grows in Jandía, from Gran Valle to Barranco de Escobones.

Distribution: Endemic to F.

Fig. 6.80 *Euphorbietum handiensis*. Gran Valle. Jandía. Fuerteventura

Lycio intricati-Euphorbietum balsamiferae
This is an endemic summer deciduous crassicaule scrub or shrubland of Fuerteventura, potentially found in arid and hyperarid Inframediterranean belts. Its distribution is nowadays greatly reduced and it mainly grows in some ravines in Jandía and Betancuria massifs and the N of the island.

Besides *Euphorbia balsamifera*, in this scrub *Euphorbia regis-jubae*, *Helianthemum canariense* and *Kleinia neriifolia* stand out.

It corresponds to the terminal community of the hyperarid-arid desertic Inframediterranean series of *Euphorbia balsamifera* in Fuerteventura (*Lycio intricati-Euphorbio balsamiferae sigmetum*). It also behaves as edaphoxerophilous within the upper-semiarid Inframediterranean, which is the climatophilous territory of *Euphorbia canariensis* shrubland (*Kleinio neriifoliae-Euphorbietum canariensis*).

The high presence of some species of the arid nitrophilous scrub *Chenoleoideo-Suaedetum mollis* like *Launaea arborescens* and *Lycium intricatum* is due to the severe anthropic disturbance in its potential territory.

Distribution: Endemic to F.

Neochamaeleo pulverulentae-Euphorbietum balsamiferae (Fig. 6.81)
This is an oligospecific endemic summer deciduous crassicaule desertic scrub or shrubland of La Gomera, potentially growing in arid and hyperarid Inframediterranean belts. The hyperarid belt occupies a short and narrow SSW coastal strip between Punta del Becerro and Punta de Iguala; the arid belt extends in an arc at low elevations on the E, S and W slopes, from San Sebastian in the E, to Taguluche in the W.

It is a community with few species, where besides *Euphorbia balsamifera, Kleinia neriifolia, Euphorbia berthelotii, Neochamaelea pulverulenta, Campylanthus salsoloides* and *Launaea arborescens* are common.

It corresponds to the terminal community of the hyperarid-arid desertic Inframediterranean series of *Euphorbia balsamifera* on La Gomera (*Neochamaeleo pulverulentae-Euphorbio balsamiferae sigmetum*).

An *Euphorbia canariensis* facies is shown on arid slopes of the S sector of the island, in contact with *E. canariensis* shrubland. The *Plocama pendula* facies is common in the SE between Punta Gorda and Punta Gaviota.

The *Launaeo arborescentis-Schizogynetum sericeae* is widespread in disturbed areas of the potential *E. balsamifera* scrub.

Distribution: Endemic to G.

Fig. 6.81 *Neochamaeleo pulverulentae-Euphorbietum balsamiferae*. Punta Llana. La Gomera

Rubio fruticosae-Euphorbietum balsamiferae

This is an oligospecific endemic summer deciduous crassicaule desertic scrub or shrubland on El Hierro, which potentially grows in the arid Inframediterranean belt. This potential area covers a relatively narrow strip on the W, SW, S and SE slopes.

It is characterized by *Euphorbia balsamífera, Euphorbia lamarckii, Kleinia neriifolia, Micromeria hyssopifolia*, *Rubia fruticosa* and *Schizogyne sericea*.

It corresponds to the terminal community of the arid Inframediterranean series of *Euphorbia balsamifera* on El Hierro (*Rubio fruticosae-Euphorbio balsamiferae sigmetum*).

The community is currently relegated to the SW end of the island, showing great floristic poverty when compared to its vicarious associations on other Canary Islands. This scarce representation is probably determined by the substrate's youth, mainly recent lava flows and lapilli and consequently very little well-developed soil. On the lava flow surface, lichenic vegetation predominates, and in fissures some elements of the *E. balsamifera* scrub signal its territory, although only in a fragmentary way. The arenicolous community *Schizogyno sericeae-Euphorbietum wildpretii* is particularly dominant on volcanic sand and lapilli. *E. balsamifera* scrub is scarcer to the SE, along the narrowest band of its potential territory, as a good portion of it is occupied by the permanent community of the coastal halophytic belt *Frankenio ericifoliae-Astydamietum latifoliae.*

The *Kleinia neriifolia* facies and *Periploca laevigata* facies, characterized by anemochorous plants, are particularly noticeable at very windy places.

Distribution: Endemic to H.

Aeonio-Euphorbienion canariensis

Nanophanerophytic crassicaule shrubby potential communities characterized by cactiform *Euphorbia canariensis*, typical of the upper-arid and lower-semiarid Infra- and Thermomediterranean belts, and edaphophilous shrubland of *Euphorbia atropurpurea* typical of semiarid and lower-dry Thermomediterranean belts. Distributed throughout the Canary Islands except Lanzarote.

Aeonio percarnei-Euphorbietum canariensis (Fig. 6.82)

This is a xeric crassicaule endemic shrubland of Gran Canaria that, when optimally developed, has the physiognomy of a cactiform spurge shrubland ("cardonal"; *Euphorbia canariensis*: "cardón"). It grows on lithosols or poorly developed soils, typical of the upper-arid and lower-semiarid Inframediterranean belts.

Euphorbia canariensis is the most characteristic species. Besides, *Periploca laevigata, Aeonium percarneum, Asparagus umbellatus, Echium decaisnei, Euphorbia regis-jubae, Kleinia neriifolia, Periploca laevigata, Rubia fruticosa* and others are characteristic.

It is the terminal community of (A) the upper-arid upper-Inframediterranean and lower-semiarid Inframediterranean climatophilous series; (B) the upper-semiarid Infra- and Thermomediterranean edapho-xerophilous rupicolous series; and (C) the lower-arid Inframediterranean edaphohygrophilous series on lava flows ("malpaíses"); of the cardón in Gran Canaria (*Aeonio percarnei–Euphorbio canariensis sigmetum*).

Its climatophilous area encircles the island directly above the area of the *Euphorbietum balsamiferae* and it extends to where it reaches the upper semiarid territory at heights about 250–350 m a.s.l. in the N and 800–950 m in the S. There, it contacts with the area of the climatophilous series of wild olive woodland (*Pistacio lentisci–Oleo cerasiformis sigmetum*). However, frequently on the SW island slopes, because of the prevailing salic substrates, the altitudinal succession between these two communities is not clear and a transition area can be observed between the *E. canariensis* shrubland series and the Canary pine series (*Pino canariensis sigmetum*), in its most thermophilous variant, favored on this substrate.

E. canariensis shrubland has retreated greatly and is now confined to steep rocky slopes and scarps. The best remnants are sheltered on the W slope, mainly in the SW quadrant. *Euphorbietum regis-jubae* shrubland is widely spread within the climatophilous area of the association. At the lower elevation levels of the community there is a noticeable increase in *E. balsamifera*, which marks the change to the *Euphorbietum balsamiferae*.

The descent of *E. canariensis* shrubland to desertic Inframediterranean territories is directly related to lava-flow geomorphology. The largely solid lava flows produce a shield effect decreasing the evaporation rate from the underlying soil, and maintain a favourable water balance. This allows the edaphohygrophilous descent of the *Aeonio percarnei-Euphorbietum canariensis* into the climatophilous domain of the *Euphorbietum balsamiferae*. Such a descent also occurs on debris at the foot of bare cliffs that collect run-on water.

The subass. *pistacietosum* corresponds to *E. canariensis* shrubland enriched with thermophilous elements, generally found growing as edaphophilous on ridges within the climatophilous area of thermophilous woodland (*Pistacio lentisci–Oleetum cerasiformis*).

The *Euphorbia balsamifera* facies develops in particularly xeric locations within the area of the association, while the *Euphorbietum regis-jubae* is widespread in disturbed places within the association's potential area.

Distribution: Endemic to C.

Fig. 6.82 *Aeonio percarnei-Euphorbietum canariensis*. Güigüí. Gran Canaria

Aeonio valverdensis-Euphorbietum canariensis

This is a xeric crassicaule endemic shrubland of El Hierro that, when optimally developed, has the physiognomy of a cactiform spurge shrubland. It is typical of the upper-arid and lower-semiarid Inframediterranean belts.

The main species apart from *Euphorbia canariensis* are *Aeonium valverdense, Euphorbia lamarckii, Kleinia neriifolia, Periploca laevigata* and *Rubia fruticosa*, among others.

It corresponds to the terminal community of the upper-arid upper-Inframediterranean and the lower-semiarid Inframediterranean climatophilous series of *E. canariensis* on El Hierro (*Aeonio valverdensis-Euphorbio canariensis sigmetum*).

Its potential area surrounds the island on the W, SW, S and SE slopes, above the area of *E. balsamifera scrub* (*Rubio-Euphorbietum balsamiferae*), and above sea-level on the remaining slopes. This area is limited upwardly by the upper-semiarid xeric Inframediterranean bioclimatic belt, potentially a domain of juniper woodland (*Rubio-Juniperetum canariensis*). The *Aeonio valverdensis-Euphorbietum canariensis* penetrates up the spurs into this latter belt as an edaphoxerophilous community.

The subass. *typicum* appears today fragmentarily, mainly along a strip from the NE extreme of the island to Las Playas. It is precisely in this strip, but near the coast, where the subass. *euphorbietosum balsamiferae* is commonest. This latter subassociation, that grows in more saline and xeric environments, is characterized by the presence of *E. balsamifera*; it is ocasionally dominant and then determines the physiognomy of the scrubland. On the SE coast, in the contact zone with the desertic *E. balsamifera* scrub (*Rubio-Euphorbietum balsamiferae*), a halo-xerophile *E. balsamifera* facies can be identified.

A large amount of the remaining potential area of the association, as occurs on the slopes of El Julan, formed by recent lava flows with no or scarce soil development (lithosols), can be considered sparsely populated by the community's most aggressive and drought resistant anemochorous elements. It therefore colonizes these territories in a fragmentary and incipient way, although the physiognomy does not retain the typical appearance of *E. canariensis* shrubland. In these situations a *Kleinia neriifolia* facies stands out, with the striking absence of *E. canariensis* itself and an abundance of *Aeonium hierrense*, *A. valverdense* and *Kleinia neriifolia*.

The *Schizogyno sericeae-Euphorbietum wildpretii* grows on lapilli and volcanic sands in its potential climatophilous area. A *Periploca laevigata* facies develops on stony debris and partially consolidated colluvial substrate.

The perennial *Cenchro-Hyparrhenietum sinaicae* grassland, which has been favored by goat grazing, also occupies a great area and may physiognomically characterize the landscape with its striking straw coloration most of the year.

Distribution: Endemic to H.

Echio breviramis-Euphorbietum canariensis (Fig. 6.83)
This is a xeric crassicaule endemic shrubland of La Palma that, when optimally developed, has the physiognomy of a cactiform spurge shrubland. It grows on lithosols or poorly developed soils, typical of upper-arid and lower-semiarid Inframediterranean belts.

Besides *Euphorbia canariensis*, its characteristic species are *Aeonium davidbramwellii, Convolvulus floridus*, *Echium brevirame*, *Euphorbia lamarckii*, *Kleinia neriifolia*, *Lavandula canariensis*, *Periploca laevigata* and *Rubia fruticosa*.

It corresponds to the terminal community of upper-arid upper-Inframediterranean and lower-semiarid Inframediterranean climatophilous series and upper-semiarid Infra- and Thermomediterranean edaphophilous-rupicolous series, of *E. canariensis* on La Palma (*Echio breviramis-Euphorbio canariensis sigmetum*).

Its potential area forms an almost continuous climatophilous strip through the NW and SW island slope, from the locality of Punta de Juan Adalid (Garafía), in the

NW, to El Time, in the SW, above the potential area of *Euphorbia balsamifera* scrub and below the potential area of juniper woodland. In the N and NE of the island, it has a fragmented distribution as edapho-xerophilous within the potential area of juniper woodland. In the E, between Punta Salinas and Punta Ganados, it extends from sea level to juniper potential area, continuing southward to Montaña del Azufre along a narrow strip above the lapilli vegetation and the area of potential *E. balsamifera* scrub. Additionally, where edaphoxerophilous, it locally occurs around the whole island perimeter on spurs, steep walls and malpaíses in the upper-semiarid Inframediterranean belt.

Nowadays *E. canariensis* shrubland is rare and clumps are to be found throughout its potential area, although often fragmentary. Noteworthy are those of Martín Luis and Barranco Seco (Puntallana), and Barranco del Humo (Breña Alta).

The *E. balsamifera* facies grows near the coast under dryer and saltier conditions. The widespread *Periploca laevigata* facies preferably grows on loose stony soils or lapilli.

Distribution: Endemic to P.

Fig. 6.83 *Echio breviramis-Euphorbietum canariensis*. Santa Lucía. La Palma

Euphorbietum atropurpureae (Fig. 6.84)
This is a xeric crassicaule endemic shrubland of Tenerife, with a physiognomy of dendroid spurge, dominated by *Euphorbia atropurpurea*. It occurs over the S and SW of the island on the rocky slopes of the upper mid-lands (700–1100 m), from Güímar Valley to Masca. It also grows in Teno, but at a lower elevation.

Euphorbia atropurpurea, E. lamarckii, Aeonium haworthii and *Retama rhodorhizoides* are characteristic species.

It preferably grows on ledges and rocky steep slopes in the semiarid and lower-dry Thermomediterranean belts, within the climatophilous area of juniper woodland (*Junipero canariensis-Oleetum cerasiformis*).

It corresponds to the terminal community of the semiarid-dry Thermomediterranean edaphoxerophilous series of *E. atropurpurea* in Tenerife (*Euphorbio atropurpureae sigmetum*).

Besides the typical subass. *euphorbietosum atropurpureae,* also the subass. *juniperetosum canariensis* is recognized. It shows an ecotonic character with juniper woodland.

The *Aeonium haworthii* facies is common on the cliffs of Cumbre de Baracán (Teno). The *Euphorbietum lamarckii* represents the main substitutional shrubland of the association.

Distribution: Endemic to T.

Fig. 6.84 *Euphorbietum atropurpureae*. Arguayo. Tenerife

Euphorbietum berthelotii-canariensis (Fig. 6.85)
This is a xeric crassicaule endemic shrubland of La Gomera that, when optimally developed, has the physiognomy of a cactiform spurge shrubland. It is typical of the upper-arid and lower-semiarid Inframediterranean belts.

The main species apart from *Euphorbia canariensis* are *Kleinia neriifolia, Euphorbia berthelotii, Rubia fruticosa*, *Periploca laevigata* and *Plocama pendula*, among others.

It corresponds to the terminal community of the upper-arid upper-Inframediterranean and lower-semiarid Inframediterranean climatophilous, and lower-semiarid and dry (wc) Infra-Thermomediterranean edaphoxerophilous series of *E canariensis* on La Gomera (*Euphorbio berthelotii- canariensis sigmetum*).

Its potential area rings the island; in the southern half above the climatophilous area of *Euphorbia balsamifera* scrub (*Neochamaeleo pulverulentae-Euphorbietum balsamiferae*), and in the northern half from near sea level, above the *Frankenio-Astydamietum latifoliae* or *Euphorbietum aphyllae* (potential edaphophilous community of windward promontories and cliffs), until it reaches the potential area of juniper woodland (*Brachypodio-Juniperetum canariensis*). It can enter the climatophilous area of the latter as edaphoxerophilous potential vegetation on steep rocky places with scarce soil.

This community has substantially lost in area and nowadays can only be observed in isolated, cleared and sheltered steep locations, throughout its potential area. The best remnants of the subass. *typicum* are commoner in the E of the island. The subass. *euphorbietosum balsamiferae* (northern *E. balsamifera* scrub) is frequent in particularly xeric conditions favored by rocky substrates on the N and NE slopes; the abundance of *E. balsamifera* gives a particular physiognomy to the association.

There is a *Periploca laevigata* facies that develops on slope debris very close to the potential area of the *Brachypodio-Juniperetum canariensis*. The *Plocama pendula* facies grows on very fragmented, eroded and porous old lava substrates, particularly developed on the NE, E, SE and NW sides of the island.

The main substitutional shrubland is the *Euphorbietum berthelotii,* widely distributed over the potential area of *E. canariensis* shrubland. Its floristic composition differs little from the latter but there is an impoverishment in species and the absence of *Euphorbia canariensis* is striking. The nitrophilous shrubland *Artemisio thusculae-Rumicetum lunariae* is also common in anthropic areas, as are the nitrophilous herbaceous communities of the *Echio-Galactition*.

Distribution: Endemic to G.

Fig. 6.85 *Euphorbietum berthelotii-canariensis*. Cliffs North to Avalos. La Gomera

Kleinio neriifoliae-Euphorbietum canariensis (Fig. 6.86)
This is a crassicaule endemic shrubland of Fuerteventura. It grows on lithosols, lava flow fields, or poorly developed soils. It is typical of upper-arid and lower-semiarid Inframediterranean belts.

It is characterized by *E. canariensis* together with *Kleinia neriifolia* and *Rubia fruticosa*. The constant presence of the companion species *Launaea arborescens* indicates the aridity and disturbance in its territory.

It corresponds to the terminal community of the upper-arid upper-Inframediterranean and lower-semiarid Inframediterranean climatophilous series of *E. canariensis* on Fuerteventura (*Kleinio neriifoliae-Euphorbio canariensis sigmetum*).

Only scarce remnants of the community are to be found nowadays that, when optimally developed, has the physiognomy of a cactiform spurge shrubland, as in Jandía and Montaña Cardones, where the purest facies are located.

Distribution: Endemic to F.

Fig. 6.86 *Kleinio neriifoliae-Euphorbietum canariensis*. Cofete. Jandía. Fuerteventura. (Ph: PLP)

Periploco laevigatae-Euphorbietum canariensis (Fig. 6.87)
This is a xeric crassicaule endemic shrubland of Tenerife that, when optimally developed, has the physiognomy of a cactiform spurge shrubland. It grows on lithosols or little developed soils, typical of upper-arid and lower-semiarid Inframediterranean belts.

It is mainly characterized by *E. canariensis*, together with *Kleinia neriifolia, Rubia fruticosa, Periploca laevigata, E. lamarckii, Asparagus umbellatus,*

Convolvulus floridus, Justicia hyssopifolia, Lavandula buchii, Scilla haemorrhoidalis, among others.

It is the terminal community of (A) the upper-arid upper-Inframediterranean and lower-semiarid Inframediterranean climatophilous series; (B) the upper-semiarid Inframediterranean, semiarid and lower-dry Thermomediterranean (wc) edaphophilous rupicolous series; and (C) the lower-arid Inframediterranean edapho-hygrophilous series on lava flows, of *Euphorbia canariensis* on Tenerife (*Periploco-Euphorbio canariensis sigmetum*).

Its climatophilous area encircles the island directly above the area of *Euphorbia balsamifera* scrub (*Ceropegio-Euphorbietum balsamiferae*), at about 500 m in the S sector of the island. The upper limit coincides with the beginning of the upper-semiarid ombrotype where the climatophilous series of juniper woodland is established (*Junipero canariensis–Oleo cerasiformis sigmetum*) at elevations between 100–200 m in the N, mainly growing on cliffs and coastal lava flows. On the SW and SE slopes it rises up to about 850–900 m. This area is very narrow, especially in the N and S sector of the ring. On steep slopes of S Anaga, *E. canariensis* shrubland can reach the coast. The edaphophilous enlargement of its area widens its territory, favored by lava flows, steep slopes and cliffs. It is particularly noticeable on the ancient massifs of Anaga, Teno, and Adeje; on the latter the 'cardón' can coexist with occasional Canary pines. A good example on lava flows can be seen in 'Malpaís' de Güímar. This descent of *E. canariensis* shrubland into arid Inframediterranean territories is attributed to a shielding effect of the solid lava flows which decrease the evaporation rate from the underlying soil, maintaining a favorable water balance and allowing this edapho-hygrophilous behavior. In the South of the island, the distribution of *E. canariensis* shrubland is interrupted in many places due to the ascent of *E. balsamifera* scrub on pumice substrate.

Apart from the typical subass. *euphorbietosum canariensis*, the subass. *salsoletosum divaricatae* (halophilous *E. canariensis* shrubland) can be distinguished, typical of northern locations subject to strong saline atmosphere or an increase of soluble salt in the soil. The subass. *euphorbietosum balsamiferae*, with a physiognomy of *Euphorbia balsamifera* scrub, established on warm xeric coastal slopes of the N of the island. The subass. *jasminetosum odoratissimi* includes thermophilous elements of the *Mayteno-Juniperion,* and occupies rocky locations in contact with thermophilous woodland (*Junipero canariensis– Oleetum*) or even dry laurel forest (*Visneo mocanerae-Arbutetum canariensis*).

The *Plocama pendula* facies is particularly common on strongly weathered old lava flows, and on lapilli (Fig. 6.88).

The area of the *E. canariensis* shrubland has been affected by agriculture, urban and industrial development, and has retreated greatly. Nowadays it is mainly confined to steep rocky slopes, scarps and lava flows. The *Artemisio thusculae–Rumicetum lunariae* is the main nitrophilous substitutional shrubland, together with the *Euphorbietum lamarckii* in less disturbed places. The *Cenchro ciliaris–Hyparrhenietum sinaicae* substitutional perennial herbaceous layer represents a later stage than the former and is also widely spread, not only in the area of this series but also in the *E. balsamifera* series and *Junipero turbinata* subsp. *canariensis* series.

Distribution: Endemic to T.

Fig. 6.87 *Periploco laevigatae-Euphorbietum canariensis*. San Andrés. Anaga. Tenerife

Fig. 6.88 *Plocama pendula* facies. Lava flows of the Arafo volcano. Tenerife

Euphorbienion aphyllae
Chamaephytic crassicaule potential (edaphoxero-anemogenic) scrublands characterized by the succulent *Euphorbia aphylla*, with slender pencil-stems, typical of very windy rocky littoral territories of the arid and semiarid Inframediterranean belts. Distributed in the central Canary Islands (C, T, G).

Astydamio latifoliae-Euphorbietum aphyllae
This is a xeric crassicaule endemic scrubland growing on the coastal rocky cliffs and promontories swept by strong marine trade-wind of the north and northwest of Gran Canaria.

It corresponds to the terminal community of the arid and lower-semiarid Inframediterranean edaphoxero-anemogenic littoral vegetation series of *E. aphyllae* on Gran Canaria (*Astydamio latifoliae-Euphorbio aphyllae sigmetum*), which mainly develops within the climatophilous territory of the *Euphorbietum balsamiferae*, and occasionally penetrates into the NW cliffs within the lower-semiarid ombrotype, potential territory of the climatophilous the *Aeonio-Euphorbietum canariensis*.

Euphorbia aphylla, Ceropegia fusca, E. balsamifera, Kleinia neriifolia, Rubia fruticosa, and typical plants of the rocky saline belt (*Frankenio-Astydamion latifoliae*) like *Astydamia latifolia*, *Frankenia capitata* and *Limonium pectinatum*, are characteristic.

Distribution: Endemic C.

Ceropegio dichotomae-Euphorbietum aphyllae (Fig. 6.89)
This is a xeric crassicaule endemic edaphoxero-anemogenic scrubland endemic to the northern cliffs of Teno (Tenerife), swept by strong marine winds. It grows between 50 and 350 m a.s.l. within the climatophilous territory of *E. canariensis* shrubland (*Periploco-Euphorbietum canariensis*). It always faces the sea and bears the saline air which makes the coastal rocky vivacious halophytes of the *Frankenio-Astydamion latifoliae* common in the community.

It corresponds to the terminal community of the semiarid Inframediterranean edaphoxero-anemogenic littoral vegetation series of *E. aphyllae* on Tenerife (*Ceropegio dichotomae-Euphorbio aphyllae sigmetum*).

Characteristic species are *Argyranthemum coronopifolium*, *Ceropegia dichotoma*, *Cheirolophus burchardii* and *Euphorbia aphylla*.

There is a *Euphorbia atropurpurea* variant of sciophilous cooler places, as well as a more general aerophilous *Astydamia latifolia* and *Limonium pectinatum* variant.

Distribution: Endemic to T.

Fig. 6.89 *Ceropegio dichotomae-Euphorbietum aphyllae*. El Fraile. Teno. Tenerife

Euphorbietum aphyllae (Fig. 6.90)
This is a xeric crassicaule endemic edaphoxero-anemogenic scrubland growing on the coastal rocky cliffs and promontories of La Gomera constantly swept by north-east marine trade-winds. It is a chamaephytic or nanophanerophytic community that grows within the climatophilous territory of *Euphorbia canariensis* shrubland (*Euphorbietum berthelotii-canariensis).*

It corresponds to the terminal community of the semiarid Inframediterranean edaphoxero-anemogenic littoral vegetation series of *E. aphylla* on La Gomera (*Euphorbio aphyllae sigmetum).*

E. aphylla, Aeonium lindleyi subsp. *viscatum, Ceropegia dichotoma* subsp. *krainzii* are characteristic, and *E. balsamifera, Kleinia neriifolia, Neochamaelea pulverulenta, Plocama pendula* and *Rubia fruticosa* are common. Also plants of the coastal rocky saline belt (*Frankenio-Astydamion latifoliae*) penetrate into the community.

There is an *E. balsamifera* facies in the northeast sector of the island, between Punta Llana and Agulo. Also a *Plocama pendula* facies grows in broken-up ancient basaltic substrates.

Distribution: Endemic to G.

Fig. 6.90 *Euphorbietum aphyllae*. Coast of Vallehermoso. La Gomera

Euphorbion regisjubo-lamarckii
Potential and seral dendroid *Euphorbia spp.* shrubby communities from the hyperarid to lower-dry Inframediterranean, and shrublands of *Plocama pendula* growing in colluvia and ramblas within the same belts. Endemic to the Canaries.

Euphorbienion regisjubo-lamarckii
Potential and seral dendroid *Euphorbia spp.* shrubby communities (*E. lamarckii, E. rejis-jubae* and *E.* berthelotii) from the hyperarid to lower-dry Inframediterranean. Throughout the Canary Islands.

Euphorbietum berthelotii (Fig. 6.91)
This is a xeric crassicaule endemic shrubland of La Gomera, with a physiognomy of dendroid spurge shrubland, dominated by *Euphorbia berthelotii*, an island endemic of great colonizing ability in disturbed places, on soils with a slightly compacted deep clay layer.

It corresponds to a substitutional shrubby stage of *E. canariensis* shrubland (*Euphorbietum berthelotii-canariensis*) and juniper woodland (*Rubio-Juniperetum canariensis*). It preferably develops on the southern slopes, within arid, semiarid and the dry Infra- and Thermomediterranean, in a strip extending from Puntallana (in the East) to Alojera (in the West).

Besides *E. berthelotii*, *Kleinia neriifolia*, *Neochamaelea pulverulenta*, *Plocama pendula* and *Rubia fruticosa* subsp. *melanocarpa* grow in the community. *Argyranthemum frutescens* subsp. *parviflorum* joints the community at its Eastern extreme. The presence of *Launaea arborescens* and *Schizogyne serícea* is evidence of a more degraded stage, within arid and semiarid ombrotypes.

Distribution: Endemic to G.

Fig. 6.91 *Euphorbietum berthelotii*. Barranco de la Villa. La Gomera

Euphorbietum lamarckii (Fig. 6.92)
This is a xeric crassicaule shrubland, with a physiognomy of dendroid spurge shrubland, dominated by *Euphorbia lamarckii.*

It constitutes a substitutional shrubby stage of diverse potential communities within the hyperarid to dry Infra- and Thermomediterranean belts, below the cloud level, and shows greater development on southern and western slopes.

Besides *E. lamarckii*, *Kleinia neriifolia* and *Periploca laevigata* are frequent with some interspersed *Plocama pendula*. Furthermore, several plants of the potential communities of those belts and others of the *Forsskaoleo angustifoliae-Rumicetalia lunariae* are included among the companion species.

This is a widespread community which generally constitutes an advanced seral stage in the recovery of the potential communities of the aforementioned belts, and varies in appearance. For this reason it has been sometimes considered as facies within the terminal communities of *Euphorbia* scrublands and shrublands, *Juniperus* woodland, and *Palm* groves, even within substitutional communities of the *Forsskaoleo-Rumicetalia*, depending on the prevalent companion species of the *Euphorbietum lamarckii.* These are the previously identified facies:

Tenerife:

Ceropegio fuscae-Euphorbietum balsamiferae facies of *Euphorbia lamarckii*
Euphorbietum atropurpureae facies of *Euphorbia lamarckii*
Periploco laevigatae-Euphorbietum canariensis facies of *Euphorbia lamarckii*
Junipero canariensis-Oleetum cerasiformis facies of *Euphorbia lamarckii*
Periploco laevigatae-Phoenicetum canariensis facies of *Euphorbia lamarckii*
Rhamno crenulatae-Hypericetum canariensis facies of *Euphorbia lamarckii*

Artemisio thusculae-Rumicion lunariae facies of *Euphorbia lamarckii*
Launaeo arborescentis-Schizogynetum sericeae facies of Euphorbia lamarcki

La Palma:
Echio breviramis-Euphorbietum balsamiferae facies of *Euphorbia lamarckii*
Echio breviramis-Euphorbietum canariensis facies of *Euphorbia lamarckii*
Rhamno crenulatae-Juniperetum canariensis facies of *Euphorbia lamarckii*
Rhamno crenulatae-Hypericetum canariensis facies of *Euphorbia lamarckii*
Artemisio thusculae-Rumicion lunariae facies of *Euphorbia lamarckii*

El Hierro:
Rubio fruticosae-Euphorbietum balsamiferae facies of *Euphorbia lamarckii*
Aeonio valverdensis-Euphorbietum canariensis facies of *Euphorbia lamarckii*
Rubio fruticosae-Juniperetum canariensis facies of *Euphorbia lamarckii*
Rhamno crenulatae-Hypericetum canariensis facies of *Euphorbia lamarckii*
Artemisio thusculae-Rumicion lunariae facies of *Euphorbia lamarckii*

Distribution: Endemic to T, P, H.

Fig. 6.92 *Euphorbietum lamarckii*. Arafo. Tenerife

Euphorbietum regis-jubae (Fig. 6.93)
This is a xeric crassicaule shrubland, with a physiognomy of dendroid spurge shrubland, dominated by *Euphorbia regis-jubae*.

It constitutes a substitutional shrubby stage of diverse potential communities within the hyperarid to dry Infra- and Thermomediterranean belts, below the cloud level, and shows better development on south and western slopes.

Besides *E. regis-jubae*, *Kleinia neriifolia* and *Periploca laevigata* are frequent with some interspersed *Plocama pendula*. Furthermore, several plants of the potential communities of those belts and others of the *Forsskaoleo angustifoliae-Rumicetalia lunariae* are among the companion species.

This is a widespread community which generally constitutes an advanced seral stage in the recovery of the potential communities of the aforementioned belts, and shows a varied appearance. For this reason, sometimes it has been considered as facies within the terminal communities of *Euphorbia*, *Juniperus* woodland, and *Palm* grove community, and even within substitutional communities of the *Forsskaoleo-Rumicetalia*, depending on the prevalent companion species making up each *Euphorbietum regis-jubae*. These are the facies:

Gran Canaria:
Euphorbietum balsamiferae facies of *Euphorbia regis-jubae*.
Aeonio percarnei-Euphorbietum canariensis facies of *Euphorbia regis-jubae*.
Pistacio lentisci-Oleetum cerasiformis facies of *Euphorbia regis-jubae*.
Periploco laevigatae-Phoenicetum canariensis facies of *Euphorbia regis-jubae*.
Rhamno crenulatae-Hypericetum canariensis facies of *Euphorbia regis-jubae*.
Artemisio thusculae-Rumicion lunariae facies of *Euphorbia regis-jubae*.

Lanzarote:
Asterisco intermedii-Euphorbietum balsamiferae
- subass. *euphorbietosum balsamiferae* facies of *Euphorbia regis-jubae*.
- subass. *aeonietosum lancerottensis* facies of *Euphorbia regis-jubae*.
Convolvulo lopezsocasi-Oleetum cerasiformis facies of *Euphorbia regis-jubae*.

Distribution: Endemic to L, C.

Fig. 6.93 *Euphorbietum regis-jubae*. La Santa. Lanzarote

Euphorbio berthelotii-Schizogynetum sericeae (Fig. 6.94)
This is a nanophanerophytic shrubland which grows in colluvia and coastal debris, in the S and W of La Gomera.

It corresponds to the terminal community of the arid to semiarid Inframediterranean edaphoxerophyllous colluvial series of *Euphorbia berthelotii* and *Schizogyne sericea* on La Gomera.

Euphorbia berthelotii, *Kleinia neriifolia* and *Schizogyne sericea* are the characteristic species.

Distribution: Endemic to G.

Fig. 6.94 *Euphorbio berthelotii-Schizogynetum sericeae*. Quiebracanillas. Valle Gran Rey. La Gomera

Kleinio-Asparagetum pastoriani (Fig. 6.95)
This is a xeric crassicaule endemic shrubland of Fuerteventura, with a physiognomy of dendroid spurge shrubland.

It is a substitutional stage of *Euphorbia canariensis* shrubland (*Kleinio neriifoliae-Euphorbietum canariensis*) in places where the decrease in anthropic action has allowed a slow but steady recovery of the plant landscape. Thus, it is distributed below the potential area of *Olea cerasiformis* woodland and above the potential area of *E. balsamifera* scrub, although it descends further over malpaíses.

Asparagus pastorianus, *E. regis-jubae*, *Helianthemum canariense* and *Kleinia neriifolia* are the most characteristic species. Other species of wide ecological range like *Fagonia cretica*, *Launaea arborescens*, *Lotus lancerottensis* and *Lycium intricatum*, more typical of the widespread nitrophilous shrublands of the island, are constant and sometimes co-dominant in this community.

The *Euphorbia regis-jubae* facies is the typical one. A *Kleinia neriifolia* facies develops at very windy places.

Distribution: Endemic to F.

Fig. 6.95 *Kleinio neriifoliae-Asparagetum pastoriani*. Malpaís de Lajares. Fuerteventura

Schizogyno sericeae-Euphorbietum wildpretii (Fig. 6.96)

This is a nanophanerophytic shrubland which grows in volcanic ashes, colluvia, debris and semi-weathered malpaises sharing a sandy soil texture. It is mainly established within arid and semiarid Inframediterranean belts.

It corresponds to the terminal community of the arid and lower-semiarid Inframediterranean edaphoxerophilous series of *Schizogyne sericea* and *Euphorbia lamarckii* subsp. *wildpretii*, over volcanic ashes, on La Palma and El Hierro (*Schizogyno sericeae-Euphorbio wildpretii sigmetum*).

E. lamarckii, *Kleinia neriifolia*, *Rumex lunaria* and *Schizogyne sericea* are the most frequent species.

The subass. *schizogynetosum sericeae* characterizes this community on El Hierro, and it extends all around the island's perimeter, preferentially within the arid ombrotype. The subass. *ceballosietosum fruticosae* differentiated by *Ceballosia fruticosa, Rubia fruticosa* and *Argyranthemum hierrense* grows also on El Hierro, it is more water demanding, and settles on the North coast within the semiarid ombrotype.

On La Palma the subass. *echietosum breviramis* grows, characterized by the island endemic *Echium brevirame.* It is widespread along the SE and SW coasts, particularly in the SE and S tip of the Neopalma edifice. The *Schizogyne sericea* facies is particularly abundant on volcanic ashes.

Distribution: Endemic to P and H.

Fig. 6.96 *Schizogyno sericeae-Euphorbietum wildpretii*. Llanos de la Irama. El Hierro

Plocamenion pendulae

Shrublands of *Plocama pendula* growing in colluvia and ramblas of the the hyper-arid to lower-semiarid Inframediterranean belts. Canary Islands.

Plocametum pendulae (Fig. 6.97)

This is a community characterized by the microphanerophyte *Plocama pendula*, typical of alluvial deposits along ramblas (seasonal stream beds) within the territories of a desertic bioclimate (of the climatophilous dominance of *Euphorbia balsamifera* scrub), and to a lesser extent within the xeric bioclimate and lower-semiarid ombrotype (of the climatophilous dominance of *E. canariensis* shrubland). These streams only carry water as a result of periodic rain or sometimes flash floods. The community outstands in the arid landscape because of its greenery throughout the year. *P. pendula* has a strong root development, able to reach deep water tables.

It corresponds to the western Canary terminal community of the alluvial arid and lower-semiarid Inframediterranean edaphohygrophilous series in ramblas of *P. pendula* (*Plocamo pendulae sigmetum*).

In general, a mosaic formed by the *Plocametum pendulae*, the hemicryptophytic grassland *Cenchro-Hyparrhenietum sinaicae* and the halo-nitrophilous scrub *Launaeo arborescentis-Schizogynetum sericeae* are observed along these ramblas. On rocky outcrops in these ramblas and on the banks with more soil, the ecotone between *Euphorbia balsamifera* scrub and *Plocametum pendulae* is visible, which has been described as a subassociation of *E. balsamifera* scrub.

The subass. *euphorbietosum lamarckii* for Tenerife is the *typical* subassociation. The subass. *euphorbietosum regis-jubae* is representative on Gran Canaria, and subass. *euphorbietosum berthelotii* on La Gomera.

Distribution: Endemic to C, T, G.

Fig. 6.97 *Plocametum pendulae*. Barranco de Tirajana. Gran Canaria

Periplocenion laevigatae
Shrublands of *Periploca laevigatae* growing on screes and stony places in the semi-arid and dry Inframediterranean belts. Canary Islands.

***Periploca laevigata* community**
This is a permanent community of Infra- and Thermomediterranean screes, stony places and collapsing cliffs, physiognomically characterized by the nanophanerophytic twiner *Periploca levigata.* It is particularly developed on El Hierro, on the screes of cliffs at Tibataje, Sabinosa and Las Playas. On La Gomera it also appears at similar sites forming a mosaic with the *Euphorbio berthelotii-Schizogynetum sericeae*. On La Palma and Tenerife it has been also reported. Until now this community has remained hidden behind several facies considered in this text under the following syntaxa:

- *Aeonio valverdensis-Euphorbietum canariensis* facies of *Periploca laevigata* (H)
- *Brachypodio arbusculae-Juniperetum canariensis* facies of *Periploca laevigata* (G)

- *Echio breviramis-Euphorbietum canariensis* facies of *Periploca laevigata* (P)
- *Euphorbio berthelotii-canariensis* facies of *Periploca laevigata* (G)
- *Junipero canariensis-Oleetum cerasiformis* facies of *Periploca laevigata* (T)
- *Rhamno crenulatae-Juniperetum canariensis* facies of *Periploca laevigata* (P)
- *Rubio fruticosae-Juniperetum canariensis* facies of *Periploca laevigata* (H)

Further study is required.
Distribution: G, P, H.

6.2.7.2 Thermo-Sclerophyllous Woodland, Scrublands and Shrublands of the Mid-Lands

Rhamno crenulatae-Oleetea cerasiformis
Phytosociological class that includes potential thermo-sclerophyllous woodlands of *Olea* and *Juniperus*, and potential Canary palm-tree community, as well as associated scrub and shrublands: potential and seral broom shrubland of *Retama rhodorhizoides,* potential and seral shrubland of *Rhamnus crenulata* and *Hypericum canariense,* and seral scrub on eroded soils mainly dominated by *Micromeria* and *Cistus.* All of them growing within the (arid) semiarid to dry Infra- to Thermomediterranean belts. Endemic to the Canarian-Madeiran biogeographical subregion.

Rhamno crenulatae-Oleetalia cerasiformis
Includes potential thermo-sclerophyllous woodlands of *Olea cerasiformis* and *Juniperus turbinata* subsp. *canariensis* of the Canaries and Madeira and potential Canary palm community, as well as associated scrub and shrublands: potential and seral broom shrublands of *Retama rhodorhizoides*, and potential and seral shrubland of *Rhamnus crenulata* and *Hypericum canariensis* (Table 6.2).

Table 6.2 Characteristic species of the vegetation of *Rhamno crenulatae-Oleetea cerasiformis*

Rhamno crenulatae-Oleetea cerasiformis				
Scientific name	**Family**	**Common name**	**Biotype**	**Distribution**
TREES				
Dracaena draco subsp *draco*	Dracaenaceae	"drago"	M	H,P,G,T,C
Juniperus turbinata subsp. *canariensis*	Cupressaceae	"sabina"	M	H,P,G,T,C
Maytenus canariensis	Celastraceae	"peralillo"	N/M	H,P,G,T,C,F
Olea cerasiformis	Oleaceae	"acebuche"	M	H,P,G,T,C,F,L
Phoenix canariensis	Arecaceae	"palmera canaria"	M	H,P,G,T,C,F,L
Pistacia atlantica	Anacardiaceae	"almácigo"	M	P,G,T,C,F
Pistacia lentiscus	Anacardiaceae	"lentisco"	M	G,T,C,F,L
Sideroxylon canariensis	Sapotaceae	"marmolán"	M	H,P,G,T,C,F
SCRUB and SHRUBS				
Anagyris latifolia	Fabaceae	"oro de risco	N	P,G,T,C
Argyranthemum callichryson	Asteraceae	"margarita gomera amarilla"	Ch/N	G
Argyranthemum escarrei	Asteraceae	"magarza de La Aldea"	Ch/N	C
Asparagus plocamoides	Asparagaceae	"esparragón colgante"	N	P,G,T,C
Asparagus scoparius	Asparagaceae	"esparragón raboburro"	N	H,P,G,T,C,F
Asteriscus sericeus	Asteraceae	"jorado"	Ch/N	F
Bupleurum handiense	Apiaceae	"anís de Jandía"	N	F,L
Bupleurum salicifolium subsp. *aciphyllum*	Apiaceae	"anís de risco"	N	H,P,G,T,C
Bupleurum salicifolium subsp. *salicifolium*	Apiaceae	"anís salvaje"	N	H,G
Bystropogon plumosus	Lamiaceae	"poleo peludo"	N	T
Carlina salicifolia	Asteraceae	"malpica cabezote"	Ch/N	H,G,T,C,F,L
Cistus monspeliensis subsp. *canariensis*	Cistaceae	"jaguarzo"	N	H,P,G,T,C
Convolvulus lopezsocasi	Convolvulaceae	"corregüelón de Famara"	N/Li	L
Convolvulus scoparius	Convolvulaceae	"leñanoel"	N	H,P,G,T,C
Dorycnium eriophthalmum	Fabaceae	"trébol de risco blanco"	N	H,P,G,T,C
Echium aculeatum	Boraginaceae	"ajinajo"	N	H,P,G,T
Echium decaisnei subsp. *decaisnei*	Boraginaceae	"taginaste blanco"	N	C
Echium decaisnei subsp. *purpuriense*	Boraginaceae	"taginaste blanco oriental"	N	F,L
Echium giganteum	Boraginaceae	"taginaste gigante"	N	T
Echium hierrense	Boraginaceae	"ajinajo herreño"	N	H

(continued)

Table 6.2 (continued)

Rhamno crenulatae-Oleetea cerasiformis				
Scientific name	**Family**	**Common name**	**Biotype**	**Distribution**
Echium strictum	Boraginaceae	"taginaste chico canario"	Ch/N	H,P,G,T,C
Erysimum bicolor	Brassicaceae	"alhelí de medianía"	Ch	H,P,G,T,C,F
Euphorbia atropurpurea	Euphorbiaceae	"tabaiba majorera"	N	T
Globularia salicina	Globulariaceae	"mosquera común"	N	H,P,G,T,C
Gymnosporia cryptopetala	Celastraceae	"arto"	N	F,L
Hypericum canariense	Hypericaceae	"granadillo canario"	N	H,P,G,T,C
Jasminum odoratissimum	Oleaceae	"jazmín silvestre"	N	H,P,G,T,C,F
Justicia hyssopifolia	Acanthaceae	"mataprieta"	N	G,T
Marcetella moquiniana	Rosaceae	"palosangre"	N	G,T,C
Micromeria hyssopifolia var. *hyssopifolia*	Lamiaceae	"tomillo de pinar"	Ch	H,T
Micromeria varia subsp. *gomerensis*	Lamiaceae	"tomillo salvaje gomero"	Ch	G
Micromeria varia subsp. *rupestris*	Lamiaceae	"tomillo salvaje majorero"	Ch	F,L
Micromeria varia subsp. *varia*	Lamiaceae	"tomillo salvaje común"	Ch	G,T
Osyris lanceolata	Santalaceae	"bayón canario"	N	P,G,T
Paronychia canariensis	Caryophyllaceae	"nevadilla canaria"	Ch	H,P,G,T,C
Retama rhodorhizoides	Fabaceae	"retama blanca canaria"	N	H,P,G,T,C,F
Rhamnus crenulata	Rhamnaceae	"espinero"	N	H,P,G,T,C,F,L
Rubia fruticosa subsp. *fruticosa*	Rubiaceae	"tasaigo"	N	H,P,G,T,C,F,L
Ruta pinnata	Rutaceae	"ruda canaria"	N	P,T
Sideritis brevicaulis	Lamiaceae	"chajorra de Teno"	C/N	T
Sideritis pumila	Lamiaceae	"salviarrisco"	N	F,L
Spartocytisus filipes	Fabaceae	"retama fina"	N	H,P,G,T
Teline osyrioides subsp. *osyrioides*	Fabaceae	"retamón de Teno"	N	T
Teline osyrioides subsp. *sericea*	Fabaceae	"retamón del Sur"	N	T
Teucrium heterophyllum	Lamiaceae	"jocama"	N	H,P,G,T,C
HERBS				
Androcymbium hierrense	Colchicaceae	"cebollín estrellado"	G	H,P,G
Brachypodium arbuscula	Poaceae	"pajonazo"	H	H,G,T
Tamus edulis	Dioscoreaceae	"norsa"	G-Li	H,P,G,T,C
Vicia cirrhosa	Fabaceae	"chicharilla canaria florida"	H-Li	H,P,G,T,C

(continued)

Table 6.2 (continued)

Rhamno crenulatae-Oleetea cerasiformis				
Scientific name	**Family**	**Common name**	**Biotype**	**Distribution**
Transgressive species with significant presence in some communities of the class				
TREES				
Heberdenia excelsa	Myrsinaceae	"aderno"	M	H,P,G,T,C,F
SCRUB and SHRUBS				
Asparagus umbellatus	Asparagaceae	"esparraguera común"	N/Li	H,P,G,T,C,F
Cistus symphytifolius	Cistaceae	"amagante de pinar"	N	P,T
Echium brevirame	Boraginaceae	"arrebol"	N	P
Euphorbia berhelotii	Euphorbiaceae	"tabaiba picuda"	N	G
Euphorbia lamarckii subsp. *wildpretii*	Euphorbiaceae	"tabaiba amarga"	N	H,G,P,T
Euphorbia regis-jubae	Euphorbiaceae	"tabaiba salvaje"	N	P,G,C,F,L
Helianthemum canariensis	Cistaceae	"jarilla turmera"	Ch	H,P,G,T,C,F,L
Micromeria herpyllomorpha	Lamiaceae	"tomillo palmero"	Ch	P
Periploca laevigata	Apocynaceae	"cornical"	N/Li	H,P,G,T,C,F,L
Spergularia fimbriata	Caryophyllaceae	"romerillo manso"	Ch	T,C,F,L

Common names according to Machado and Morera (2005)

Mayteno canariensis-Juniperion canariensis

Thermo-sclerophyllous woodlands of *Olea cerasiformis* and *Juniperus turbinata* subsp. *canariensis,* and potential and seral shrubland of *Rhamnus* and *Hypericum,* in the semiarid to dry Infra- and Thermomediterranean belts. Endemic to the Canary Islands.

Woodlands

Brachypodio arbusculae-Juniperetum canariensis (Fig. 6.98)

This is a xeric open juniper woodland endemic to La Gomera. Its climatophilous area rings the island above the climatophilous area of *Euphorbia canariensis* shrubland (*Euphorbio berthelotii-canariensis*), potential area of *E. aphylla* scrubland (in the NW) and below that of dry laurel forest (*Visneo mocanerae-Arbutetum canariensis*) on the slopes under the influence of the t-w.c., and below the climatophilous area of pine forest (*Cisto gomerae–Pinetum canariensis*) on parts of the southern slopes. It can also develop upwards as an edaphoxerophilous community.

It corresponds to the terminal community of the upper-semiarid and dry Inframediterranean (wc), and upper-semiarid and lower-dry Thermomediterranean (wc) climatophilous series of juniper (*Juniperus turbinata* subsp. *canariensis*) woodland on La Gomera (*Brachypodio arbusculae-Junipero canariensis sigmetum*).

Apart from *Juniperus turbinata* subsp. *canariensis*, the most characteristic species are *Olea cerasiformis*, *Pistacia atlantica*, *Rubia fruticosa*, *Euphorbia berthel-*

otii, *Kleinia neriifolia*, *Jasminum odoratissimum*, *Asparagus umbellatus*, *Rhamnus crenulata* and *Brachypodium arbuscula.*

Besides the subass. *typicum*, the subass. *ericetosum arboreae* (humid juniper woodland) grows in contact with the dry laurel forest. A *Periploca laevigata* facies can be distinguished on slope debris in the Hermigua Valley and a *Plocama pendula* facies develops over Miocene pyroclasts at several places on the island.

La Gomera's juniper woodland is today largely an open community, but impoverished as a result of human action. Its biotope was the chosen territory in which to establish rural settlements, grow crops and graze animals, because of its favorable orographic and climatic conditions. The juniper woodland is nowadays dispersed into several patches on the island, but some important nuclei still persist, like those of Hermigua, Agulo, Vallehermoso and Tazo.

Within its potential area, the most characteristic substitutional communities are rockrose scrub (*Micromerio gomerensis-Cistetum canariensis*) and *E. berthelotii* shrubland (*Euphorbietum berthelotii*), both widely spread over the island's S slope, and *Artemisio thusculae–Rumicetum lunariae* in anthropic places. The *Rhamno crenulatae–Hypericetum canariensis* is the substitutional shrubland found mostly in disturbed places within the potential area of the humid juniper woodland (subass. *ericetosum arboreae*), which shows a notable presence of the Canary endemic *Spartocytisus filipes*.

Distribution: Endemic to G.

Fig. 6.98 *Brachypodio arbusculae-Juniperetum canariensis*. Vallehermoso. La Gomera

Convolvulo lopezsocasii-Oleetum cerasiformis
This is an endemic association of Lanzarote, made up of the thermo-sclerophyllous climatophilous wild olive woodland. Its potential climatophilous area extends over a small surface above about 430 m in the N and 450 m in the S, over the highest mountains of the island, especially on the Famara massif.

It corresponds to the terminal community of the upper-semiarid Infra- and Thermomediterranean climatophilous series of the wild olive (*Olea cerasiformis*) in Lanzarote (*Convolvulo lopezsocasi-Oleo cerasiformis sigmetum*).

The most characteristic species are *O. cerasiformis, Convolvulus lopezsocasi, Gymnosporia cryptopetala*, *Phillyrea angustifolia*, *Pistacia lentiscus* and *Rhamnus crenulata*. These occur in fissures and ledges with shallow soil on the most inaccessible NW-facing cliffs of the Famara massif, constituting remnants of the community. Today the community has almost disappeared because of human activity.

Besides the typical subass. *oleetosum cerasiformis*, the subass. *euphorbietosum balsamiferae* has been differentiated, growing on rocks and walls of the summit. With a northern exposure the abundance of epiphytic lichens is striking, since it is favored by the moisture bearing NE trade-winds.

Distribution: Endemic to L.

Junipero canariensis-Oleetum cerasiformis (Fig. 6.99)
This association includes the thermo-sclerophyllous woodland of Tenerife, mainly juniper (*Juniperus turbinata* subsp. *canariensis*) woodland and its less abundant physiognomic facies of wild olive (*Olea cerasiformis*) woodland, and mastic (*Pistacia atlantica*) woodland. The structure and physiognomy are those of open xeric woodland, which generally thrives on developed shallow soils.

Its climatophilous area rings the island directly above the climatophilous area of *Euphorbia canariensis* shrubland (*Periploco-Euphorbietum canariensis*), reaching the potential area of dry laurel forest (*Visneo mocanerae-Arbutetum canariensis*) at about 300 m a.s.l. in the N, except on the NW facing slopes of Tacoronte-Santa Úrsula at about 400 m. On the SW and SE slopes it borders the area of pine forest (*Sideritido solutae-Pinetum canariensis*), except in the areas where clouds overflow the crest of the ridge, and in the local NE-facing area of Güímar where it borders the area of laurel forest. The upper potential natural limit in these SW and SE sectors varies considerably from the theoretical climatophilous position. The presence of large areas dominated by salic lava flows makes the pine forest (*Sideritido-Pinetum canariensis*) descend into the territories of the upper-semiarid Thermomediterranean, the lower-dry Thermomediterranean (wc) and even the top of the lower-semiarid Thermomediterranean. In any case, a good proportion of these areas, principally that of the upper-semiarid Thermomediterranean, are covered by the *Sideritido-Pinetum canariensis* subass. *cistetosum canariensis*, which occupies the mixed territory of contact between juniper and pine forest, where some dispersed remnants are still to be found.

It corresponds to the terminal community of the upper-semiarid Inframediterranean, semiarid Thermomediterranean and lower-dry Thermomediterranean (wc) of the climatophilous series of juniper (*Juniperus turbi-*

nata subsp. *canariensis*) and wild olive (*Olea cerasiformis*) woodland on Tenerife (*Junipero canariensis-Oleo cerasiformis sigmetum*).

The tree species which physiognomically characterize the woodland are *Juniperus turbinata* subsp. *canariensis*, *Olea cerasiformis* and *Pistacia atlantica*. Several shrubs of the mid-lands like *Convolvulus floridus*, *Hypericum canariense*, *Jasminum odoratissimum* and *Rhamnus crenulata* also occur, as well as others that are more xerophytic and widespread like *Euphorbia lamarckii*, *Kleinia neriifolia* and *Rubia fruticosa* subsp. *fruticosa*.

Besides the typical subass. *juniperetosum canariensis*, the most common, the subass. *ericetosum arboreae* is recognizable in more humid environments with more humic soil, in the transition to laurel forest. It shows a greater floristic diversity by incorporating plants like *Erica arborea*, *Ilex canariensis*, *Morella faya* and *Visnea mocanera*.

Some facies can also be distinguished. The *Pinus canariensis* facies characterizes some salic outcrops and recent lava flows in the W of the island, where the pines are interspersed within juniper woodland. The *Pistacia atlantica* facies corresponds to the scarce remnants of the *Pistacia atlantica* woodland, characteristic of places with a shallower water table. This facies in the North of Teno also includes trees of *Heberdenia excelsa*. The *Periploca laevigata* facies displays the abundance of this species on screes and stony places in the territory of the association.

The current representation of the community has dramatically retreated because of the transformation of its territory by farming and urban development. Only scattered remnants exist and those of Anaga, especially Afur and Mesa del Sabinal, are noteworthy.

Among the most representative substitutional communities are rockrose scrub (*Cistetum symphytifolio-canariensis*) on denuded soils; white broom shrubland (*Echio aculeati-Retametum rhodorrhizoidis*) on dry slope debris and abandoned fields, especially in the Teno massif; the nitrophilous shrubland *Artemisio thusculae-Rumicetum lunariae* on disturbed places; the bitter spurge shrubland *Euphorbietum lamarckii*; and the perennial grassland of the *Cenchro-Hyparrhenietum sinaicae*. Deforested zones near the dry laurel forest (*Visneo mocanerae-Arbutetum canariensis*) potential area are occupied by the *Rhamno crenulatae-Hypericetum canariensis*.

Note: For now we include the *Pistacio atlanticae-Heberdenietum excelsae* Nezadal & Wells 2009, an association given for North Teno, within the *Junipero canariensis-Oleetum cerasiformis* facies of *Pistacia atlantica*, and the *Rhamno-Hypericetum canariensis* facies of *Heberdenia excelsa*. The local presence of *Heberdenia excelsa* there, a typical plant of the *Pruno-Lauretea,* could be explained by the shallower water table typical of the territory of these facies.

Distribution: Endemic to T.

Fig. 6.99 *Junipero canariensis-Oleetum cerasiformis*. Afur. Anaga. Tenerife

Micromerio rupestris-Oleetum cerasiformis (Fig. 6.100)
This is an endemic association of Fuerteventura, made up by thermo-sclerophyllous climatophilous wild olive woodland, which grows on shallow well-structured soils. Its climatophilous area is located directly above the area of *Euphorbia* shrubland (*Kleinio neriifoliae-Euphorbietum canariensis*), about 450 m in the N and 500 m in the S, and caps the island summits except in the Jandía massif where it would reach up to the potential area of the *Echio handiensis-Visneetum mocanerae*, at about 700 m.

It corresponds to the terminal community of the semiarid Infra- and Thermomediterranean climatophilous series of the wild olive (*Olea cerasiformis*) on Fuerteventura (*Micromerio rupestris-Oleo cerasiformis*).

The most characteristic species are *Olea cerasiformis* and *Micromeria varia* subsp. *rupestris*. Sporadically other typical species like *Pistacia atlantica* and *Rhamnus crenulata* are to be found. Limited to Jandía scarps are *Jasminum odoratissimum, Maytenus canariensis*, *Phillyrea angustifolia, Pistacia lentiscus* and *Bupleurum handiense.*

It has almost completely disappeared because of wood harvesting, farming and grazing, especially after the Conquest. Thus, its current distribution is reduced to the presence of isolated dwarf individuals of the characteristic species on the summits of the Betancuria massif and scarps of Jandía. On the trade-winds-facing slopes, a humid wild olive woodland transitional to dry laurel forest may have been developed, with participation of the driest tolerant species of the potential dry laurel forest of that island.

Note: Provisionally, we consider the *Bupleuro handiensis-Maytenetum canariensis* described for Jandía comprised within the *Micromerio rupestris-Oleetum cerassiformis* Rodriguez et al. 2000, the typical thermo-sclerophyllous woodland in Fuerteventura. Some endemic plants like *Gymnosporia cryptopetala, Bupleurum handiense* and *Sideritis pumila* typical of the dwarf relictual thermo-sclerophyllous woodland area of Jandía in Fuerteventura, which could be considered as differential, also grow on the Famara massif in Lanzarote, where the *Convolvulus lopezsocasi-Oleetum cerasiformis* grows.

Distribution: Endemic to F.

Fig. 6.100 *Micromerio rupestris-Oleetum cerasiformis*. Betancuria Massif. Fuerteventura

Pistacio lentisci-Oleetum cerasiformis (Fig. 6.101)
This is an endemic association of Gran Canaria, which includes stands of thermo-sclerophyllous juniper, wild olive, and mastic woodlands. Its climatophilous area rings the island directly above the area of *Euphorbia* shrubland (*Aeonio percarnei-Euphorbietum canariensis)*, reaching the area of dry laurel forest (*Visneo-Arbutetum canariensis*) at near 900 m a.s.l. in the NW, 500 m in the N and 600–700 m in the E. In the SW part of this ring, territory where the salic substrate is dominant, it has little potential presence relegated to basaltic outcrops; it borders upwards with the potential area of pine forest over such rocks. On the salic substrates of this sector, the pine forest *Pinetum canariensis* subass. *juniperetosum canariensis* is dominant.

It is an association of variable physiognomy depending on the dominant tree: wild olive woodland with dominance of *Olea europaea* subsp. *cerasiformis*; mastic-tree woodland with dominance of *Pistacia atlantica*, and mastic-bush thicket with dominance of *Pistacia lentiscus*. Nowadays, juniper (*Juniperus turbinata* subsp. *canariensis*) woodland (“sabinar”) is not found, as only isolated trees remain.

It corresponds to the terminal community of the upper-semiarid Inframediterranean, semiarid and lower-dry Thermomediterranean belts (wc), which is the climatophilous series of mastic-bush (*Pistacia lentiscus*) and wild olive (*Olea europaea* subsp. *cerasiformis*) in Gran Canaria (*Pistacio lentisci-Oleo cerasiformis sigmetum*).

In its floristic cohort, the shrubs of the *Mayteno-Juniperion canariensis* are frequent, like *Asparagus plocamoides* and *Bupleurum salicifolium* subsp. *aciphyllum*. Besides, widespread shrubs of the *Kleinio-Euphorbietea canariensis* are frequent, like *Asparagus umbellatus*, *Euphorbia regis-jubae*, *Kleinia neriifolia*, *Periploca laevigata* and *Rubia fruticosa*.

The current representation of the community is very scarce due to the radical transformation of its potential territory, mainly by farming and urban development. The best preserved remnants are the stands of wild olive scattered over the whole NE *e.g.*, Barranco de los Cernícalos (Telde); mastic-bushes, also relegated to the NE e.g., Caldera de Bandama – Monte Lentiscal; and the mastic-tree clumps spread mainly over the NW, where they show a good regeneration at the bottom of valleys, especially in La Aldea.

Among the substitution communities with a widespread area are rockrose scrub (*Euphorbio-Cistetum canariensis*), the nitrophilous shrubland *Artemisio-Rumicetum lunariae*, the bitter spurge shrubland (*Euphorbietum regis-jubae*), elsewhere as a recovering stage of the deforested woodland, and also the perennial grassland of the *Cenchro-Hyparrhenietum sinaicae*. In the N, in adjacent areas to dry laurel forest *Rhamno-Hypericetum canariensis* appears.

Distribution: Endemic to C.

Fig. 6.101 *Pistacio lentisci-Oleetum cerasiformis*. Barranco de Guayadeque. Gran Canaria

Rhamno crenulatae-Juniperetum canariensis (Fig. 6.102)
This is a xeric open juniper woodland endemic to La Palma. Its potential climatophilous area surrounds the island above the *Euphorbio canariensis* shrubland (*Echio breviramae-Euphorbietum canariensis*) potential area and below the dry laurel forest (*Visneo mocanerae-Arbutetum canariensis*) potential area on the N, NE and E slopes, and the pine forest (*Loto hillebrandii-Pinetum canariensis*) potential area in the SE, S, SW and partially NW. Within the potential area of the series that follows in elevation upon the *Rhamno crenulatae-Juniperetum canariensis* this community can behave as an edaphoxerophilous community.

It corresponds to the terminal community of the upper-semiarid to subhumid Inframediterranean (wc), upper-semiarid and lower-dry Thermomediterranean (wc), climatophilous series of juniper (*Juniperus turbinata* subsp. *canariensis*) on La Palma (*Rhamno crenulatae-Junipero canariensis sigmetum*).

The most characteristic species are *Juniperus turbinata* subsp. *canariensis, Euphorbia lamarckii*, *Kleinia neriifolia*, *Olea cerasiformis*, *Rhamnus crenulata* and *Rubia fruticosa*.

The *Periploca laevigata* facies which grows at stony, windy, disturbed places is widespread on the main promontories of the island.

The current representation of the community is very scarce and phragmented, mainly due to the farmed and urbanized areas within its potential territory. It is almost impossible to find optimally representative stands of this type of woodland. Within its potential area it is frequent to find some isolated individuals of juniper growing amid *Euphorbietum lamarckii,* or the *Cisto-Micromerietalia* or *Pegano-Salsoletea* substitutional scrublands. At elevations near the laurel forest, the mixture between juniper and the *Hypericum canariense* shrubland (*Rhamno crenulatae-Hypericetum canariensis*) is common.

Distribution: Endemic to P.

Fig. 6.102 *Rhamno crenulatae-Juniperetum canariensis*. Barranco de San Juan. Los Sauces. La Palma

Rubio fruticosae-Juniperetum canariensis (Fig. 6.103)
This is a xeric open juniper woodland endemic to El Hierro. Its climatophilous area surrounds the island above the potential area of *Euphorbia canariensis* shrubland (*Aeonio valverdensis-Euphorbietum canariensis*), and below the potential area of dry laurel forest (*Visneo mocanerae-Arbutetum canariensis*) in the North of the island and at the summits of south slopes under the influence of overflowing clouds, and below the pine forest area (*Bystropogono ferrensis-Pinetum canariensis*) on the South slope. It can also develop upwards as an edaphoxerophilous community.

It corresponds to the terminal community of the upper-semiarid Infra- and Thermomediterranean climatophilous series of the juniper (*Juniperus turbinata* subsp. *canariensis*) on El Hierro (*Rubio fruticosae-Junipereto canariensis sigmetum*).

Apart from *J. turbinata* subsp. *canariensis* and *Olea cerasiformis* the most characteristic species, among others, are *Asparagus umbellatus*, *Euphorbia lamarckii* (in southern juniper woodland), *Kleinia neriifolia*, *Jasminum odoratissimum*, *Rhamnus crenulata* and *Rubia fruticosa*.

Although the juniper tree is widespread within the band described as the potential area of juniper woodland, only a few more or less preserved stands of this juniper woodland as such persist. In spite of this small representation they are of the greatest importance in the Canary Islands because of their scarcity: with the exception of La Gomera they have greatly retreated in the rest of the islands. The stands of El Julan and La Dehesa are good samples of typical Canary juniper woodland. They are floristically very poor and apart from the juniper as arboreal element, the widespread shrubs of *Euphorbia* communities (*Kleinio-Euphorbietea canariensis*) are dominant in the characteristic floristic combination. On the northern slopes, especially those most benefitted by the humid influence of the NE trade-winds, the juniper woodland is enriched with the most xeric elements from the laurel forest, giving rise to the presence of the subass. *ericetosum arboreae* (humid juniper woodland). The subass. *pinetosum canariensis* is mainly found on the wide southern strip of the island corresponding to the dry Infra- and Thermomediterranean belts.

The *Periploca laevigata* facies grows mainly on screes. A very scarce *Rubia fruticosa* facies has been also distinguished.

Among the most frequent communities growing in the potential area of the juniper woodland are the *Euphorbietum lamarckii, Micromerio-Cistetum canariensis*, *Echio- Micromerietum hyssopifoliae*, *Rhamno-Hypericetum canariensis* and the *Echio hierrensis-Retametum rhodorhizoidis*. Communities of the *Pegano-Salsoletea* on abandoned cultivated ground are also common.

Distribution: Endemic to H.

Fig. 6.103 *Rubio fruticosae-Juniperetum canariensis*. La Dehesa. El Hierro

Shrublands

Rhamno crenulatae-Hypericetum canariensis (Fig. 6.104)
This is a shrubby community, with some thorny species, which constitutes a xerophytic mantle around dry laurel forest (*Visneo-Arbutetum canariensis*) and also expands into northern territories of dry to humid variants of thermo-sclerophyllous woodlands (*Mayteno-Juniperion canariensis*).

It is a species-rich community, comprising *Asparagus scoparius*, *Bupleurum salicifolium*, *Echium spp.*, *Erysimum bicolor*, *Hypericum canariense*, *Jasminum odoratissimum*, *Rhamnus crenulata* and *Spartocytisus filipes,* among others.

Apart from being a mantle community, the association shows serial character developing as substitutional shrubland in disturbed areas of the upper two thirds of the dry laurel forest potential territory, and that of the juniper woodland, mainly in the dry Infra- and Thermomediterranean zones facing the trade-winds and some humid microclimates to the leeward. Nowadays it is widespread and the spring yellow bloom of *Hypericum canariensis* stands out in the landscape of the windward mid-lands of the western islands.

The *Pistacia atlantica* facies represents a successional stage of *Juniper* woodland particularly developed at humid places of the E of Anaga and the North of Teno in Tenerife, and the E of La Gomera. A *Heberdenia* facies is common in the foothills around the Hermigua ravine in La Gomera, and the North of Teno (see "note" within *Junipero canariensis-Oleetum cerasiformis*).

Distribution: C, T, G, P, H.

Fig. 6.104 *Rhamno crenulatae-Hypericetum canariensis*. Interián. Tenerife

Phoenicion canariensis

Potential colluvial or seasonally-hygrophilous ravine community of Canary palm in the arid to dry Infra- to Thermomediterranean belts. Endemic to the Canary Islands.

Periploco laevigatae-Phoenicetum canariensis (Fig. 6.105)

This is the Canary palm tree (*Phoenix canariensis*) community which develops on seasonally-humid old colluvial soils, around tamarisk-lined watercourse edges on the eastern islands or the Canary willow community on western islands, and also around springs and dripping water crevices, mainly within the climatophilous territory of *Euphorbia* shrublands, and *Juniper* thermo-sclerophyllous woodland.

It corresponds to the terminal community of the arid to semiarid Inframediterranean and semiarid to dry Thermomediterranean edaphohygrophilous colluvial and riparian series of the Canary palm (*Periploco laevigatae-Phoenico canariensis sigmetum*).

The most characteristic species are *Phoenix canariensis, Periploca laevigata, Dracaena draco* and *Plocama pendula.*

Natural Canary palm stands are rather scarce on some islands, although on Gran Canaria and La Gomera their abundance is outstanding. *Phoenix dactylifera* introduced on the islands nowadays grows together with the Canary palm, and hybrids have developed. Sometimes *Dracaena draco* is present on colluvia where palms grow; this suggests that both meso-phanerophytes occurred together with other shrubs and forest twiners.

Distribution: Endemic to F, C, T, G, P.

Fig. 6.105 *Periploco laevigatae-Phoenicetum canariensis*. Rambla de Castro. Tenerife

Retamion rhodorhizoidis

Potential and seral colluvial broom shrubland of *Retama rhodorhizoides* of the semi-arid to dry Infra- and Thermomediterranean belts. Endemic to the Canary Islands.

Echio aculeati-Retametum rhodorhizoidis (Fig. 6.106)

This is a nanophanerophytic shrubland endemic to Tenerife that grows in dry slope debris mainly within the climatophilous territory of juniper woodland (*Junipero-Oleetum cerasiformis*).

It corresponds to the terminal community of the upper-semiarid and dry Inframediterranean (wc), and semiarid and lower-dry Thermomediterranean (wc) edaphoxerophilous colluvial series of white broom (*Retama rhodorhizoides*) on Tenerife (*Echio aculeati-Retamo rhodorhizoides sigmetum*). In a secondary way, it also expands as substitutional community favored by grazing, and in old cultivated lands.

The association is physiognomically characterized by *Retama rhodorhizoides*, together with *Echium aculeatum* and *Euphorbia lamarckii*, which are also accompanied by widespread species from the xeric belts of the island.

It particularly develops in the Teno massif (Carrizales-Masca-Los Gigantes) and Santiago del Teide Valley. Also, to a lesser extent, it grows at the edges of the Güímar Valley: at the foot of La Ladera, and between Araya and Igueste de Candelaria.

Both in Masca and Ladera de Güímar, a change in this community towards the substitutional *Euphorbietum lamarckii* shrubland is observed, at disturbed places, as is the change to the *Euphorbietum atropurpureae* on undisturbed rocky substrate.

Distribution: Endemic to T.

Fig. 6.106 *Echio aculeati-Retametum rhodorhizoidis*. Igueste de Candelaria. Tenerife

Echio decaisnei-Retametum rhodorhizoidis

This is a nanophanerophytic broom shrubland endemic to Gran Canaria that develops on loose substrates, incipient soils on pyroclastic cones, or colluvial deposits within the climatophilous territory of wild olive woodland (*Pistacio lentisci-Oleetum cerasiformis*), in the NE sector of Gran Canaria.

It corresponds to the terminal community of the semiarid to dry Infra- and Thermomediterranean edaphophilous colluvial series of white broom (*Retama rhodorhizoides*) on Gran Canaria (*Echio decaisnei-Retamo rhodorhizoides sigmetum*). It can also expand to grazed terrrain.

It is characterized by *Retama rhodorhizoides*, which is dominant among a cohort that includes xeric species of the *Euphorbia* communities like *Euphorbia regis-jubae*, *Kleinia neriifolia* and *Rubia fruticosa* as well as others of wider distribution like *Echium decaisnei*.

Distribution: Endemic to C.

Echio hierrensis-Retametum rhodorhizoidis

This is a broom shrubland endemic to El Hierro that grows in dry slope debris within the climatophilous territory of juniper woodland (*Brachypodio arbusculae-Juniperetum canariensis*).

It corresponds to the terminal community of the upper-semiarid and dry Inframediterranean, and upper-semiarid and lower-dry Thermomediterranean (wc) edaphoxerophilous colluvial series of white broom (*Retama rhodorhizoides*) on El Hierro (*Echio hierrensis-Retamo rhodorhizoides sigmetum*).

It is mainly characterized by *R. rhodorhizoides* as the dominant species within a floristic cohort of widespread species of the island's most thermophilous and xeric belts, like *Echium aculeatum*, *E. hierrensis*, *Euphorbia lamarckii subsp. wildpretii*, *Kleinia neriifolia* and *Rubia fruticosa.*

This white broom shrubland currently occurs in the N sector of Riscos de Tibataje, below La Peña, Riscos de Isora and Los Herreños, where besides alternating with the remains of the mentioned arboreal vegetation, it alternates with other scrub, mainly rockrose on decapitated soils (*Micromerio-Cistetum canariensis*), or communities of the *Pegano-Salsoletea* on abandoned cultivated ground.

Distribution: Endemic to H.

Euphorbio berthelotii-Retametum rhodorhizoidis (Fig. 6.107)
This is a nanophanerophytic shrubland endemic to La Gomera that grows on dry slope debris mainly within the climatophilous territory of juniper woodland (*Brachypodio arbusculae-Juniperetum canariensis*), and sometimes within that of *E. canariensis* shrubland (*Euphorbio berthelotii-canariensis*).

It corresponds to the terminal community of the semiarid and dry Inframediterranean, and upper-semiarid and lower-dry Thermomediterranean (wc) edaphoxerophilous colluvial series of white broom (*Retama rhodorhizoides*) on La Gomera (*Euphorbio berthelotii-Retamo rhodorhizoidis sigmetum).*

The association is physiognomically characterized by *R. rhodorhizoides*, accompanied by widespread species from the xeric belts of the island, like *Echium aculeatum*, *Euphorbia berthelotii*, *Kleinia neriifolia*, *Micromeria varia* subsp. *varia*, *Neochamaelea pulverulenta* and *Rubia fruticosa.*

The community is widespread particularly over the NW sector of the island (Arguamul-Alojera-Taguluche) and more localized in the W (Valle Gran Rey), N (Hermigua Valley) and S (Santiago Valley). Its expansion in the NW is largely anthropic, mainly due to fires and grazing.

Distribution: Endemic to G.

Fig. 6.107 *Euphorbio berthelotii-Retametum rhodorhizoidis*. Tazo. La Gomera

Euphorbio wildpretii-Retametum rhodorhizoidis (Fig. 6.108)
This is a nanophanerophytic broom shrubland endemic to La Palma that grows on dry slope colluvial deposits, old alluvial soils, malpaises and lapillis mainly within the climatophilous territory of juniper woodland (*Rhamno crenulatae-Juniperetum canariensis*), partly within that of *E. canariensis* shrubland (*Echio breviramis-Euphorbietum canariensis*).

It corresponds to the terminal community of the semiarid and dry Inframediterranean, and semiarid and lower-dry Thermomediterranean (wc) edaphoxerophilous colluvial series of white broom (*Retama rhodorhizoides*) on La Palma (*Euphorbio wildpretii-Retamo rhodorhizoides sigmetum*).

It is characterized by *R. rhodorhizoides*, *Euphorbia lamarckii* subsp. *wildpretii*, *Echium brevirame* and *Kleinia neriifolia.*

It is particularly widespread over the geologically young southern half of the island, where an abundance of fractured lava flow substrates favour its development. In part of the community's current area its expansion could have been favoured by fires and grazing.

Distribution: Endemic to P.

Fig. 6.108 *Euphorbio wildpretii-Retametum rhodorhizoidis*. El Espigón. Bajamar. La Palma

Scrublands

Micromerio hyssopifoliae-Cistetalia canariensis
Permanent rupicolous or seral scrublands growing on eroded stony or decapitated soils mainly dominated by *Micromeria* and *Cistus* species, typical of the semiarid to dry Infra- to Thermomediterranean belts, mainly in potential territory of *Euphorbia canariensis* shrubland, thermo-sclerophyllous woodland, and thermophilous pine forest, with scarce penetration into lower territories of thermophilous laurel forest. A good proportion of these scrublands contain summer deciduous plants. The characteristic autumn to spring dark-green color of those dominated by *Cistus* disappears in summer when the leaves of its most characteristic species fall. Endemic to Canary Islands.

Micromerio hyssopifoliae-Cistion canariensis
Single Alliance.

Carlino salicifoliae-Echietum famarae
This is a chamaephytc and nanophanerophytic seral scrub that grows on cliffs and ledges with scarce soil, on the N-NW facing slopes of the Jandía massif within the semiarid Thermomediterranean belt, just below the area influenced by trade-wind clouds.

Echium decaisnei subsp. *purpuriense* (=*E. famarae*), *Carlina salicifolia, Andryala pinnatifida* subsp. *buchiana, Bituminaria bituminosa* and *Ferula lancerottensis* are its characteristic species. Widespread companion species of the *Kleinia-Euphorbietea, Rhamno-Oleetea* and *Aeonietea* stand out.

The community is subject to intensive grazing except in the most inaccessible places of the massif, reaching its optimum within the potential territory of thermo-sclerophyllous woodland.

Distribution: Endemic to F.

Cistetum symphytifolio-canariensis (Fig. 6.109)
This is a chamaephytic-nanophanerophytic association that is summer deciduous and endemic to Tenerife. It grows on eroded stony or decapitated soils, with scarce soil covering the parent rock discontinuously, mainly in the upper level of the climatophilous area of *Euphorbia canariensis* shrubland (*Periploco-Euphorbietum canariensis*), potential juniper woodland (*Junipero-Oleetum cerasiformis*), the lower level area of dry laurel forest (*Visneo-Arbutetum canariensis*) and the lower level area of pine forest (*Sideritido solutae-Pinetum canariensis*). It is much more developed on the mid-elevation southern slopes.

It is mainly characterized by *Cistus monspeliensis* subsp. *canariensis*, *Cistus symphytifolius*, *Euphorbia lamarckii* and *Micromeria hyssopifolia*, together with some other scrub plants of the xeric and pluviseasonal territory.

As nitrogen increases in the soil by grazing, the community evolves towards the nitrophilous *Pegano-Salsoletea*, mainly the *Artemisio-Rumicetum lunariae*.

At the most impoverished places a *Micromeria hyssopifolia* facies thrives.

Distribution: Endemic to T.

Fig. 6.109 *Cistetum symphytifolio-canariensis*. Lomo Morín. Los Silos. Tenerife

Echio aculeati-Micromerietum hyssopifoliae (Fig. 6.110)
This is a chamaephytic-nanophanerophytic scrub, a permanent community of lapilli and volcanic ashes within the Thermo- and Mesomediterranean belts on El Hierro, territories of the climatophilous potential of juniper woodland and pine forest. It also has a nitrophilous behavior, spreading secondarily to abandoned cultivated lands and stabilized edges of farm and forestry tracks.

Echium aculeatum and *Micromeria hyssopifolia* are its characteristic species. Also, the isolated presence of some plants belonging to the potential communities of those territories is common (Canary pine, Canary juniper, dendroid spurges, etc.).

In its most typical appearance it has a high presence of *Echium aculeatum*, and it extends from lower-dry up to subhumid territories on S slopes that are not or only little influenced by overflowing trade-wind clouds towards the south.

The chamaephytic cushionlike *Micromeria hyssopifolia* facies, less than 40 cm in height, grows abundantly on the ridges of the island's summit, between 1200 and 1500 m, influenced by frequent trade-wind clouds.

The subass. *rubietosum fruticosae* is the most thermophilic, growing in the territory of juniper woodland or the transition to pine forest. Typical plants of the *Kleinio-Euphorbietea* and *Forsskaoleo-Rumicetalia* are differential: *Rubia fruticosa, Kleinia neriifolia, Euphorbia lamarckii* and *Artemisia thuscula.* Young individuals of *Juniperus turbinata* subsp. *canariensis* are common, showing a recovery of the natural landscape.

Distribution: Endemic to H.

Fig. 6.110 *Echio aculeati-Micromerietum hyssopifoliae*. El Tomillar. El Hierro. (Ph: PLP)

Echio breviramis-Micromerietum herpyllomorphae (Fig. 6.111)
This is a chamaephytic-nanophanerophytic scrub, a permanent community of lapilli and volcanic ashes within the Infra- and Thermomediterranean belts on La Palma. It is especially well developed in the island's southern tip in the municipality of Fuencaliente. Furthermore, it behaves as a nitrophilous community secondarily spreading on loose earthy pyroclasts, abandoned cultivated lands, road edges and cuttings, and screes in the above-mentioned belts.

Echium brevirame and *Micromeria herpyllomorpha* are its characteristic species. Dynamic species of the lower zone like *Rumex lunaria* and *Kleinia* neriifolia, grasses like *Aristida adscensionis* and *Hyparrhenia sinaica,* and *Tolpis laciniata* and *Bituminaria bituminosa* are the commonest companion species.

Distribution: Endemic to P.

Fig. 6.111 *Echio breviramis-Micromerietum herpyllomorphae*. Cumbre Vieja. La Palma

Euphorbio regis-jubae-Cistetum canariensis (Fig. 6.112)
This is a non- or scarcely nitrophilous chamaephytic or nanophanerophytic drought-deciduous scrub of Gran Canaria. It grows on eroded stony or decapitated soils, mainly within the climatophilous area of thermo-sclerophyllous woodland (*Pistacio-Oleetum cerasiformis*), thermophile pine forest (*Pinetum canariensis* subass. *typicum*), and to a lesser extent at the upper level of the potential area of *E. canariensis* shrubland (*Aeonio percarnei-Euphorbietum canariensis*), from the NW slope, all

along the W, until the SE slopes of the island. Generally it is an open scrub with a maximum coverage of 75%.

Besides *Cistus monspeliensis* subsp. *canariensis* and *Euphorbia regis-jubae,* it is characterized by xeric and thermophilous plants like *Echium onosmifolium*, *Lavandula minutolii* and *Micromeria varia* subsp. *meridialis*. The presence of *Ononis angustissima* and *Salvia canariensis* indicates a certain degree of nitrification, mainly by grazing. Then, the scrub tends to change towards *Pegano-Salsoletea* communities, mainly to the *Artemisio thusculae-Rumicion lunariae* (*Rumex lunaria, Artemisia thuscula* or *Argyranthemum spp*. facies).

Micromeria lanata and *Echium decaisnei* facies constitute an impoverished stunted appearance of the community.

Distribution: Endemic to C.

Fig. 6.112 *Micromerio gomerensis-Cistetum canariensis*. Imada. La Gomera

Micromerio gomerensis-Cistetum canariensis

This is a scarcely or non-nitrophilous chamaephytic-nanophanerophytic summer-deciduous association, endemic to Gomera. It grows on eroded stony or decapitated soils, mainly within the climatophilous area of juniper woodland (*Brachypodio arbusculae-Juniperetum canariensis*) and pine forest (*Cisto gomerae-Pinetum canariensis*), although it can also be found in the upper level of cardón shrubland (*Euphorbio berthelotii-canariensis*) and even within the potential territory of laurel forest (*Pruno-Lauretalia*). It is more abundant in the S sector of the island.

It is mainly characterized by *Micromeria varia* subsp. *gomerensis*, *Cistus monspeliensis* subsp. *canariensis* and some widespread species of the *Kleinio-Euphorbietea canariensis* such as *Euphorbia berthelotii* and *Echium aculeatum*.

A chamaephytic *Micromeria varia* subsp. *gomerensis* facies is recognized.

The subass. *adenocarpetosum foliolosi* grows in the coldest area of the association, which corresponds to potential territories of laurel forest and pine forest. *Adenocarpus foliolosus* and *Chamaecytisus proliferus* subsp. *angustifolius* are its differential taxa.

Distribution: Endemic to G.

Micromerio herpyllomorphae-Cistetum canariensis

This is a non- or scarcely nitrophilous chamaephytic or nanophanerophytic drought-deciduous scrub of La Palma, growing on stony or decapitated soils. It is scattered within the climatophilous territory of juniper woodland (*Rhamno-Juniperetum* canariensis) encircling the island, and that of the most xerophytic pine forest (*Loto-Pinetum canariensis* subas. *cistetosum symphytifolii*) of the island's western slope.

It is mainly characterized by *Micromeria herpyllomorpha* and *Cistus monspeliensis* subsp. *canariensis*, and some widespread species of the *Kleinio-Euphorbietea canariensis* are common companions.

Distribution: Endemic to P.

Micromerio hyssopifoliae-Cistetum canariensis

This is a non- or scarcely nitrophilous chamaephytic or nanophanerophytic drought-deciduous scrub of El Hierro, growing on stony or decapitated soils, growing mainly within the climatophilous territory of juniper woodland (*Rubio-Juniperetum canariensis*) or dry laurel forest (*Visneo-Arbutetum canariensis*).

It is characterized by *Cistus monspeliensis* subsp. *canariensis*, *Echium hierrense, Euphorbia lamarckii* and *Micromeria hyssopifolia*, among other mainly xeric plants.

The subass. *hypericetosum canariensis* constitutes the humid rockrose community of the north slope of the island, *Hypericum canariense* and *Pericallis murrayii* being differential.

Distribution: Endemic to H.

Micromerio variae-Globularietum salicinae (Fig. 6.113)

This is a low to medium-sized scrub that constitutes a substitutional stage of dry and subhumid laurel forest (*Visneo-Arbutetum canariensis* and *Lauro-Perseetum indicae*), and humid juniper woodland on decapitated stony soils. It mainly grows on ridges, outcrops and steep eroded slopes.

Globularia salicina and *Micromeria varia* are its characteristic species.

On Tenerife is particularly present in Anaga and Teno. On La Palma it mainly grows in the NE. It has a similar general appearance as rockrose scrub but is a brighter green.

Distribution: Endemic to T, P.

Fig. 6.113 *Micromerio variae-Globularietum salicinae*. Barranco Huertas de San Andrés. Anaga. Tenerife

Spergulario fimbriatae-Helianthemetum canariensis
This is a chamaephytic dwarf community formed by stunted chamaephytes growing on exposed windy and strongly grazed, much eroded soils showing calcic and petrocalcic superficial horizons (Petrocalcids and Haplocalcids). It constitutes a substitutional stage of potential *Euphorbia* communities and thermo-sclerophyllous woodland, that has been favored by traditional clearing for fuelwood and decapitation of soils

Helianthemum canariense, *Micromeria varia* subsp. *rupestris* and *Spergularia fimbriata* are dominant, and the geophyte *Allium subvillosum* is usually abundant.

Distribution: Endemic to L, F?.

6.2.7.3 Laurel Forest

Pruno hixae-Lauretea novocanariensis (Figs. 6.114, 6.115 and 6.116)
Large phytosociological class that includes: 1. Potential laurel forest of the dry to humid Infra- to Mesomediterranean belts (c and ces); 2. Seral microforest of *Morella faya* and *Erica arborea* and mantle or seral legume shrublands of the same belts. Legume shrublands can extend into the pine forest area within the Mesomediterranean belts (wc); 3. Riparian potential willow forest and mantle and seral bramble ticket. Class endemic to Canarian-Madeiran subregion (Table 6.3).

Potential Climatophilous Forest

Pruno hixae-Lauretalia novocanariensis
Potential laurel forest of the dry to humid Infra- to Mesomediterranean belts (c and ces). Endemic to Canarian-Madeiran subregion.

Visneo mocanerae-Apollonion barbujanae
Potential laurel forest of the dry to subhumid Infra- to Thermomediterranean belts (c). Endemic to Canarian-Madeiran subregion.

Echio handiensis-Visneetum mocanerae
This is a relictual eastern Canary mesophytic heminebulous forest, related to the western dry laurel forest (*Visneo-Arbutetum canariensis*), thermophilous and resistant to relatively xeric conditions. Its current fragments grow between 700 and 807 m a.s.l. on the summit of the Jandía massif (Fuerteventura), on very steep cliffs exposed windward to the NE trade-winds.

Visnea mocanera and *Heberdenia excelsa* are its characteristic dwarf tree species, here greatly reduced in size and density by the limiting environmental conditions and goat grazing. *Echium handiense* and *Ononis christii* are the characteristic shrubs.

It corresponds to the terminal community of the upper-semiarid (and dry) Thermomediterranean belts (c), of the climatophilous series of the subxeric forest of *V. mocanera* and *H. excelsa* on Fuerteventura (*Echio handiensis-Visneo mocanerae sigmetum*).

The community area is extraordinarily reduced as a consequence of felling and grazing over the centuries. Only fragments are to be found on ledges and in crevices of the Jandía cliffs. Some companion plants of humid laurel forest areas like *Smilax aspera, Andryala pinnatifida, Ranunculus cortusifolius* and *Polypodium macaronesicum* are common, together with a rich epiphytic moss and lichen flora.

Distribution: F.

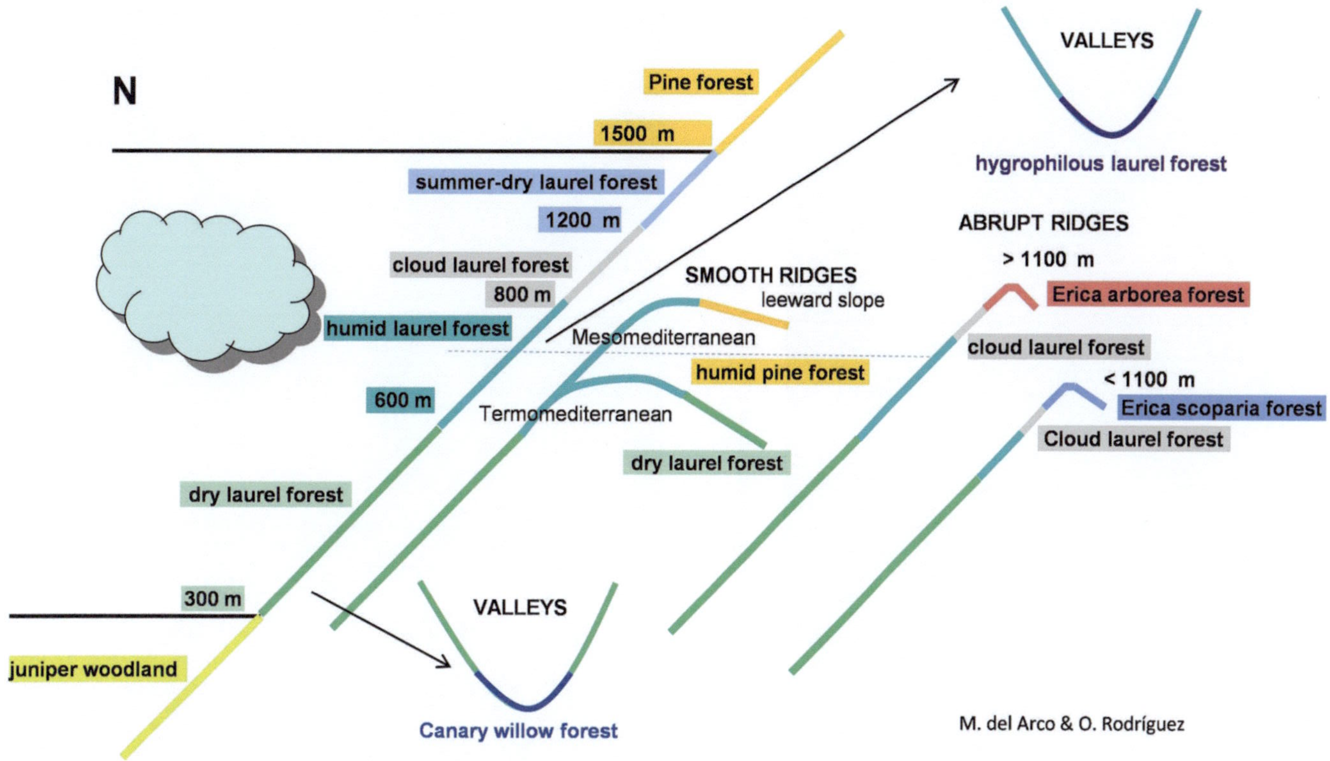

Fig. 6.114 General distribution pattern of potential communities of laurel forest, and contacts with other communities

Fig. 6.115 *Pruno hixae-Lauretea novocanariensis*. A view of the laurisilva of Anaga from Pico del Inglés. Tenerife

Fig. 6.116 *Pruno hixae-Lauretea novocanariensis*. El Cedro. La Gomera. (Ph: AFL)

Table 6.3 Characteristic species of the vegetation of *Pruno hixae-Lauretea novocanariensis*

Pruno hixae-Lauretea novocanariensis				
Scientific name	**Family**	**Common name**	**Biotype**	**Distribution**
Pruno hixae-Lauretalia novocanariensis				
TREES				
Apollonias barbujana subsp. barbujana	Lauraceae	"barbuzano"	M	H,P,G,T,C,F
Apollonias barbujana subsp. *ceballosi*	Lauraceae	"barbuzano negro"	M	G
Arbutus canariensis	Ericaceae	"madroñero canario"	M	H,P,G,T,C
Erica platycodon	Ericaceae	"tejo canario"	M	H,G,T
Euphorbia mellifera	Euphorbiaceae	"tabaiba de monteverde"	M	P,G,T
Heberdenia excelsa	Primulaceae	"aderno"	M	H,P,G,T,C,F
Ilex canariensis	Aquifoliaceae	"acebiño"	M	H,P,G,T,C
Ilex perado subsp.*lopezlilloi*	Aquifoliaceae	"naranjero salvaje gomero"	M	G
Ilex perado subsp. *platyphylla*	Aquifoliaceae	"naranjero salvaje"	M	P,G,T
Laurus novocanariensis	Lauraceae	"loro"	M	H,P,G,T,C,F,L?
Morella faya	Myricaceae	"faya canaria"	M	H,P,G,T,C,F,L
Morella rivas-martinezii	Myricaceae	"faya herreña"	M	H,P,G
Ocotea foetens	Lauraceae	"til"	M	H,P,G,T,C
Persea indica	Lauraceae	"viñátigo"	M	H,P,G,T,C
Picconia excelsa	Oleaceae	"paloblanco"	M	H,P,G,T,C,F
Pleiomeris canariensis	Primulaceae	"delfino"	M	P,G,T,C
Prunus lusitanica subsp. *hixa*	Rosaceae	"hija"	M	H,P,G,T,C
Rhamnus glandulosa	Rhamnaceae	"sanguino"	M	P,G,T,C
Sambucus palmensis	Adoxaceae	"saúco canario"	M	P,G,T,C
Sideroxylon canariensis	Sapotaceae	"marmolán"	M	H,P,G,T,C,F
Viburnum rigidum	Adoxaceae	"follao canario"	M	H,P,G,T,C
Visnea mocanera	Pentaphylacaceae	"mocán"	M	H,P,G,T,C,F
SHRUBS				
Asparagus fallax	Asparagaceae	"esparraguera de monteverde"	N-Li	G,T
Bencomia caudata	Rosaceae	"bencomia de monte"	N	P,T,C
Cedronella canariensis	Lamiaceae	"algaritope"	N	H,P,G,T,C
Crambe strigosa	Brassicaceae	"colderrisco canaria"	Ch/N	P,G,T
Hedera canariensis	Araliaceae	"yedra canaria"	N/M-Li	H,P,G,T,C
Hypericum glandulosum	Hypericaceae	"malfurada de monte"	N	P,G,T,C,F
Hypericum grandifolium	Hypericaceae	"malfurada grande"	N	H,P,G,T,C,F,L

(continued)

Table 6.3 (continued)

Pruno hixae-Lauretea novocanariensis				
Scientific name	**Family**	**Common name**	**Biotype**	**Distribution**
Isoplexis canariensis	Plantaginaceae	"crestagallo"	N	P,G,T
Isoplexis chalcantha	Plantaginaceae	"crestagallo de Doramas"	N	C
Ixanthus viscosus	Gentianaceae	"reina del monte"	N	H,P,G,T,C
Phyllis nobla	Rubiaceae	"capitana"	N	H,P,G,T,C
Semele androgyna	Asparagaceae	"gibalbera"	N-Li	H,P,G,T,C
Smilax aspera subsp.*mauritanica*	Smilacaceae	"zarzaparrilla común"	N-Li	H,P,G,T,C,F
Smilax canariensis	Smilacaceae	"zarzaparrilla canaria"	N-Li	P,G,T,C,F
HERBS				
Arisarum simorrhinum	Araceae	"zumillo candil"	G	T,C,F,L
Carex canariensis	Cyperaceae	"cuchillera canaria"	H	H,P,G,T,C
Carex perrauderiana	Cyperaceae	"cuchillera ancha"	H	P,G,T,F
Gennaria diphylla	Orchidaceae	"orquídea de dos hojas"	G	H,P,G,T,C,L
Luzula canariensis	Juncaceae	"lúzula canaria"	H	G,T
FERNS (no characteristic of the class but with a significant presence in its communities)				
Asplenium anceps	Aspleniaceae	"culantrillo calambur"	H	H,P,G,T
Asplenium hemionitis	Aspleniaceae	"hierba candil"	H	H,P,G,T,C,F,L
Asplenium onopteris	Aspleniaceae	"culantrillo negro de monte"	H	H,P,G,T,C,F,L
Athyrium filix-femina	Athyriaceae	"helecho hembra"	H	H,P,G,T,C
Blechnum spicant	Blechnaceae	"helecho peine"	H	G,T
Culcita macrocarpa	Dicksoniaceae	"helecho colchonero"	H	T
Cystopteris fragilis	Athyriaceae	"helecho de manantial"	H	H,P,G,T,C
Cystopteris viridula	Athyriaceae	-	H	P,G,T,C
Davallia canariensis	Davalliaceae	"helecho batatilla"	H/C	H,P,G,T,C,F,L
Diplazium caudatum	Athyriaceae	"helecho de monte"	H	P,G,T,C
Dryopteris aemula	Dryopteridaceae	–	H	G
Dryopteris affinis subsp. *affinis*	Dryopteridaceae	–	H	G,T
Dryopteris guanchica	Dryopteridaceae	"helecho penco dentado"	H	H,G,T

(continued)

Table 6.3 (continued)

Pruno hixae-Lauretea novocanariensis				
Scientific name	**Family**	**Common name**	**Biotype**	**Distribution**
Dryopteris oligodonta	Dryopteridaceae	"helecho penco"	H	H,P,G,T,C
Hymenophyllum tunbrigense	Hymenophyllaceae	"helechilla"	H	G,T,C
Hymenophyllum willsonii	Hymenophyllaceae	–	H	G,T,C
Polypodium macaronesicum	Polypodiaceae	"Polipodio del país"	H/C	H,P,G,T,C,F,L
Polystichum setiferum	Dryopteridaceae	"píjara macho"	H	H,P,G,T
Pteridium aquilinum	Hypolepidaceae	"helechera"	G	H,P,G,T,C,L
Pteris incompleta	Pteridaceae	"rasguño dentado"	H	P,G,T,C
Vandenboschia speciosa	Hymenophyllaceae	"helecho de cristal"	H	H,P,G,T,C
Woodwardia radicans	Blechnaceae	"píjara"	H	G,T
Andryalo pinnatifidae-Ericetalia canariensis				
TREES				
Erica arborea	Ericaceae	"brezo"	M	H,P,G,T,C,F,L
SHRUBS				
*Adenocarpus foliolosus*var. *foliolosus*	Fabaceae	"codeso de monte"	N	H,P,G,T,C
*Adenocarpus foliolosus*var. *villosus*	Fabaceae	"codeso de monte"	N	P,C
Andryala pinnatifida subsp. *pinnatifida*	Asteraceae	"estornudera común"	C	H,P?,G,T,C
Bystropogon canariensis	Lamiaceae	"poleo de monte"	N	H,P,G,T,C
Chamaecytisus proliferus subsp. *proliferus* var. *canariae*	Fabaceae	"tagasaste grancanario"	N	C
Chamaecytisus proliferus subsp. *proliferus* var. *hierrensis*	Fabaceae	"tagasaste herreño"	N	H
Chamaecytisus proliferus subsp. *proliferus* var. *palmensis*	Fabaceae	"tagasaste palmero"	N	P
Chamaecytisus proliferus subsp. *proliferus* var. *proliferus*	Fabaceae	"tagasaste tinerfeño"	N	T
Convolvulus canariensis	Convolvulaceae	"corregüelón de monte"	N/M-Li	H,P,G,T,C
Daphne gnidium	Thymelaeaceae	"torvisco"	N	
Gesnouinia arborea	Urticaceae	"estrelladera"	N	H,P,G,T,C
Rubia fruticosa subsp. *periclymenum*	Rubiaceae	"tasaigo trepador"	N-Li	G,T,C
Rubia peregrina subsp. *agostinhoi*	Rubiaceae	"ruiviño"	N-Li	H,P,G,T,C,F,L

(continued)

Table 6.3 (continued)

Pruno hixae-Lauretea novocanariensis				
Scientific name	**Family**	**Common name**	**Biotype**	**Distribution**
Sideritis canariensis	Lamiaceae	"chajorra de monte"	N	H,P,T
Sideritis macrostachys	Lamiaceae	"chajorra reina"	N	T
Teline canariensis	Fabaceae	"retamón canario"	N	T,C
Teline pallida subsp. *pallida*	Fabaceae	"gildana de Anaga"	N	T
Teline rosmarinifolia subsp. *eurifolia*	Fabaceae	"gildana de Faneque"	N	C
Teline splendens	Fabaceae	"herdanera"	N	P
Teline stenopetala subsp. *microphylla*	Fabaceae	"gacia menuda"	N	H,G
Teline stenopetala subsp. *stenopetala*	Fabaceae	"gacia de monte"	N	P,T,C
HERBS				
Canarina canariensis	Campanulaceae	"bicacarera"	N-Li	H,P,G,T,C
Cryptotaenia elegans	Apiaceae	"perejil de monteverde"	H	P,G,T
Dracunculus canariensis	Araceae	"taraguntía"	G	H,P,G,T,C,F
Rubo bollei-Salicetalia canariensis				
TREES				
Salix canariensis	Salicaceae	"sauce canario"	M	H,P,G,T,C
SHRUBS				
Rubus bollei	Rubiaceae	"zarza de monte"	N-Li	H,P,G,T,C,F
Rubus palmensis	Rubiaceae	"zarza palmera"	N-Li	P,T,C
Rubus ulmifolius	Rubiaceae	"zarza común"	N-Li	H,P,G,T,C,F
Asparagus asparagoides	Asparagaceae	"alicacán"	N-Li	H,P,G,T,C,L
Urtica morifolia	Urticaceae	"ortigón"	C/N	H,P,G,T,C
Transgressive species with significant presence in some communities of the class				
Chamaecytisus proliferus subsp. *angustifolius* (in *Andryalo-Ericetalia*)	Fabaceae	"escobón de pinar"	N	H?,G,T
Micromeria benthamii (in *Andryalo-Ericetalia*)				
Pericallis murrayi (in *Pruno-Lauretalia*)				
Species of *TRIFOLIO-GERANIETEA SANGUINEI* characteristic of the laurel forest fringe				
Geranium reuteri	Geraniaceae	"patagallo canario"	H	H,P,G,T,C
Myosotis discolor subsp. *canariensis*	Boraginaceae	"nomeolvides canaria"	H	H,P,G,T,C

(continued)

Table 6.3 (continued)

Pruno hixae-Lauretea novocanariensis				
Scientific name	**Family**	**Common name**	**Biotype**	**Distribution**
Myosotis latifolia	Boraginaceae	"nomeolvides de monte"	H	H,P,G,T,C
Pericallis appendiculata var. *appendiculata*	Asteraceae	"alamillo de monteverde"	C	H,P,G,T,C
Pericallis appendiculata var. *preauxiana*	Asteraceae	"alamillo de Doramas"	C	C
Pericallis cruenta	Asteraceae	"tusilago morado"	H	H,G,T,C
Pericallis steetzii	Asteraceae	"arcila"	H	G
Pericallis tussilaginis	Asteraceae	"tusilago canario"	H	T,C
Pericallis webbii	Asteraceae	"flor de mayo"	H	C
Ranunculus cortusifolius	Ranunculaceae	"morgallana"	H/G	H,P,G,T,C,F,L
Scrophularia smithii	Scrophulariaceae	"fistulera"	H	H,P,G,T,C
Viola anagae	Violaceae	"violeta de Anaga"	G/H	T
Viola riviniana	Violaceae	"violeta de monte"	H/C	H,P,G,T

Common names according to Machado and Morera (2005)

Visneo mocanerae-Arbutetum canariensis (Fig. 6.117)

This is a medium-sized, dense, western Canary mesophytic heminebulous forest, the "dry laurel forest", the most thermophilous and resistant to the xeric conditions of the climatophilous communities of laurel forest. It grows at the lowest levels affected by NE t-w.c. (300–600 m), but on NW sheltered slopes influenced by clouds it reaches higher altitudes, and also grows in the upper elevations of some southern Thermomediterranean slopes subject to overflowing clouds (Fig. 6.118).

Its climatophilous area extends over the slopes under the influence of NE trade-winds, above the potential area of thermo-sclerophyllous woodland. It reaches up to the area of dominance of the subhumid ombrotype, at around 600 m, where potential humid laurel forest (*Lauro novocanariensis-Perseetum indicae*) starts to occur. This area can be interrupted over salic substrate by pine forest, as on Tenerife at mid-elevations in Garachico-Los Realejos, the lower part of Ladera de Tigaiga, and Orotava-Santa Úrsula between 300 and 750 m (*Morello fayae-Pinetum canariensis* subass. *arbutetosum canariensis*), Roque de los Pinos, Anaga (subass. *cistetosum chinamadensis*), and Vallehermoso, La Gomera, over salic outcrops (*Cisto gomerae-Pinetum canariensis* subass. *juniperetosum canariensis*).

The trees *Visnea mocanera, Apollonias barbujana*, *Picconia excelsa*, *Ilex canariensis, Heberdenia excelsa, Morella faya* and *Erica arborea* are common in the association; *Visnea mocanera,* although characteristic, is not so common on La Palma and Gran Canaria. The shrubs *Hypericum canariense, Jasminum odoratissimum, Daphne gnidium* and *Spartocytisus filipes*, among others, are frequent at forest edges and in substitutional shrubland.

It corresponds to the terminal community of the dry Inframediterranean (c), upper-semiarid (c) and dry Thermomediterranean (c), and subhumid Thermomediterranean (oc) climatophilous series of the subxeric forest of *Arbutus canariensis* and *Visnea mocanera*, with *Apollonias barbujana* and *Picconia excelsa* (*Visneo mocanerae-Arbuto canariensis sigmetum*).

Nowadays the association is found fragmented over steep areas like ravine cornices and high ledges in the mid-elevations of the N, as in the Hermigua and Vallehermoso basins (La Gomera), and NE and E sectors of La Palma. Also in some well preserved nuclei as in Jinama (El Hierro) and Monte del Agua (Teno) and Anaga – Vueltas de Taganana, Monte Aguirre, etc. (Tenerife).

The substitutional shrubland *Rhamno-Hypericetum canariensis* is common in the lower two thirds of the potential area of the association, and the *Morello fayae-Ericetum arboreae* in the upper third. The nitrophilous shrubland *Artemisio thusculae-Rumicetum lunariae* is also widespread.

Distribution: C, T, G, P, H.

Fig. 6.117 *Visneo mocanerae-Arbutetum canariensis*. Jinama. El Hierro

Fig. 6.118 *Visneo mocanerae-Arbutetum canariensis*, on Thermomediterranean slopes subject to overflowing clouds. Monte Aguirre. Tenerife

Ixantho viscosae-Laurion novocanariensis
Potential laurel forest of the upper-dry to humid Infra- to Mesomediterranean belts (c and ces). Endemic to Canarian-Madeiran subregion.

Lauro novocanariensis-Perseetum indicae (Figs. 6.119 and 6.120)
This is a tall, dense, western Canary mesophytic forest, the "humid laurel forest", floristically diverse at its optimum. Its climatophilous area is on the northern slopes under the influence of NE t-w.c., from 600 m a.s.l., above the potential area of dry laurel forest (*Visneo-Arbutetum canariensis*) up to about 1250 m a.s.l., where it reaches the climatophilous area of cold (and summer dry) laurel forest (*Pericallido murrayii-Morelletum fayae* -T, P, H, C- and *Violo rivinianae-Morelletum fayae* -G). It establishes in subhumid Infra-, dry Thermo- and dry subhumid and humid Mesomediterranean belts (c). The salic substrates reduce its potential area to the benefit of salic pine forest, and on lapilli cones it is substituted by humid pine forest (several subassociations of the western islands' pine forests). There are many variations to the aforementioned distributional pattern depending on island altitude, local topography and substrate:

On Gran Canaria, its potential area is reduced in the NW of the island on salic bedrock of the Tamadaba massif where salic pine forest (*Micromerio pineolentis-Pinetum canariensis*) develops.

Fig. 6.119 *Lauro novocanariensis-Perseetum indicae*. Vueltas de Taganana. Anaga. Tenerife

Fig. 6.120 *Persea indica* canopy. Vueltas de Taganana. Anaga. Tenerife

On Tenerife its climatophilous territory extends along the N slope of the island, NE slopes of S Anaga ravines and windward slopes of Güímar in the SE, above the potential area of the *Visneo-Arbutetum canariensis*. In the N, on the crests of the Anaga ridge, its upper limit coincides with the anemo-nebulous community *Ilici-Ericetum platycodonis*. There is a considerable reduction in the potential area of the *Lauro-Perseetum* over the salic lava flows in Garachico-Los Realejos (deriving from the recent volcanic activity of Mount Teide a few thousand years ago) and salic territories of Orotava-Santa Úrsula (belonging to the second Canary volcanic cycle < 5.1 Ma). There occurs the *Morello fayae-Pinetum canariensis* subass. *typicum*, between 750–1500 m a.s.l.

On La Gomera two plant communities are common on the ridge crests within the potential area of this association: the *Ilici canariensis-Ericetum platycodonis* at altitudes below 1100 m, that is year-round under the influence of clouds; and the *Micromerio lepidae-Ericetum arboreae* especially above 1100 m a.s.l., which is drier due to the wind effect and lower frequency of t-w.c. at those altitudes in summer.

As a consequence of a lesser influence of clouds at the leeward slopes of ridges where the clouds overflow, the humid laurel forest (*Lauro-Perseetum indicae*) of windward Thermomediterranean territories changes to dry laurel forest (*Visneo-Arbutetum canariensis*) to the leeward and that of Mesomediterranean territories changes to humid pine forest there (several subassociations of the islands' pine forests).

There is a dominance of broad-leaved trees with shiny perennial leaves. The most common laurifolious trees making up the forest matrix are *Laurus novocanariensis, Ilex canariensis* and *Morella faya*. Besides these, *Heberdenia excelsa* stands out among the trees as well as *Ilex perado* subsp. *platyphylla* and *Persea indica,* which although present, are more common in edaphohygrophilous positions. *Apollonias barbujana* and *Picconia excelsa* are present at lower elevation. *Prunus lusitanica* subsp. *hixa* is particularly abundant in Anaga – T. Furthermore, the needle-leaved *Erica arborea* is common, probably because of old anthropic disturbance of the forest. Among the small trees are *Viburnum rigidum, Rhamnus glandulosa* and *Sambucus palmensis* – the latter very rare. Accompanying them are a number of shrubs, lianas, herbs, and ferns.

It corresponds to the terminal community of the subhumid Infra-, dry Thermo- and dry subhumid and humid Mesomediterranean belts (c), the western Canary climatophilous series of the forest of *Laurus novocanariensis* and *Persea indica* (*Lauro novocanariensis-Perseo indicae sigmetum*).

Several facies can be distinguished: a *Persea indica* facies, especially in edaphohygrophilous environments; a *Morella faya* facies in disturbed places in the recovering phase; an *Erica arborea* facies on windy crests.

The subass. *ocoteetosum foetentis*, described from La Gomera (Garajonay N.P. and Teselinde and Enchereda), grows on very cloudy slopes below the crest ridges with *Erica* communities (*Ilici canariensis-Ericetum platycodonis* and *Micromerio lepidae-Ericetum arboreae*) at variable elevations between 800 and 1250 m or higher depending on local topography). This cloud forest receives abundant fog-drip and hosts a rich epiphytic lichen and moss flora, and has a high coverage of ferns in the understorey. Some endangered species like *Ilex perado* subsp. *lopezlilloi* and *Euphorbia mellifera* grow here. There are two variants: (a) the most humid, in general below 1100 m, with abundant clouds all year round. *Ocotea foetens* and *Ilex perado* subsp. *platyphylla* are present. *Aichryson laxum, Davallia canariensis* and *Polypodium macaronesicum* are frequent as epiphytes, and locally *Hymenophyllum tunbrigense. Pericallis appendiculata* and the ferns *Woodwardia radicans, Polystichum setiferum, Dryopteris guanchica* and *Dryopteris affinis* are abundant in the understorey; (b) above 1100 m, with less clouds in summer but still abundant fog-drip, the latter tree species are not present and *Laurus novocanariensis, Morella faya, Ilex canariensis* and *Erica arborea* are protagonists. This variant differs from the *Violo riviniana-Morelletum fayae* because of the increased presence of ferns and epiphytes and the greater abundance of water-demanding species like *Cryptotaenia elegans, Hedera canariensis, Polystichum setiferum* and *Dryopteris guanchica*.

The main anthropic substitutional community *Morello-Ericetum arboreae* occurs spread over a much wider area, interspersed with the broom shrubland (*Telino-Adenocarpion*) which is found in drier and sunny rocky sites as a mantle community at the fringes of the forest and also expands as a substitutional stage. Several broom shrublands are to be noticed with some special island distribution, characterized by several species of *Teline*, island varieties or subspecies of *Chamaecytisus proliferus* of the laurel forest area, and *Adenocarpus foliolosus* broom scrubland. Interspersed bramble shrubland *Rubio periclymeni–Rubetum* and bracken facies of *Pteridium aquilinum* are common in damper disturbed places. The perennial herbaceous community *Piptathero miliacei–Foeniculetum vulgaris* also grows on deforested humid soils as do the nitrophilous meadows of the *Galactito tomentosae-Trachynietum distachyae*.

This forest has considerably retreated over the centuries due mainly to tree-felling, expansion of cultivated land and forestry. For centuries the secondary forest has supplied sticks from saplings and new sprouting stems as supports for the plants of the mono-cultures (tomato and banana) of the islands.

On El Hierro most of the potential area of the forest has been transformed by tree felling and plowing of flat zones, particularly at Nizdafe. Moreover, the presence of lava flows and recent lapilli in part of the area reduces its chances of establishment. Its current presence is reduced to a few localities in the N of the island, arising fragmentarily within the substitutional *Erica-Morella* thicket (*Fayo-Ericetum arboreae*), or in edaphohygrophilous situations within the dry laurel forest, mainly on the heights of the scarps of Frontera and El Derrabado. Some of the association's characteristic tree species are missing, particularly *Persea indica*, that is virtually absent from the island. On La Palma it is well represented and in the deep valleys and ravines of the N and NE excellent stands survive, some as impressive as that which justified establishing the Los Tilos Biosphere Reserve. La Gomera hosts the best remnants, and the forest shows an excellent state of conservation, since the vast majority of its potential territory belongs to the Garajonay National Park. In Tenerife several nuclei still persist and they are now included within the Canary Natural

Reserve Network. The purest and most extensive relics are in Anaga and Teno; they are protected by law and form part of the Anaga Rural Park (and its Integral Reserves) and Teno Rural Park. On Gran Canaria this community is now practically absent despite recent recuperation efforts. Only small impoverished relicts persist on some steep slopes and in some humid valleys. Even the typical substitutional *Morello–Ericetum arboreae* has a greatly reduced presence there.

Distribution: C, T, G, P, H.

Pericallido murrayii-Morelletum fayae (Fig. 6.121)
This is a dense, mid-height western Canary mesophytic forest, the "cold, and summer-dry laurel forest" characteristic of the highest elevations of laurel forest. It potentially grows in the dry to humid Mesomediterranean belts (ces), above the area of humid laurel forest (*Lauro-Perseetum indicae*) between 1250 (1300) and 1500 m a.s.l., a territory with a continentality index higher than 12.

The community is dominated in its optimum state by *Morella faya* together with the most cold-tolerant species of the laurel forest, among them *Ilex canariensis, Erica arborea* and *Morella rivas-martinezii*. At its lowest elevation *Laurus novocanariensis* also grows. Several species of *Pericallis* grow on it: *P. murrayi, P. webbii, P. cruenta,* and *P. papyracea* that have been used as insular differential species.

It corresponds to the terminal community of the dry to subhumid Mesomediterranean belts (ces) of the western Canary climatophilous series of *Morella faya* (*Pericallido murrayii-Morello fayae sigmetum*).

This *Morella faya* forest has been strongly transformed through forest exploitation and nowadays, except on El Hierro, it is not easily recognizable. In its natural area the substitutional *Morello-Ericetum arboreae,* substitutional humid pine forest with *Erica* and *Morella*, and plantations of Canary pine and Monterrey pine (*Pinus radiata*) are dominant.

On Gran Canaria the community (nowadays extinct) has been provisionally named as subass. *pericallidetosum webbii*. It would occupy a narrow strip in the North limited below by humid laurel forest (*Lauro-Perseetum indicae*) or by stands of humid pine forest (*Pinetum canariensis* subas. *ericetosum arboreae*) over lapilli, and limited above by typical pine forest (*Pinetum canariensis*). This strip would be dotted or interrupted by pine forest fragments over recent lapilli and salic outcrops. Apart from bioclimatic data, the upper territories of the current distribution of the broom shrubland *Chamaecytiso canariae-Adenocarpetum foliolosi* determine its territory within the potential area of laurel forest.

On Tenerife it is represented by the subass. *pericallidetosum cruentae* with a potential area in the North over basaltic substrate, bordering below with humid laurel forest and above with the pine forest domain (*Sideritido-Pinetum canariensis*).

The related forest of La Gomera has been defined as *Violo rivinianae-Morelletum fayae*.

On La Palma the subass. *pericallidetosum papyraceae* occurs. Its potential area is located on the North and East slope, limited below by humid laurel forest and above by pine forest (*Loto-Pinetum canariensis*). Additionally, on the northern ridge of Cumbre Nueva, in leeward areas with overflowing clouds, it fringes on the humid pine forest (*Loto-Pinetum canariensis* subass. *ericetosum arboreae*) and towards the South on the *Loto-Pinetum canariensis* subass. *adenocarpetosum foliolosi*.

On El Hierro it is represented by the subass. *typicum* (= *pericallidetosum murrayii*), with its potential area on the island's summit, bordering northwards with humid laurel forest (*Lauro-Perseetum* indicae) and southwards with pine forest (*Bystropogono ferrensis-Pinetum canariensis*).

Distribution: C (potential), T, G, P, H.

Fig. 6.121 *Pericallido murrayii-Morelletum fayae*. Garajonay. La Gomera

Violo rivinianae-Morelletum fayae

This is a dense, middle high mesophytic forest (10–15 m) endemic to La Gomera, whose climatophilous area starts at 1250–1300 m a.s.l. and caps the island. It potentially occurs above the humid laurel forest (*Lauro-Perseetum indicae*) on the N slopes, and above pine forest (*Cisto gomerae-Pinetum canariensis*) on the dry S slope still influenced by clouds.

At its optimum the community is dominated by *Morella faya,* accompanied by the most cold-tolerant species of the laurel forest, *Ilex canariensis* and *Erica arborea*, along with some *Laurus novocanariensis* mainly at low elevations.

It corresponds to the terminal community of the subhumid Mesomediterranean (ces), climatophilous series of the *Morella faya* forest (*Violo rivinianae-Morello fayae sigmetum*).

The altitudinal transition from *Lauro-Perseetum indicae* is driven by the predominant absence of clouds in summer at the elevations where the community grows, which results in an increase in solar radiation, temperature, and evaporation, causing an increase in continentality index (Ic) values to > 12.

The community, although transformed over the years, was until 2012 in a relatively acceptable state of conservation in the northern half of its climatophilous area, where it shared the territory with the Canary tree-heath crest community (*Micromerio lepidae-Ericetum arboreae*). But its southern area had been much more transformed by human use, and the substitutional shrublands *Morello-Ericetum arboreae* and *Micromerio gomerensis-Cistetum canariensis* filled the territory, together with some patches of *Pinus canariensis* and *Pinus radiata* plantation, which were being eradicated. Unfortunately, a wildfire occurred in the southern sector of Garajonay N.P. in the summer of 2012, changing this vegetation pattern and considerably diminishing the community area, now relegated to scarce northern areas.

Potential Edaphophilous Forests

Crest Ridge Communities

Ilici canariensis-Ericetum platycodonis (Fig. 6.122)
This is a dense, low to middle high, anemo-nebulous and ombrophilous western Canary mesophytic forest that grows on spurs and ridge crests located between 600 and 1100 m, very windy and subject to clouds all year round (even in summer). This forest captures a great deal of water by condensation on its leaves and then fog-drip, as witnessed by the lush bryophyte and lichen cover and richness below the canopy. But paradoxically the wind action causes dryness at the canopy top, which lends it an aero-xerophile character promoting the dominance of the needle-leaved tree *Erica platycodon.*

Besides the characteristic tree *Erica platycodon, Ilex canariensis, Laurus novocanariensis, Morella faya, Prunus lusitanica* subsp. *hixa* (on Tenerife) and *Viburnum rigidum* are frequent. Also common on the thick moder-humus layer are the ferns *Asplenium onopteris, Blechnum spicant, Dryopteris oligodonta, Pteridium aquilinum* and *Woodwardia radicans.*

It corresponds to the terminal community of the subhumid-humid Thermo- and Mesomemediterranean anemogenic and nebulous edaphophilous series of *Ilex canariensis* and *Erica platycodon* on Tenerife and La Gomera (*Ilici canariensis-Erico platycodonis sigmetum*).

On Tenerife it is characteristic of the summits of Anaga, and on La Gomera along the Carbonero-Inchereda-Juel ridge in the NE sector, and Epina in the NW, giving rise to narrow strips all along the ridges.

Distribution: Endemic to T, G.

Fig. 6.122 *Ilici canariensis-Ericetum platycodonis*. Vueltas de Taganana. Tenerife

Micromerio lepidae-Ericetum arboreae (Fig. 6.123)
This is a dense, low to middle high, anemogenic, seasonally cloudy and ombrophilous mesophytic forest that grows on the windy ridges at the top of La Gomera. These ridges, especially above 1100 m a.s.l., are periodically subject to t-w.c, causing alternating extremely humid periods with rather dry spells in summer. The forest can also grow at lower elevations with a more steady presence of clouds, but on relatively smooth summits where fog-drip is lower than those abrupt crests where the *Ilici-Ericetum platycodonis* occurs.

Erica arborea is the dominant species of the tree layer, which shows xeromorphic characteristics in line with the drying wind action. Other species in the community are *Morella faya, Laurus novocanariensis* and *Ilex canariensis*. *Juniperus cedrus* subsp. *cedrus* is locally frequent. The moss layer is highly developed, there are abundant epiphytic lichens, and several ferns in the understory. Together with the carpet of bryophytes (*Scapania gracilis*, *Isothecium myosuroides*, *Antitrichia curtipendula*, etc.) and lichens (*Cladina* spp., *Cladonia* spp. and *Peltigera* spp.) on the soil, the cushion-like specimens of *Micromeria lepida* subsp. *lepida* are striking as well as the abundance and constancy of *Viola riviniana, Galium scabrum* and *Aichryson spp.* among others.

It corresponds to the terminal community of the subhumid-humid Thermo- and Mesomediterranean anemogenic and seasonally-nebulous edaphophilous series of *Erica arborea* on La Gomera (*Micromerio lepidae-Erico arboreae sigmetum*).

Distribution: Endemic to G.

Fig. 6.123 *Micromerio lepidae-Ericetum arboreae*. Cumbre de Tajaqué. La Gomera

Ravine-Bed Laurel Forest

Diplazio caudati-Ocoteetum foetentis (Fig. 6.124)
This is a tall, dense, western Canary meso-hydrophytic, dark and nebulous forest, the "hydrophilous laurel forest". It grows on well developed soils subject to abundant fog-drip, streams, moist ravine beds and basins with wet soils almost all year round, within the general climatophilous area of humid laurel forest (*Lauro-Perseetum indicae*) and the upper third part of that of dry laurel forest (*Visneo-Arbutetum canariensis*).

Ilex perado subsp. *platyphylla*, *Laurus novocanariensis* and *Ocotea foetens* are the characteristic trees, which form a dense canopy generating a dark understory. The hygrophilous ferns *Culcita macrocarpa*, *Diplazium caudatum* and *Vandenboschia speciosa* are also characteristic. The moist dark environment favors the growth of young *Ocotea* and the ferns.

It corresponds to the terminal community of the subhumid-humid Infra- to Mesomediterranean edaphohygrophilous and nebulous series of the forest of *Ocotea foetens* and *Ilex perado* subsp. *platyphylla* with *Diplazium caudatum* and *Vandenboschia speciosa* (*Diplazio caudati-Ocoteo foetentis sigmetum*).

Felling of trees and drying of streams have caused a serious retreat of the *Diplazio caudati-Ocoteetum foetentis*, as a result of which the humid laurel forest (*Lauro-Perseetum indicae*) invaded its area. Felling and thinning-out of the forest also has allowed the penetration of the Canary willow community (*Rubo-Salicetum canariensis*) onto ravines beds with enough edaphic moisture.

The community nowadays is greatly reduced in area and to be found on the windward side of Tenerife, La Gomera and La Palma. On Tenerife there are some good remnants in ravine heads of the eastern part of Anaga Rural Park. On Gomera there are outstanding localities like El Cedro basin, La Palmita, Atrabancos, Barranco de Liria and the La Meseta forest above Macayo, probably the best remnant on the island. On La Palma it is common in ravines of the NE sector within the Los Tiles Biosphere Reserve. On Gran Canaria only a few small remnants persist, almost limited to Tilos de Moya. It is not present on El Hierro except probably in the humid basins of the Meseta de Nizdafe in the past, where some individuals of *Ocotea foetens* now grow at the locality today known as Garoé (local common name for the species).

Distribution: C, T, G, P, H (as potential).

Fig. 6.124 *Diplazio caudati-Ocoteetum foetentis*. Los Tilos. La Palma

Willow Forest and Humid Mantle Community

Rubo bollei-Salicetalia canariensis
Riparian potential willow forest of *Salix canariensis* and humid mantle and seral bramble thicket. Endemic to Canarian-Madeiran subregion.

Salicion canariensis
Riparian potential willow forest. Endemic to Canarian-Madeiran subregion.

Rubo ulmifolii-Salicetum canariensis (Fig. 6.125)
This is a hydrophytic and heliophilous azonal community that grows in streams, humid ravine beds with running water during a large part of the year, and natural oozing crevices.

The winter-deciduous *Salix canariensis* and *Morella faya* are the characteristic trees, and *Rubus ulmifolius* and *Ageratina adenophora* the most common shrubs.

It corresponds to the terminal community of the Infra- to Supramediterranean, xeric to pluviseasonal, western Canary edaphohygrophilous series of the Canary willow (*S. canariensis*) (*Rubo-Salici canariensis sigmetum*).

It naturally grows in ravine beds at the lowest elevation of the potential territory of dry laurel forest where light can penetrate and favor its development. Its further expansion within the rest of the laurel forest area is favored by felling of the largest hygro-tolerant trees that normally limit the light penetrating the canopy.

Apart from the typical subassociation, the subass. *ericetosum arboreae* is distinguished. It includes the willow stands growing in disturbed territories of laurel forest, and shows a greater participation of their typical constituent elements. The subass. *polygonetosum salicifolii* constitutes a highly nitrophilous stage of the community.

The subass. *woodwardietosum radicantis* was described for willows on sunny steep slopes with abundant water within the laurel forest territory; it shows an abundant presence of the fern *Woodwardia radicans*. This could be an anthropogenic formation due to felling of laurel forest that was quickly replaced by willows.

Distribution: Endemic to C, T, G, P.

Fig. 6.125 *Rubo ulmifolii-Salicetum canariensis*. Barranco del Infierno. Tenerife

Rubio periclymeni-Rubion ulmifolii
Seral bramble thicket and mantle communities of hygrophilous laurel forest and Canary willow community. Canarian-Madeiran subregion.

Rubio periclymeni-Rubetum ulmifolii (Fig. 6.126)
This is a western Canary lianoid and prickly bramble shrubland, which constitutes the mantle of the Canary willow community (*Rubo-Salicetum canariensis*) and the humid mantle of mesophytic laurel forest, and it also substitutes those two communities. It also grows on humid steep slopes, anthropic cliff edges and walls of agricultural fields.

Rubus ulmifolius, Rubia fruticosa subsp. *periclymenum, Pteridium aquilinum* and *Hypericum grandifolium* are characteristic, among other plants mainly belonging to the laurel forest shrub and herb cohort.

Its *Pteridium aquilinum* facies is common on humid slopes and abandoned agricultural lands.

Distribution: C, T, G, P, H.

Fig. 6.126 *Rubio periclymeni-Rubetum ulmifolii*. Taganana. Tenerife

Fayal-Brezal. Legume Scrublands and Shrublands of Laurel Forest and Pine Forest

Andryalo pinnatifidae- Ericetalia arboreae
Seral microforest of *Morella faya* and *Erica arborea* and mantle or seral legume shrublands of the dry to humid Infra- to Mesomediterranean belts (c). Legume shrublands can extend into pine forest areas within the Mesomediterranean belts (wc). Endemic to the Canary Islands.

Morello fayae-Ericion arboreae
Seral microforest of *Morella faya* and *Erica arborea* of the dry to humid Infra- to Mesomediterranean belts (c). Canary Islands.

Morello fayae-Ericetum arboreae (Fig. 6.127)
This is a shrubby or sub-arboreal seral western Canary community widely distributed as a consequence of the anthropic destruction of laurel forest. It has the physiognomy of a tree-heath or wax myrtle – tree-heath ("fayal-brezal") thicket of variable cover but generally it is dense. It constitutes a substitutional stage of all tree communities of laurel forest, particularly cold and summer-dry laurel forest (*Pericallido-Morelletum fayae* and *Violo rivinianae-Morelletum fayae*), humid laurel forest (*Lauro-Perseetum indicae*) and the lowest territory of dry laurel forest (*Visneo-Arbutetum canariensis*).

Morella faya, Erica arborea, Ilex canariensis, Daphne gnidium and *Hypericum grandifolium* among others, are characteristic species.

Nowadays it is the dominant community of the potential area of the laurel forest, although on Gran Canaria due to major landscape transformations it is scarce.

Apart from the type subass. *ericetosum arboreae*, the subass. *telinetosum canariensis*, transitional to the Canary broom community *Telinetum canariensis,* has been described from Tenerife; the *Chamaecytisus proliferus* subsp. *proliferus* (a laurel forest tree-lucerne) facies is well represented at the foothills of Monte del Agua facing Las Portelas (Tenerife). The *Ulex europaeus* facies is well represented in marginal, disturbed laurel forest environments in the North of Tenerife.

Distribution: C, T, G, P, H.

Fig. 6.127 *Morello fayae-Ericetum arboreae*. El Bailadero. Anaga. Tenerife

Telino canariensis-Adenocarpion foliolosi (Fig. 6.128)
Mantle or seral legume scrub and shrublands of the dry to humid Infra- to Mesomediterranean belts (c) of potential laurel forest territory, and dry to humid Mesomediterranean belts (wc) of potential pine forest territory.

***Adenocarpus foliolosus* community**
This is a shrub community dominated by *Adenocarpus foliolosus* in the potential areas of laurel and pine forest; its distribution may be related to fires. It is still a community with an uncertain definition within the *Telino-Adenocarpion foliolosi*, identified on several western islands.

Distribution: T, G, P.

Fig. 6.128 *Telino canariensis-Adenocarpion foliolosi.* Above Taganana. Tenerife

Adenocarpo foliolosi-Chamaecytisetum angustifolii
This is a nano-microphanerophytic community with a preference for ledges, cornices and rocky biotopes in the dry-subhumid Thermo- and Mesomediterranean belts of the S sector of La Gomera, within the potential territory of pine forest and cold and summer-dry laurel forest.

It is floristically characterized by *Chamaecytisus proliferus* subsp. *angustifolius* (usually dominant), *Adenocarpus foliolosus* and to a small extent by *Sideritis lotsyi* and *Bystropogon origanifolius.* Apart from these, some dynamic plants belonging to the *Cisto-Micromerietalia* and *Pruno-Lauretalia* are frequently present.

Its expansion as a substitutional community in this S sector has probably been favored by deforestation and fires, extending its area to the upper territories of juniper woodland and also to high territories of humid laurel forest on the central plateau of the island, and those of dry laurel forest in overflowing cloud areas. Today the community is also widely spread over abandoned agricultural areas.

Distribution: Endemic to G.

Chamaecytisetum angustifolii (Fig. 6.129)
This is a nano-microphanerophytic shrubland that mainly grows within the dry Mesomediterranean belt, the potential climatophilous area of pine forest, preferentially on colluvia, stony areas and lava flows.

Apart from *Chamaecytisus proliferus* subsp. *angustifolius,* several species (some of them local endemics) can be considered characteristic, like *Bystropogon canariensis* var. *smithianus, Descurainia lemsii, D. gonzalezii, Sideritis oroteneriffae, S. soluta* subsp. *soluta* and *Echium virescens*.

At its highest altitudes it mixes with plants of summit broom shrubland, and at its lowest altitudes with widespread plants of the *Kleinio-Euphorbietea*, like *Euphorbia lamarckii*. *Bystropogon origanifolius* var. *origanifolius* and *Cistus symphytifolius* occur in the potential natural pine forest area. *Erica arborea* is common on N-facing slopes, and in the Güímar valley above laurel forest territory.

It seems that this shrubland was widespread in the past within the aforementioned belt but afforestation mainly with *Pinus canariensis* and *P. radiata* have reduced its area.

Distribution: Endemic to T.

Fig. 6.129 *Chamaecytisetum angustifolii*. Fuente de Joco. Tenerife

Chamaecytiso canariae-Adenocarpetum villosi

This is a shrubby legume community distributed over the upper-dry and subhumid Thermo- and Mesomediterranean belts of the NW sector of Gran Canaria. It is mainly a substitutional stage of laurel and pine forest on the salic massif of Tamadaba and north facing edges of pine forest of the island's summits in contact with clouds.

Adenocarpus foliolosus var. *villosus* and *Chamaecytisus proliferus* subsp. *proliferus* var. *canariae* are its characteristic taxa, and *Teline microphylla* is also present.

Distribution: Endemic to C.

Lotetum campylocladi (Fig. 6.130)
This is a chamaephytic oligospecific community typical of Canary pine forest that mainly grows in burned places, along dirt roads and the slopes of their cuttings, and in disturbed soils raked after collecting pine needle litter.

It is characterized by the massive presence of the procumbent chamaephyte (occasionally hemicryptophyte) *Lotus campylocladus* (subsp. *campylocladus* – T, and subsp. *hillebrandii* – P, H), which is rare in the purest pine forest stands. In summer and fall sometimes it is unnoticed because of loss of vegetative parts, but in spring it is striking because *L. campylocladus* gives rise to spectacular blooms.

Distribution: T, P, H.

Fig. 6.130 *Lotetum campylocladi*. Bejenado skirts. La Palma

Micromerio benthamii-Telinetum microphyllae
This is a broom shrub community widely distributed in the upper-dry and subhumid Mesomediterranean belts of the summits of Gran Canaria. It is largely distributed as a substitutional stage of laurel forest in the NE sector of the island, and also within

the potential area of the pine forest. It preferably grows on shallow soils, rocky places and sunny slopes.

Teline microphylla, an island endemic, is the dominant species in the community. Besides, *Andryala pinnatifida* subsp. *pinnatifida*, *Adenocarpus foliolosus* and *Micromeria benthamii*, among others, are companion in the subass. *typicum*. In the central part of the island the *Micromeria lanata* facies is recognizable.

The subass. *sideritidetosum dasygnaphalae,* whose differential taxa are *Argyranthemum adauctum* subsp. *canariense*, *Erysimum albescens* and *Sideritis dasygnaphala,* is a substitutional community of pine forest in the coolest area of the summit. In contrast, at the warmer SW slope the subass. *euphorbietosum regis-jubae* occurs; it has some *Kleinio-Euphorbietea canariensis* species as differentials, like *Aeonium percarneum*, *Euphorbia regis-jubae* and *Kleinia neriifolia*.

This association has increased its area through fire and grazing.

Distribution: Endemic to C.

Telinetum canariensis (Fig. 6.131)
This is a broom shrub community, endemic to Tenerife that mainly grows in dry-subhumid Thermomediterranean and subhumid Mesomediterranean areas (c). It constitutes the dry mantle of the laurel forest, on rocky outcrops and cliffs, from Anaga to Teno. It also expands as a substitutional community through laurel forest disturbance, mainly by fires and grazing, particularly at sunny spots of Anaga where it is widespread.

Teline canariensis is the main characteristic species giving the physiognomy to the community, and locally *T. pallida* subsp. *pallida.* Other common, widely distributed plants include *Erica arboreae, Artemisia thuscula, Rubia fruticosa, Micromeria varia* and *Asphodelus ramosus* as companion species.

Besides the subass. *telinetosum canariensis* that represents the typical broom shrub, two more subassociations are recognized:

The subass. *telinetosum pallidae*, typical of phonolitic outcrops of Anaga surrounded by the dominant basalts of the territory. The two local differentiating endemics are *Teline pallida* subsp. *pallida* and *Cistus chinamadensis*.

The subass. *ericetosum platycodonis* indicates the contact with the ridge crest forest of *Erica platycodon* (*Ilici-Ericetum platycodonis*), which is its differential species.

Distribution: Endemic to T.

Fig. 6.131 *Telinetum canariensis*. Head of Barranco de las Huertas de San Andres. Anaga. Tenerife

6.2.7.4 Pine Forest, Summit Broom Shrublands, Violets and Other Summit Scrub

Chamaecytiso-Pinetea canariensis
Large phytosociological class that includes potential *Pinus canariensis* forest, potential *Juniperus cedrus* woodland, potential Meso- to Oromediterranean broom shrublands, permanent and seral summit scrublands of La Palma and Tenerife, and the permanent community of *Viola cheiranthifolia* on screes and lapilli of the Teide stratocone. They grow in dry to humid (Thermo-) Meso- to Supramediterranean, and (arid) semiarid to dry Oromediterranean belts. Endemic to Western Canary subprovince (Table 6.4).

Chamaecytiso-Pinetalia canariensis
Potential *Pinus canariensis* forest of the dry to humid Thermo- to Mesomediterranean belts (wc), and potential *Juniperus cedrus* woodland of the dry to humid Meso- and Supramediterranean belts (wc). Endemic to Western Canary subprovince.

Cisto symphytifolii-Pinion canariensis
Potential *Pinus canariensis* forest of the dry to humid Thermo- to Mesomediterranean belts (wc). Distribution: C, T, G, P, H.

Table 6.4 Characteristic species of the vegetation of *Chamaecytiso-Pinetea canariensis*

Chamaecytiso-Pinetea canariensis				
Scientific name	**Family**	**Common name**	**Biotype**	**Distribution**
Chamaecytiso-Pinetalia canariensis				
TREES				
Juniperus cedrus	Cupressaceae	"cedro canario"	M	P,G,T,C
Pinus canariensis	Pinaceae	"pino canario"	M	H,P,G,T,C
SCRUB and SHRUBS				
Argyranthemum adauctum subsp. *dugourii*	Asteraceae	"magarza tinerfeña de pinar"	Ch/N	T
Bystropogon origanifolius var. *canariae*	Lamiaceae	"poleo de pinar grancanario"	N	C
Bystropogon origanifolius var. *ferrensis*	Lamiaceae	"poleo de pinar herreño"	N	H
Bystropogon origanifolius var. *origanifolius*	Lamiaceae	"poleo de pinar tinerfeño y gomero"	N	G,T
Bystropogon origanifolius var. *palmensis*	Lamiaceae	"poleo de pinar palmero"	N	P
Chamaecytisus proliferus subsp. *angustifolius*	Fabaceae	"escobón"	N	H?,G,T
Chamaecytisus proliferus subsp. *calderae*	Fabaceae	"tagasaste blanco"	N	P
Chamaecytisus proliferus subsp. *meridionalis*	Fabaceae	"escobón blanco del sur"	N	C
Cistus asper	Cistaceae	"amagante del Risco de Las Playas"	N	H
Cistus chinamadensis subsp. chinamadensis	Cistaceae	"amagante de Chinamada"	N	T
Cistus chinamadensis subsp. *gomerae*	Cistaceae	"amagante gomero"	N	G
Cistus chinamadensis subsp. *ombriosus*	Cistaceae	"amagante de Mansilva"	N	H
Cistus horrens	Cistaceae	"amagante de Paleotamarán"	N	C
Cistus ocreatus	Cistaceae	"amagante de Tamadaba"	N	C
Cistus osbeckiifolius subsp. *osbeckiifolius*	Cistaceae	"amagante de cumbre"	N	T
Cistus osbeckiifolius subsp. tomentosus	Cistaceae	"amagante de Tágara"	N	T
Cistus symphytifolius var. canus	Cistaceae	"amagante de Cumbre Vieja"	N	P
Cistus symphytifolius var. *symphytifolius*	Cistaceae	"amagante del Norte"	N	T,P
Cistus symphytifolius var. *villosus*	Cistaceae	"amagante del Sur"	N	T

(continued)

Table 6.4 (continued)

Chamaecytiso-Pinetea canariensis				
Scientific name	**Family**	**Common name**	**Biotype**	**Distribution**
Descurainia lemsii	Brassicaceae	"pajonera de Ayosa"	Ch	T
Lotus campylocladus subsp. *campylocladus*	Fabaceae	"corazoncillo de Tenerife"	Ch	T
Lotus campylocladus subsp. *hillebrandii*	Fabaceae	"corazoncillo de La Palma"	Ch	H?,P
Lotus holosericeus	Fabaceae	"corazoncillo plateado"	Ch	C
Lotus pyranthus	Fabaceae	"pico de fuego"	Ch	P
Lotus spartioides	Fabaceae	"corazoncillo de Gran Canaria"	Ch	C
Micromeria benthamii	Lamiaceae	"tomillo de Bentham"	Ch	C
Micromeria lanata	Lamiaceae	"tomillo blanco"	Ch	C
Micromeria varia subsp. *meridionalis*	Lamiaceae	"tomillo salvaje"	Ch	C
Micromeria pineolens	Lamiaceae	"tomillón de Tamadaba"	N	C
Phillyrea angustifolila subsp. *canariensis*	Oleaceae	"olivillo"	N	C
Sideritis oroteneriffae	Lamiaceae	"chajorra de cumbre"	Ch/N	T
Sideritis soluta	Lamiaceae	"chajorra sureña"	N	T
Teline stenopetala subsp. *sericea*	Fabaceae	"gacia blanca"	N	P
SPARTOCYTISETALIA SUPRANUBII				
SCRUB and SHRUBS				
Adenocarpus viscosus subsp. *spartioides*	Fabaceae	"codeso palmero"	N	P
Adenocarpus viscosus subsp. *viscosus*	Fabaceae	"codeso de cumbre"	N	G,T
Andryala pinnatifida subsp. *teydensis*	Asteraceae	"estornudera del Teide"	Ch	T
Argyranthemum tenerifae	Asteraceae	"margarita del Teide"	Ch	T
Bencomia exstipulata	Rosaceae	"bencomia de cumbre"	N	P,T
Bufonia paniculata	Caryophyllaceae	"escobilla estrellada"	T	P,G,T,C
Carlina xeranthemoides	Asteraceae	"malpica de cumbre"	Ch	T
Cheirolophus teydis	Asteraceae	"cabezón de la cumbre"	N	P,T
Descurainia bourgeauana	Brassicaceae	"pajonera de cumbre"	Ch/N	P,T
Descurainia gilva	Brassicaceae	"pajonera palmera"	Ch	P
Descurainia gonzalezii	Brassicaceae	"pajonera de pinar"	Ch	T
Echium auberianum	Boraginaceae	"tajinaste picante"	Ch	T
Echium wildpretii subsp. *trichosiphon*	Boraginaceae	"tajinaste rosado"	Ch/N	P
Echium wildpretii	Boraginaceae	"tajinaste rojo"	Ch/N	T
subsp. *wildpretii*				

(continued)

Table 6.4 (continued)

Chamaecytiso-Pinetea canariensis				
Scientific name	**Family**	**Common name**	**Biotype**	**Distribution**
Erigeron calderae	Asteraceae	"pompón de Las Cañadas"	H	T
Erysimum scoparium	Brassicaceae	"alhelí de cumbre"	Ch	P,T
Genista benehoavensis	Fabaceae	"retamón palmero"	N	P
Helianthemum juliae	Cistaceae	"jarilla de Las Cañadas"	Ch	T
Lactuca palmensis	Asteraceae	"lechuga palmera"	H	P
Laphangium teydeum	Asteraceae	"borriza del teide"	T/H	T
Micromeria herpyllomorpha	Lamiaceae	"tomillo palmero"	Ch	P
Micromeria lachnophylla	Lamiaceae	"tomillo del Teide"	Ch	T
Micromeria lasiophylla subsp. *lasiophylla*	Lamiaceae	"tomillo de cumbre tinerfeño"	Ch	T
Micromeria lasiophylla subsp. *palmensis*	Lamiaceae	"tomillo de cumbre palmero"	Ch	P
Nepeta teydea	Lamiaceae	"tonática"	Ch	P,T
Pimpinella cumbrae	Apiaceae	"perejil de cumbre"	Ch	P,T
Plantago webbii	Plantaginaceae	*"pinillo blanco"*	Ch	P,T,C
Polycarpaea tenuis	Caryophyllaceae	"pataconejo de cumbre"	Ch	P,T
Pterocephalus lasiospermus	Dipsacaceae	"rosalito de cumbre"	N	T
Pterocephalus porphyranthus	Dipsacaceae	"rosalito palmero"	N	P
Scrophularia glabrata	Scrophulariaceae	*"fistulera de cumbre"*	Ch	P,T
Sideritis eriocephala	Lamiaceae	"chajorra del Teide"	Ch	T
Silene nocteolens	Caryophyllaceae	"canutillo del Teide"	Ch	T
Spartocytisus supranubius	Fabaceae	"retama del Teide"	N	P,T
Stemmacantha cynaroides	Asteraceae	"cardo de plata"	Ch	T
Teline stenopetala subsp. *spachiana*	Fabaceae	"escobonillo"	N	T,C
Tolpis webbii	Asteraceae	"lechuguilla de cumbre"	Ch	T
Viola palmensis	Violaceae	"violeta de La Palma"	Ch	P
HERBS				
Arrenatherum calderae	Poaceae	"mazorrilla del Teide"	H	P,T
Todaroa montana	Apiaceae	"cañaheja blanca"	H	H,P,G,T,C
Viola cheiranthifolia	Violaceae	"violeta del Teide"	H	T
Species of TRIFOLIO-GERANIETEA SANGUINEI characteristic of the pine forest fringe				
Pimpinella dendrotragium	Apiaceae	"perejil cabruno"	Ch	P,T
Rumex maderensis	Polygonaceae	"aceda de Madeira"	Ch	H,P,G,T,C

Bystropogono ferrensis-Pinetum canariensis (Fig. 6.132)
This is an association endemic to El Hierro, which represents the Thermo- and Mesomediterranean pine forest. Its climatophilous area is restricted to the southern slope, above the climatophilous area of juniper woodland (*Rubio frutisocae-Juniperetum canariensis*), extending from about 700–900 m reaching up to the semi-nebulous summit vegetation of laurel forest or that of the lapilli cones.

This is an oligospecific pine forest, even poorer than that of the other islands. *Pinus canariensis* is the characteristic tree. *Cistus asper*, the equivalent species to *Cistus symphytifolius* in the pine forest of Tenerife and La Palma, is locally restricted to edaphoxerophilous pine forest on the steep slopes of Las Playas. The native *Chamaecytisus proliferus*, typical on other islands, does not appear. *Bystropogon origanifolius var. ferrensis* spreads on rocky places within the potential area of pine forest.

It corresponds to the terminal community of the upper-dry – subhumid Thermomediterranean (wc) and dry – subhumid Mesomediterranean (wc) climatophilous series of the *Pinus canariensis* forest on El Hierro (*Bystropogono ferrensis-Pino canariensis sigmetum*).

The subass. *typicum* is the pine forest that stretches from El Crés to Mercader. The subass. *ericetosum arboreae* includes the moist pine forest with *E. arborea* and *Morella faya* in areas with mist overflowing from the northern slope, above all on the moist lapilli favorable to pine development.

In addition to these communities it is worth noting the wide area covered by juniper woodland with pine on the southern slopes, mainly in the lower-dry Thermomediterranean belt, within the climatophilous area of juniper woodland (*Rubio-Juniperetum canariensis* subass. *pinetosum canariensis*).

Distribution: Endemic to H.

Fig. 6.132 *Bystropogono ferrensis-Pinetum canariensis*. El Morcillo. El Hierro

Cisto gomerae-Pinetum canariensis (Fig. 6.133)
This is an association endemic to La Gomera, whose climatophilous area extends along a narrow southern strip lying between 1000–1200 (1250) m, between the potential area of juniper woodland (*Brachypodio-Juniperetum canariensis*) and the summit laurel forest (*Ixantho-Laurion novocanariensis*). It additionally appears with a salic edaphophilous character in Garabato (Vallehermoso), in contact with juniper woodland and dry laurel forest (*Visneo-Arbutetum canariensis*).

The most representative taxa of the typical subassociation apart from *Pinus canariensis* are *Bystropogon origanifolius* var. *origanifolius*, *Chamaecytisus proliferus* subsp. *angustifolius*, and *Cistus chinamadensis* subsp. *gomerae*.

Its structure and physiognomy correspond to an open forest, the terminal community of dry and subhumid Mesomediterranean (wc) climatophilous series; and the terminal community of upper-semiarid and lower-dry Thermomediterranean (wc), and upper-semiarid to subhumid Thermo- and Mesomediterranean (c) salic edaphophilous series of *Pinus canariensis* on La Gomera (*Cisto gomerae-Pino canariensis sigmetum*).

The only representative stand of the subassociation *typicum* is to be found in Imada. Although located on a salic substrate, it is within a typical bioclimatic belt of climatophilous pine forest (dry pluviseasonal lower-Mesomediterranean). There are some xeric elements present, such as *Kleinia neriifolia, Euphorbia berthelotii, Echium aculeatum* and *Cistus monspeliensis* subsp. *canariensis*. These xerothermophilous plants are indicators of thermophilous pine forest, similarly to those of the other Islands' southern low elevation pine forest. Some humid elements like *Erica arborea* are to be found, due to the influence of nearby southward overflowing trade-wind clouds. Probably in the narrow potential pine-forest strip, dry and humid pine forests may have formed a mosaic depending on the exposure.

The Gomera pine forest has become almost extinct and only a few remnants persist. The retreat of pine forest was so strong, and probably occurred so soon after the conquest of the island by the Spaniards, that many authors have assumed the pre-conquest absence of wild pine forest. However, molecular analysis and archeological evidence support the existence of autochthonous *Pinus canariensis* on La Gomera. Bioclimatic belts suitable for its development, the abundant presence of its substitutional and mantle community *Adenocarpo foliolosi-Chamaecytisetum angustifolii*, and old isolated individuals linked to salic outcrops and fragmentary little stands, all support our thesis of accepting the potentiality of pine forest on the island.

The subass. *juniperetosum canariensis* (Fig. 6.134) potentially develops on salic outcrops within the Thermo- and Mesomediterranean belts in the general climatophilous area of juniper woodland (*Brachypodio-Juniperetum canariensis*) and laurel forest (*Pruno-Lauretalia novocanariensis*), and constitutes a salic edaphophilous subseries: the *Cisto gomerae-Pino canariensis* faciation of *Juniperus turbinata* subsp. *canariensis.* Its differential taxa are *Brachypodium arbuscula, Erica arborea, Hypericum canariense, Ilex canariensis, Juniperus turbinata* subsp. *canariensis, Morella faya, Spartocytisus filipes* and *Visnea mocanera.* Although there are many salic outcrops in the potential area, only some have isolated pines or pine stands. For instance, on the central salic plugs of the island (Agando, Ojila and La Zarzita), characteristic plants of the pine forest are present like *Cistus chinamadensis* subsp. *gomerae* and *Chamaecytisus proliferus* subsp. *angustifolius*, but *P. canariensis* itself is only found at Agando. The community, although favored by reforestation, nowadays has its best and only representation on crags at the northern locality of Garabato (Vallehermoso) at 300–500 (750) m.

Distribution: Endemic to G.

Fig. 6.133 *Cisto gomerae-Pinetum canariensis*. Roque de Imada. La Gomera

Fig. 6.134 *Cisto gomerae-Pinetum canariensis* subass. *juniperetosum canariensis*. Barranco del Ingenio. Vallehermoso. La Gomera

Loto hillebrandii-Pinetum canariensis (Fig. 6.135)

This is a pine forest endemic to La Palma. Its climatophilous area starts at about 1500 m on the N, NE and E slopes above the climatophilous area of laurel forest, and at 550 m in the NW, 650–800 m in the SW and 500–650 m on the SE slope, above the potential area of juniper woodland (*Rhamno-Juniperetum canariensis*) or white broom shrubland (*Euphorbio-Retametum rhodorhizoidis*). On the SE slopes above Santa Cruz de La Palma, partially sheltered from NE trade-winds, a combination of typical and humid pine forest, depending on local exposure, descends to very low altitudes at about 350–450 m. It reaches the area of the summit shrubland (*Genisto benehoavensis-Adenocarpetum spartioidis*) at about 2000 m, near the island's summit, at the edges of Caldera de Taburiente.

Its most representative species, besides *Pinus canariensis,* are *Bystropogon origanifolius* var. *palmensis*, *Chamaecytisus proliferus* subsp. *calderae*, *Cistus symphytifolius* and *Lotus hillebrandii*, among others. The current floristic understory poverty is probably due to introduced herbivores (rabbits, goats and Barbary sheep).

It corresponds to the terminal community of upper-dry, subhumid and humid Thermomediterranean (wc) and subhumid-humid Mesomediterranean (wc), climatophilous series of the Canary pine forest on La Palma (*Loto hillebrandii-Pino canariensis sigmetum*).

The pine forest is the commumity that shows the largest climatophilous area on the island, and despite heavy anthropic activity and countless fires has the greatest current representation, potentiated by afforestation.

The subass. *cistetosum symphytifolii* (*typicum*) includes the typical pine forest of the dry and subhumid Thermomediterranean belts, outside the influence of NE trade-wind clouds. Pine trees can grow under diverse topographical conditions, from very steep slopes with scarce soil and a low cover, to areas with a smoother topography and deep soil, forming a dense continuous forest.

The subass. *ericetosum arboreae* (humid pine forest) includes the most humid pine forest, in contact with laurel forest, with weak influence of mist. *Morella faya, Erica arborea* and *Ilex canariensis* are frequent. Also its potential area covers slopes with clouds overflowing from Mesomediterranean areas with laurel forest. It behaves edaphophilously on lapilli beds and cones within the climatophilous area of laurel forest. It expands by substitution due to previous felling of laurel forest.

The subass. *juniperetosum canariensis* includes the thermophilous pine forest in contact with Canary juniper woodland, potentially in dry-subhumid Thermomediterranean territory. Nowadays most of the area corresponds to pine forest with rockrose and dendroid spurges, as well other shrubs typical of thermo-sclerophyllous woodland (*Oleo-Rhamnetea crenulatae*).

The subass. *spartocytisetosum filipis* has been proposed for dry Infra- and Thermomediterranean territory of the SW sector of the Caldera National Park. It is floristically characterized by the presence in the underbrush of the endemic broom *Spartocytisus filipes*, together with some endemic shrubs of the coastal belts like *Kleinia neriifolia* and *Globularia salicina*, and vines such as *Periploca laevigata* and *Rubia fruticosa*.

The subass. *adenocarpetosum spartioidis* defines the summit pine forest of the subhumid-humid Mesomediterranean in contact with the legume summit shrubland.

The subass. *juniperetosum cedri* corresponds to subhumid Meso- and Supramediterranean pine forest stands of the inner cliffs of the Caldera de Taburiente, with the presence of *Juniperus cedrus* subsp. *cedrus.*

The subass. *adenocarpetosum foliolosi* represents the subhumid-humid Mesomediterranean pine forest of the southern summits of the island of recent volcanic formation.

Distribution: Endemic to P.

Fig. 6.135 *Loto hillebrandii-Pinetum canariensis*. Caldera de Taburiente. La Palma

Micromerio pineolentis-Pinetum canariensis (Fig. 6.136)
This is a mesophytic pine forest endemic to Gran Canaria, typical of salic substrates and outcrops. The canopy is dominated by *Pinus canariensis*, and this species is accompanied by other trees typical of laurel forest like *Morella faya* and *Erica arborea*. Its potential area is fragmented into several patches, located on the Tamadaba massif and surroundings, over ignimbrite tuffs and rhyolitic-trachyte lava flows and potentially on other smaller tephritic-phonolitic outcrops like those of Los Barrancos-Los Picachos, above Tentiniguada. It is found always in areas influenced by t-w.c. within the climatophilous potentiality of laurel forest. In Tamadaba it is more thermal and develops in the upper-dry Thermo- and Mesomediterranean belts (c) (climatophilous area of dry and humid laurel forest), while in the Tentiniguada area it could also have developed in the cooler upper-dry Mesomediterranean belt (ces) (climatophilous area of cold and summer-dry laurel forest).

Its characteristic species, besides *Pinus canariensis,* are *Morella faya, Erica arborea, Hypericum grandifolium*, *Cistus ochreatus, Bystropogon origanifolius* var. *canariae,* among others. *Phillyrea angustifolia* subsp. *canariensis, Micromeria pineolens* and *Hypericum canariensis* are warmer differential species in Tamadaba. Epiphytic lichenic communities of moist environments, like the *Pseudoevernietum furfuraceae* and *Usneetum atlanticae,* are conspicuous on trunks and in the canopy.

It corresponds to the terminal community of the upper-dry Thermo-, and Mesomediterranean [(c) and (ces)] edaphophilous series of the salic forest of *Pinus canariensis* on Gran Canaria (*Micromerio pineolentis-Pino canariensis sigmetum*).

The association initially included all the pine forest of the island, and later was considered synonymous with the *Pinetum canariensis*. Now it is differentiated from the latter and it is limited to salic cloudy environments.

In the humid area of Tamadaba, some marginal potential area of the association is covered by the broom shrubland *Chamaecytiso canariae-Adenocarpetum villosii*, and eroded areas by the rockrose scrub *Euphorbio regis-jubae-Cistetum canariensis*. The eastern potential area above Tentiniguada is today mainly covered by the broom shrubland *Micromerio benthami-Telinetum microphyllae*.

Distribution: Endemic to C.

Fig. 6.136 *Micromerio pineolentis-Pinetum canariensis*. Tamadaba. Gran Canaria

Morello fayae-Pinetum canariensis

This is a salic pine forest endemic to Tenerife, located on the salic lava flows of the mid-altitude lands of Garachico-Los Realejos, and in the salic territories of Orotava-Santa Úrsula. Besides these areas, it grows on some isolated salic outcrops such as Roque de los Pinos (Anaga) or La Azadilla (Los Realejos).

Apart from *Pinus canariensis*, there are characteristic species commonly linked to laurel forest like *Erica arborea*, *Ilex canariensis, Laurus novocanariensis*, *Morella faya*, *Arbutus canariensis* and *Visnea mocanera*. In the undergrowth there are several shrubs, with a dominance of those coming from laurel forest, such as *Daphne gnidium, Pteridium aquilinum* and *Hypericum grandifolium*, among others, and commonly *Cistus symphytifolius*, a typical pine forest shrub.

It occurs on salic substrates at lower altitudes than those of the climatophilous pine forest, from dry Infra- and Thermomediterranean to subhumid Mesomediterranean belts (c and ces). The vegetation of these belts, on basaltic substrates, shows the climatophilous dominance of dry laurel forest (*Visneo mocanerae-Arbutetum canariensis*), humid laurel forest (*Lauro novocanariensis-Perseetum indicae*), and cold and summer-dry laurel forest (*Pericallido murrayi-Morelletum fayae*), but on such salic substrates *Pinus canariensis* grows well and dominates over the laurel forest trees with which it grows.

It represents the terminal community of the dry Infra- and Thermomediterranean and subhumid Mesomediterranean (c and ces) edaphophilous series of the salic forest of *Pinus canariensis* and *Morella faya* on Tenerife (*Morello fayae-Pino canariensis sigmetum*).

The association corresponds to a particular plant community previously included in the Tenerife pine forest (within the *Sideritido solutae-Pinetum canariensis*), but it differs in several features: salic substrate, bioclimatic territory, and floristic composition.

Three subassociations are distinguished:

The subass. *typicum*, floristically characterized by the presence of *Pinus canariensis*, *Laurus novocanariensis*, *Morella faya*, *Ilex canariensis* and *Erica arborea*, and by the absence of trees from the more thermophilous and dry environments of dry laurel forest. It constitutes the salic pine forest of the subhumid lower-Mesomediterranean bioclimatic belt (c) which on basaltic substrates is occupied by the climatophilous community *Lauro novocanariensis-Perseetum indicae*, and at higher elevations, but without summer t-w.c., by the *Pericallido murrayi-Morelletum fayae*. It is mainly located on the salic lava flows of the upper half of the mid-altitude lands of Garachico-Los Realejos, originating from the recent volcanic activity of Mount Teide, and in the salic territories of Orotava-Santa Úrsula, belonging to the second Canary volcanic cycle, at altitudes between about 750 and 1500 m.

The subass. *arbutetosum canariensis* (Fig. 6.137) has trees of dry laurel forest like *Arbutus canariensis* and *Visnea mocanera* as differential species. The current fragmentation of the community and the extensive human use of the territory led to the practical absence of other differential trees. Additionally, in its potential territory the presence of regressive communities of dry laurel forest (*Visneo mocanerae-Arbutetum canariensis*) is characteristic, and *Cistus monspeliensis* subsp. *canariensis* is frequent. It constitutes the salic pine forest of dry Infra- and Thermomediterranean belts (c) which on basaltic substrates are occupied by the climatophilous dry laurel forest community. It is established on the salic territories spread over the mid-altitude lands of Garachico-Los Realejos, the lower part of Ladera de Tigaiga (La Azadilla -Los Realejos), and Orotava-Santa Úrsula, at altitudes between about 300 and 750 m.

The subass. *cistetosum chinamadensis* (Fig. 6.138) with *Cistus chinamadensis* subsp. *chinamadensis*, an exclusive endemism of the salic outcrop of Roque de los Pinos in Anaga, as the differential taxon of this subassociation. This is a salic pine forest that has developed in the dry Inframediterranean belt (c), in which the basaltic substrates of the surrounding territory are of the potential climatophilous dominance of dry laurel forest (*Visneo mocanerae-Arbutetum canariensis*). This subassociation is relegated to Roque de los Pinos, near the small village of Chinamada, in Anaga.

Distribution: Endemic to T.

Fig. 6.137 *Morello fayae-Pinetum canariensis* subass. *arbutetosum canariensis*. La Azadilla. Los Realejos. Tenerife

Fig. 6.138 *Morello fayae-Pinetum canariensis* subass. *cistetosum chinamadensis*. Roque de los Pinos. Chinamada. Anaga. Tenerife

Pinetum canariensis (Fig. 6.139)

This pine forest of the Thermo- and Mesomediterranean belts is an association endemic to Gran Canaria. Its climatophilous area caps the island above the laurel forest area on the N slope in the domain of the upper-Mesomediterranean thermotype (where there are frequent winter frosts and the influence of the dry NW wind prevails), and above the climatophilous area of thermophilous woodland in the South. However, the potential area of the association is larger because pine forest occurs on the salic substrate in the upper-semiarid and upper-dry Thermomediterranean belts of the SW half of the island, where the *Pinetum canariensis* subass. *juniperetosum canariensis* may be well developed. The pine forest also occurs as a potential edaphophilous community on lapilli beds and cones of the north-central part of the island, within the potential climatophilous dominance of laurel forest.

It corresponds to the terminal community of the upper-dry and subhumid Thermomediterranean, dry to subhumid Mesomediterranean belts (wc) climatophilous series, and also the terminal community of the upper-semiarid and lower-dry Thermomediterranean belts (wc), salic edaphoxerophilous series of the Canary pine forest (*P. canariensis*) on Gran Canaria (*Pino canariensis sigmetum*).

Apart from *Pinus canariensis, Bystropogon origanifolius* var. *canariae*, *Chamaecytisus proliferus* subsp. *meridionalis*, *Cistus horrens, Lotus holosericeus* and others are characteristic.

Like the other pine forests of the islands, it is an oligospecific community, probably due to the action of introduced herbivores. Nowadays it is the arboreal com-

Fig. 6.139 *Pinetum canariensis*. Near La Plata. Tirajana. Gran Canaria

munity that is best represented in Gran Canaria, favored by afforestation. The pine forests of Ojeda, Inagua and Pajonales are outstanding examples.

The *Chamaecytisus proliferus* subsp. *meridionalis* facies shows in general dense patches of this plant that reaches up to 8 m high. Other frequent species here are *Argyranthemum adauctum* subsp. *canariense*, *Bystropogon origanifolius* var. *canariae* and even *Sideritis dasygnaphala.*

There are several subassociations:

Subass. *typicum*, which according to its nomenclatural type corresponds to the thermophilous pine forest, develops in the lower-dry Thermomediterranean belt (wc) and extends to the upper-semiarid Thermomediterranean (wc) in salic territories. It is the subassociation with the largest potential area, and the most widespread distribution throughout the center and South of the island. Some elements of the *Micromerio-Cistetalia canariensis,* particularly *Cistus monspeliensis* subsp. *canariensis,* and the *Kleinio-Euphorbietalia canariensis,* particularly *Euphorbia regis-jubae* and *Kleinia neriifolia*, as well as the presence of rockrose scrub (*Euphorbio regis-jubae-Cistetum canariensis*) in disturbed areas are indicators of the territory of this subassociation. There are four outstanding areas: Tamadaba-Altavista, Inagua-Pajonales, Tauro, and Tirajana-Pilancones. Southwards, the subassociation occurs adjacent to the *Pinetum canariensis* subass. *juniperetosum canariensis* within the upper-semiarid ombrotype, which represents the transition to thermo-sclerophyllous woodland, today reduced to small rocky locations. *Juniperus turbinata* subsp. *canariensis, Olea cerasifomis, Pistacia atlantica, Bupleurum salicifolium* subsp. *aciphyllum* and *Asparagus plocamoides* are to be found among the differential species belonging to the *Mayteno-Juniperion canariensis,* whereas *Kleinia neriifolia, Ceropegia fusca* and *Kickxia scoparia* are derived from the *Kleinio-Euphorbion canariensis*.

Subass. *cistetosum horrentis* constitutes the pine forest of the dry Mesomediterranean belt, that extends above that of subass. *typicum*. Within its potential area nowadays it is only scarcely and locally represented southwards of Pico de la Bandera on the Tamadaba massif, and on the crests of Gomestén-Altavista. It is more common on the heights of Inagua, Ojeda, Pajonales, Tejeda, southwards of Pico de las Nieves and Mesa de Las Vacas-El Escobonal. Its remnants are very poor pine forest, in general on denuded soils ravaged by fires, overgrazing, and summer dryness. Only plants like *P. canariensis, Chamaecytisus proliferus* subsp. *meridionalis* or *Cistus horrens* are capable of surviving there.

Subass. *ericetosum arboreae* grows potentially on lapilli beds and cones in areas influenced by clouds within the climatophilous dominance of laurel forest, and also in finicolous areas of the NW and E, with a local southward aspect influenced by overflowing trade-wind clouds. It is characterized by the presence of some laurel forest elements, particularly *Erica arborea* and *Morella faya.* This subassociation also shows substitutional behavior in the potential area of cold and summer-dry laurel forest (*Pericallido-Morelletum fayae*).

Subass. *sideritidetosum dasygnaphalae* grows in the subhumid upper-Mesomediterranean belt, and is the coldest pine forest of the island. *Sideritis dasygnaphala* and *Erysimum albescens* are differential species.

Within the potential area of pine forests, the broom shrubland *Micromerio benthami-Telinetum microphyllae* (*Telino-Adenocarpion*) is widespread. The broom shrubland *Chamaecytiso canariae-Adenocarpetum villosii* mainly develops in the most humid and cold areas of the central part of the island, edging the pine forest area. There is a broom shrubland dominated by *Chamaecytisus proliferus* subsp. *meridionalis* that constitutes a noteworthy facies of pine forest in the dry SE sector. The decapitated soils, above all in Thermomediterranean areas, are colonized by rockrose scrub (*Euphorbio regis-jubae–Cistetum canariensis*). The *Artemisio–Rumicion lunariae* shrubland is widespread on nitrophilous soils.

Distribution: Endemic to C.

Sideritido solutae-Pinetum canariensis (Figs. 6.140, 6.141)
This is an association endemic to Tenerife which represents the Thermo- and Mesomediterranean pine forest. Its climatophilous area rings the island above the potential area of laurel forest (*Ixantho-Laurion novocanariensis*) in the North and above the potential area of thermo-sclerophyllous woodland (*Junipero-Oleetum cerasiformis*) on the SW and SE slopes, except in Güímar where it is partially in contact with the laurel forest. Its upper limit ranges between 1900 and 2100 m in the North; 2150–2250 m on the SE slope; and between 2000 and 2230 m on the SW slope; it reaches its highest altitude in the South. Its climatophilous area also extends into the Las Cañadas circus, between the south slopes of Pico Teide and the inner

Fig. 6.140 *Sideritido solutae-Pinetum canariensis*. Tamadaya. Tenerife

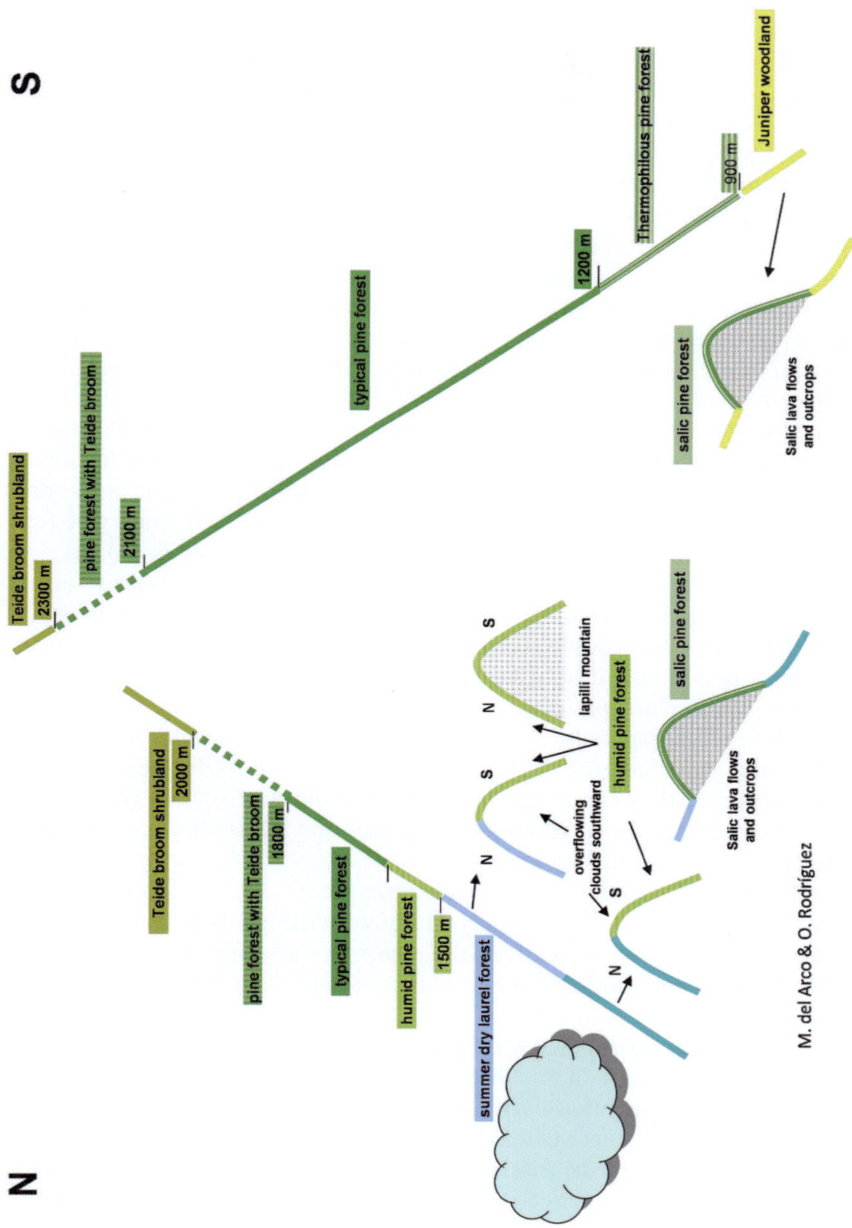

Fig. 6.141 General distribution pattern of potential communities of pine forest, and contacts with other communities

north-facing wall of the circus. Nowadays, in this area there is a reinvasion of pine forest between Montaña Guajara and Boca Tauce, while the Teide broom shrubland (*Spartocytisetum supranubii*) persists on the lava flows as edaphoxerophilous and on the sandy areas of the Ucanca plain where temperature inversion frequently causes frost and a resultant inversion of the vegetation as well.

However, the potential area of the association is larger. The salic lava flows in the SE and SW sectors of the island make the pine forest descend to the upper-semiarid and lower-dry Thermomediterranean, and even the uppermost area of lower-semiarid Thermomediterranean. This area is mainly covered by thermophilous pine forest (*Sideritido-Pinetum canariensis* subass. *cistetosum canariensis*), but it seems potentially to belong, in its mature stage, to an almost disapeared mixed community between juniper woodland and pine forest (*Sideritido-Pinetum canariensis* subass. *juniperetosum canariensis*). Also, in the North, over salic lava flows (Garachico-Los Realejos, Orotava-Santa Úrsula) and dispersed salic outcrops, the pine forest grows within the climatophilous area of laurel forest, but that is a different forest: the *Morello fayae-Pinetum canariensis.*

It corresponds to the terminal community of the upper-dry Thermomediterranean and upper-semiarid to subhumid Mesomediterranean belts (wc) climatophilous series, and also the terminal community of salic upper-semiarid and lower-dry Thermomediterranean belts (wc) edaphoxerophilous series of the Canary pine forest (*P. canariensis*) in Tenerife (*Sideritido solutae-Pino canariensis sigmetum*).

P. canariensis is the dominant characteristic tree. *Junipereus cedrus* subsp. *cedrus* may have participated in the forest but fire and wood exploitation probably relegated it to some inaccessible rocky places safe from fires. Nowadays the pine forest understory is poor, probably due to introduced herbivores. Among the shrubs and scrub, *Chamaecytisus proliferus* subsp. *angustifolius, Bystropogon origanifolius* var. *origanifolius,* especially in the western pine forest, *Argyranthemum adauctum* subsp. *dugourii, Sideritis oroteneriffae, Cistus symphytifolius* and *Sideritis soluta* are characteristic. *Lotus campylocladus* is also characteristic and gives rise to a successional community on disturbed sites, gaps and forest fringes, after fires, under intensive grazing and litter removal.

The following subassociations have been recognized:

Subass. *typicum*, especially in the dry Mesomediterranean belt, that represents the species-poorest pine forest, thinly spread along the northern slope between 1500 and 1800 m, but largely occurring along the southern slope between 1200 and 1900 m.

Subass. *cistetosum canariensis*, with a thermophilous character, extends between 900 and 1200 m, especially in the lower-dry Thermomediterranean belt, but with notable fluctuations, mainly downwards, on salic substrates. It constitutes the transition to thermo-sclerophyllous woodland (*Oleo-Rhamnetea crenulatae*) and thus the lower limit of pine forest on the leeward southern slope. It seems to have developed in the potential territory of a mixed community between juniper woodland and pine forests although it has almost disappeared. It is sometimes mentioned as *Sideritido-Pinetum canariensis* subass. *juniperetosum canariensis nom. prov.* or *pistacietosum atlanticae* nom. prov. Besides *C. monspeliensis* subsp. *canariensis, Euphorbia lamarckii* acts as differential species.

Subass. *cistetosum osbaeckiaefolii* includes the open pine forest with *C. osbaeckiaefolius,* and occasionally *Juniperus cedrus* subsp. *cedrus*, that grows over phonolitic substrate in a proportion of the dry Meso- and Supramediterranean belts of Las Cañadas (Teide). It would represent the transition towards the potential Supramediterranean *Juniperus cedrus* forest (*Adenocarpo viscosi-Juniperetum cedri)* that could have grown above the pine forest.

Subass. *ericetosum arboreae* grows in areas with variable cloud influence. *Erica arborea, Morella faya, Daphne gnidium* and *Hypericum grandifolium* are differential species. The community occurs as a potential vegetation in the contact area between pine and laurel forest, in areas with overflowing trade-wind clouds from the Mesomediterranean, and on recent lapilli beds or volcanic cones within the laurel forest climatophilous area. Besides, the community shows substitutional behavior in disturbed areas of cold and summer-dry laurel forest (*Pericallido-Morelletum fayae*).

Subass. *spartocytisetosum supranubii* is typical of the highest areas of pine forest, between 1800 and 2000 m on the North slope and 1900–2100 (2250) m on the South slope, in contact with the Supramediterranean summit broom shrubland. *Spartocytisus supranubius* and *Nepeta teydea* are differential species.

Within subass. *typicum* and *subass. spartocytisetosum supranubii* facies of *Adenocarpus viscosus* subsp. *viscosus* show as dense *Adenocarpus* scrubland on heliophile, disturbed or decapitated, grazed and burnt soils.

Despite anthropic activity and forest fires, pine forest is well represented, favored by afforestation within its potential territory. The oldest forests of Icod, Adeje, Vilaflor, Arico and Candelaria are today included within the complete pine forest ring which encircles the island, the "Corona Forestal" Nature Park.

The *Chamaecytisus proliferus* subsp. *angustifolius* broom shrubland (*Chamaecytisetum angustifolii*) is common in the territory of the forest. It grows mainly on stony ground and 'malpaíses'. The scrub *Erysimo scoparii-Pterocephaletum lasiospermi* is also frequent on decapitated soils, lithosols and lapilli beds. Rockrose scrub (*Cistetum symphytifolio-canariensis*) is also present on decapitated soils of Thermomediterranean areas.

Distribution: Endemic to T.

Juniperion cedri

Potential *Juniperus cedrus* woodland of the dry to humid Meso- and Supramediterranean belts (wc). Endemic to Western Canary subprovince (T, P).

Adenocarpo spartioidis-Juniperetum cedri

This association represents an open Canary cedar (*Juniperus cedrus* subsp. *cedrus*) forest that it is thought to have been the arboreal community on the *summit* of La Palma. Nowadays it is very sparse, more or less spread above the upper part of the pine forest belt in vegetation fragments considered relicts of ancient sparse woodland. Below these it generally grows only on rocky outcrops. The community probably had a wider abundance and distribution in the past, and was disturbed and destroyed by human activities.

The typical accompanying species of the Canary cedar on the outcrops are Canary pines and the rupicolous plants of the summit, and apart from these the typical plants of the summit broom shrubland (*Genisto benehoavensis-Adenocarpetum spartioidis*). For this reason, it is very difficult to characterize this woodland except by the sporadic presence of *J. cedrus* subsp. *cedrus*.

This community has been proposed as the terminal community of the subhumid-humid Supramediterranean edaphophilous series on lithosols of the woodland of *Juniperus cedrus* and A*denocarpus viscosus* subsp. *spartioides*, with *Echium wildpretii* subsp. *trichosiphon* and *Pterocephalus porphyranthus* on La Palma (*Adenocarpo spartioidis-Junipero cedri sigmetum*).

Distribution: Endemic to P.

Adenocarpo viscosi-Juniperetum cedri (Fig. 6.142)
This association represents a Canary cedar (*Juniperus cedrus* subsp. *cedrus*) open forest that it is thought to have been the arboreal community of the summit of Tenerife. Nowadays it is very sparse, and the cedar is found spread over Supramediterranean rocky scarps and slopes of Las Cañadas del Teide, isolated or forming fragments that are probably relicts of the ancient sparse woodland. Sometimes it is also found mixed with *Pinus canariensis* in Mesomediterranean areas or salic Supramediterranean territory (*Sideritido-Pinetum canariensis* subass. *cistetosum osbaeckiaefolius*). These fragments mainly settle on the walls of the older Pre-Caldera volcanic edifice that still remains in La Fortaleza (N of the Teide) and more continuously towards the S and SE of the Teide.

The Canary cedar woodland probably had a wider abundance and distribution in the past, and it was disturbed and destroyed by human activities, mainly felling and fires.

J. cedrus and *Adenocarpus viscosus* are the characteristic species. Besides these, rupicolus plants and plants of the summit broom shrubland (*Spartocytisus supranubius s.l.)* like *Spartocytisus supranubius, Echium wildpretii, Pterocephalus lasiospermus, Micromeria lachnophylla, Tolpis webbii,* etc. are companion species.

This community has been proposed as the terminal community of the subhumid-humid Supramediterranean edaphophilous series on lithosols of the woodland of *Juniperus cedrus* and A*denocarpus viscosus* subsp. *viscosus*, with *Echium wildpretii* subsp. *wildpretii* and *Pterocephalus lasiospermus* on Tenerife (*Adenocarpo viscosi-Junipero cedri sigmetum*).

Distribution: Endemic to T.

Fig. 6.142 *Adenocarpo viscosi-Juniperetum cedri*.La Fortaleza. Las Cañadas. Tenerife

Spartocytisetalia supranubii
Potential Meso- to Oromediterranean broom shrublands, permanent and seral summit communities of La Palma and Tenerife, and the *Viola cheiranthifolia* permanent community on the Teide stratocone. They grow in dry to humid Mesomediterranean, and (arid) semiarid to dry Oromediterranean belts. Endemic to Western Canary subprovince (T, P).

Spartocytision supranubii
Potential broom shrublands of *Spartocytisus supranubius*, *Genista benehoavensis, Adenocarpus viscosus* or *Teline stenopetala* subsp. *spachiana* in the dry- to humid Meso- to Oromediterranean belts. Endemic to Western Canary subprovince (T, P).

Descurainio bourgeauanae-Spartocytisetum supranubii (Fig. 6.143)
This is a community endemic to Tenerife confined to the NE sector of the summit broom shrubland. It extends above the climatophilous area of pine forest between 2000 and 2300 m, and a little higher in the northeastern part of Las Cañadas, an area with fall to spring frost. It colonizes preferably soft materials both salic and basaltic lapilli from recent volcanic activity after the landslide of the North wall of Las Cañadas, and pumice materials of Montaña Blanca. The territory with N aspect is colder and somewhat rainier than that of the *Spartocytisetum supranubii*.

Its floristic composition is quite similar to the *Spartocytisetum supranubii* except for its differential species *Descurainia bourgeauana.*

It corresponds to the terminal community of the dry and subhumid Meso- and Supramediterranean edaphophilous series of Teide broom shrubland on Tenerife (*Descurainio bourgeauanae-Spartocytisetum supranubii* sigmetum).

The community was intensively grazed and suffered repeated fires in the past. Now it has recovered after the creation of the Teide National Park and the banning of grazing. Today it is threatened by other introduced herbivores, mainly rabbits and mouflons that prevent recruitment and impoverish the community in favor of the expansion of less palatable species such as *Pterocephalus lasiospermus*.

The subass. *erigerontetosum calderae* grows on endorreic deposits with abundant fine fractions of soil, temporarily flooded after melting of snow. The endemism *Erigeron calderae* is the differential species.

Distribution: Endemic to T.

Fig. 6.143 *Descurainio bourgeauanae-Spartocytisetum supranubii*. Summits of the North slope of Tenerife

Genisto benehoavensis-Adenocarpetum spartioidis (Fig. 6.144)
This is the summit broom shrubland of La Palma adapted to the extreme conditions of the high mountain; it grows above the climatophilous área of pine forest (*Loto hillebrandii-Pinetum canariensis*).

The most representative species of this low and dense shrubland is *Genista benehoavensis,* and the most abundant species are *Adenocarpus viscosus* subsp. *sparti-*

oides, *Descurainia gilva, Tolpis laciniata, Erysimum scoparium*, *Festuca agustini*, *Plantago webbii, Lactuca palmensis* and *Spartocytisus supranubius.*

It corresponds to the terminal community of the subhumid-humid Supramediterranean climatophilous series of the broom shrubland of *G. benehoavensis* and *A. viscosus* subsp. *spartioides* on La Palma (*Genisto benehoavensis-Adenocarpo spartioidis sigmetum*).

G. benehoavensis was probably more abundant in the past, forming the dominant shrub with *A. viscosus* subsp. *spartioides*. Overgrazing by goats and fires were the main causes of retreat. Fire threat persists and the grazing effect of introduced wild herbivores (rabbits, goats and Barbary sheep).

Distribution: Endemic to P.

Fig. 6.144 *Genisto benehoavensis-Adenocarpetum spartioidis*. Summits of La Palma

Spartocytisetum supranubii (Fig. 6.145)
This is a summit broom shrubland endemic to Tenerife that commonly occurs above the pine forest between 2000 and 2900 m in the north and 2200–3100 m in the south, an area with a variable probable frost period between November and June. It preferentially grows on soils developed over solid lava flows and volcanic detritus of the Teide and Las Cañadas, already existing before the gravitational collapse of the old north flank of the Cañadas caldera edifice.

Spartocytisus supranubius, *Nepeta teydea, Scrophularia glabrata, Adenocarpus viscosus* subsp. *viscosus*, *Pterocephalus lasiospermus, Argyranthemum teneriffae, Tolpis webbii, Arrhenatherum calderae* and *Pimpinella cumbrae* are their commonest species.

Fig. 6.145 *Spartocytisetum supranubii*. Montaña Blanca de Los Roques. Las Cañadas. Tenerife

It corresponds to the terminal community of the dry (Meso-) Supra- and Oromediterranean climatophilous series of the Teide broom shrubland of Tenerife (*Spartocytisetum supranubii*). It also can descend to theoretically dry Mesomediterranean territory to the south of the Teide, although this is questionable because of the probable temperature inversion in Ucanca valley and surroundings (that would make it Supramediterranean) or for edaphic reasons.

As with the *Descurainio bourgeauanae-Spartocytisetum supranubii,* the community was intensively grazed and suffered repeated fires in the past. Now it has recovered after the creation of the Teide National Park and the banning of grazing. Today it is threatened by other introduced herbivores, mainly rabbits and mouflons that prevent recruitment and impoverish the community in favor of the expansion of less palatable species such as *Pterocephalus lasiospermus*.

An *Adenocarpus viscosus* facies, probably linked to past fires, is distinguished.

Apart from the subass. *typicum* with a poor floristic composition, several subassociations have been described.

The subass. *aeonietosum* corresponds to a broom shrubland that has developed at rocky places, with typical rupicolous plants of the summit like *Rhamnus integrifolia*, *Aeonium spathulatum, A. smithii, Bethencourtia palmensis, Cheilanthes pulchella, Monanthes brachycaulos,* and *Polycarpaea tenuis*. It appears to be a mixture with the *Cheilantho guanchicae-Aeonietum smithii*.

Fig. 6.146 *Spartocytisetum supranubii* subass. *echietosum wildpretii*. Cañada del Cedro. Las Cañadas. Tenerife

The subass. *echietosum wildpretii* (Fig. 6.146) grows on slope debris of Las Cañadas both on the inner slopes as well the southern external slopes, in contact with the pine forest. It also acts as pioneer vegetation at dirt road-sides with a loose substrate. *Echium wildpretii* shows a spectacular bloom in May and June.

The subass. *descurainietosum gonzalesii* grows on solid pyroclast deposits, with a particular presence of the endemic *Descurainia gonzalezii*. It mainly develops on the south external slope of the Pre-Caldera edifice.

The subass. *descurainietosum lemsii,* differentiated by *Descurainia lemsii*, grows on recent lapilli deposits of the eastern area of the *Spartocytisus supranubius* shrubland, in dry and subhumid Meso- and Supramediterranean territory spreading southwards to the pine forest area. It seems it would be better linked to the *Descurainio bourgeauanae- Spartocytisetum supranubii.*

Distribution: Endemic to T.

Telinetum spachianae (Fig. 6.147)

This is a dense legume shrubland endemic to Tenerife with a local distribution on the steep slopes of the caldera and summits of Pedro Gil (Orotava, Arafo and Güímar), between 1500 and 2000 m.

It is characterized by *Teline stenopetala* subsp. *spachiana, Sideritis oroteneriffae* and *Echium virescens,* accompanied by *Chamaecytisus proliferus* subsp. *angustifo-*

lius, Bystropogon canariensis var. *smithianus* and some isolated *Pinus canariensis* hanging on the cliffs.

It corresponds to the terminal community of the dry Meso- and Supramediterranean edaphoxerophilous series of *Teline stenopetala* subsp. *spachiana* broom shrubland on Tenerife (*Telino spachianae sigmetum*).

Distribution: Endemic to Tenerife.

Fig. 6.147 *Telinetum spachianae*. Risco del Valle. Arafo. Tenerife. (Ph: JC)

Violo cheiranthifoliae-Adenocarpetum viscosii

This is a low scrubland, a permanent anemogenous community, about 15–20 cm high, which grows on rocky phonolitic substrate mixed with pumice pyroclast in a small area at the southwest of Guajara Mountain. The territory is at an altitude of around 2700 m, in the dry upper-Supramediterranean belt, subject sometimes to strong SW wind and frequent frost.

Adenocarpus viscosus subsp. *viscosus* is the most abundant plant and serves as a refuge to *Viola cheiranthifolia*. *Erysimum scoparium, Nepeta teydea, Tolpis webbi* and *Spartocytisus supranubius* are occasionally present.

Distribution: Endemic to T.

Plantaginion webbii

Permanent and seral summit scrub, mainly of *Plantago webbii* and *Pterocephalus,* growing on lapilli, volcanic sand, screes and debris of the dry to humid Meso- to Supramediterranean belts. Endemic to Western Canary subprovince (T, P).

Plantaginenion webbii

Permanent and seral *Plantago webbii* scrub growing on lapilli, volcanic sand and debris of the subhumid to humid Meso- and Supramediterranean belts of La Palma.

Descurainio gilvae-Plantaginetum webbii (Fig. 6.148)

This is a chamaephytic-nanophanerophytic scrub with low cover, endemic to La Palma. It is a permanent community growing on lapilli, volcanic sand and debris, within the climatophilous territory of pine forest (*Loto-Pinetum canariensis*), summit Canary cedar forest (*Adenocarpo spartioidis-Juniperetum cedri*) and summit broom shrubland (*Genisto benehoavensis-Adenocarpetum spatioidis*). There, it can spread as a substitutional community to disturbed places with eroded soils.

Descurainia gilva, Plantago webbii, *Micromeria herpyllomorpha* and *Pterocephalus porphyranthus* are the most representative species.

Distribution: Endemic to P.

Fig. 6.148 *Descurainio gilvae-Plantaginetum webbii*. Cumbre Vieja. La Palma

Pterocephalenion lasiospermi

Permanent and seral summit scrub mainly of *Pterocephalus lasiospermus* and *Plantago webbii* growing on lapilli, volcanic sand, screes and debris of the dry to humid Meso- and Supramediterranean belts of Tenerife.

Arrhenathero calderae-Plantaginetum webbii
This is a pioneer hemicryptophytic-chamaephytic community, with some nanophanerophytes, endemic to Tenerife, typical of screes and gelifracted rocks; it grows at the foot of scarps on Mount Teide, walls of the Las Cañadas caldera, Roques de García and La Fortaleza.

It is established within the dry upper-Meso- and Supramediterranean belts, in the climatophilous potential area of pine forest (*Sideritido-Pinetum canariensis*), summit Canary cedar forest (*Adenocarpo viscosi-Juniperetum cedri*), and summit broom shrubland (*Spartocytisetum supranubii s.l.*).

Arrhenatherum calderae and *Plantago webbii* are the characteristic species. They are accompanied by common plants of the summit like *Pterocephalus lasiospermus, Tolpis webbii, Argyranthemum teneriffae, Scrophularia glabrata* and *Erysimum scoparium.*

Distribution: Endemic to T.

Echietum auberiani
This is an open permanent oligospecific community, with very scarce cover, that settles on semi-mobile flat pumice substrate mixed with scoria, as on Montaña Blanca, in the dry Supramediterranean belt on Tenerife.

In spring it stands out in the landscape because of the blue flowering of its characteristic species *Echium auberianum*, endemic to Las Cañadas del Teide. *Descurainia bourgeauana* and *Pterocephalus lasiospermus*, although with scarce presence, are the most frequent companion species, the others being very rare (*Erysimum scoparium, Nepeta teydea,, Scrophularia glabrata, Argyranthemum teneriffae* or *Erigeron cabrerae*).

Distribution: Endemic to T.

Erysimo scoparii-Pterocephaletum lasiospermi (Fig. 6.149)
This is a chamaephytic-nanophanerophytic permanent community endemic to Tenerife that grows on lithosols, rocky ground and lapilli of the dry-subhumid Meso- and Supramediterranean belts of the climatophilous dominance of pine forest (*Sideritido-Pinetum canariensis*), summit Canary cedar forest (*Adenocarpo viscosi-Juniperetum cedri*) and summit broom shrubland (*Spartocytisetum supranubii s.l.*). In these environments it also behaves as substitutional on road-side slopes and disturbed and eroded soils.

Descurainia bourgeauana, Erysimum scoparium, Pterocephalus lasiospermus and *Scrophularia glabrata* are characteristic species.

The *Descurainio bourgeauana* facies as prime colonizer is widespread around Izaña-El Portillo. The *Carlina xeranthemoides* facies is common in the western pine forest, and in Vilaflor.

Distribution: Endemic to T.

Violion cheiranthifoliae
Permanent community of *Viola cheiranthifolia* on screes and lapilli of the Teide stratocone, in the dry upper-Supra- and (arid) semiarid to dry Oromediterranean belts. Endemic to T.

Fig. 6.149 *Erysimo scoparii-Pterocephaletum lasiospermi*. Cañadas del Teide. Tenerife. (Ph: PLP)

Violetum cheiranthifoliae (Fig. 6.150)

This is a hemicryptophytic-chamaephytic association, a permanent community of dry upper-Supra- and (arid) semiarid to dry Oromediterranean screes around the skirts of the Teide and surrounding peaks of the Las Cañadas caldera.

It is characterized by V*iola cheiranthifolia*, an endemic hemicryptophyte that thrives on lapilli beds and pumice screes between 2400 and 3600 m, sometimes accompanied by *Argyranthemum teneriffae* and *Silene nocteolens*.

Distribution: Endemic to T.

Fig. 6.150 *Violetum cheiranthifoliae*. Mount Teide. Tenerife. (Ph: PLP)

6.3 Potential Natural Vegetation

Potential natural vegetation is the vegetation towards which natural succession will lead. It is the final equilibrium stage in vegetation succession. It may be considered as the final stable association of vegetation series (Fig. 6.151). The latter are diverse in response to the ecological characteristics of the environment. Potential natural vegetation (PNV) includes climatophilous vegetation (climatic climax) whose stability depends directly on climate. This is represented by the terminal associations of the climatophilous series. PNV also includes edaphophilous vegetation (edaphic climax); it differs from the climatic climax in response to soil type, exposure to sun, wind, drainage, etc. This is represented by the terminal associations of the edaphophilous series and by permanent communities (permaseries), which is a particular, less developed edaphophilous vegetation of some environments. Figures 6.164, 6.165, 6.166, 6.167, 6.168, 6.169 and 6.170, show potential vegetation maps of the Canaries.

We note below the terminal communities of the climatophilous and edaphophilous series, permanent communities, and some vegetation complexes.

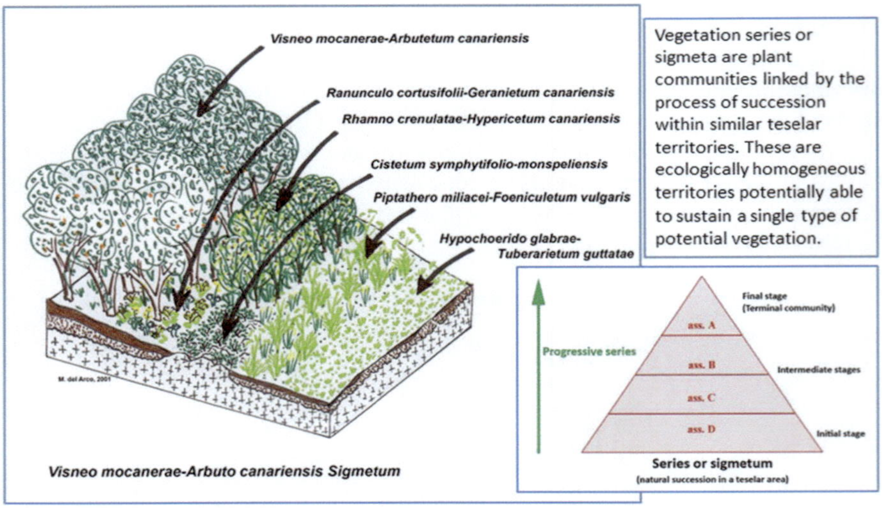

Fig. 6.151 Scheme of a vegetation series

6.3.1 Climatophilous Communities

These are plant associations representing the stable optimum into which vegetation develops under the prevailing environmental macroclimatic conditions, within the average edapho-topographic situations of the territory. Stable associations conditioned by clouds are also considered climatophilous; this occurs in the archipelago because of the action of trade-wind clouds. In general, the climatophilous associations have a wide zonal development, and can encircle the islands. Apart from their climatophilous character, their alternative edaphophilous behavior is on occasion pointed out (See footnote 1).

Sweet spurge scrublands and shrublands

These are terminal communities of the hyperarid and arid Inframediterranean climatophilous series, as well as of the lower-semiarid Inframediterraean edaphophilous series on pumice, of the *E. balsamiferae* scrublands and shrublands.

Asterisco intermedii-Euphorbietum balsamiferae (L)
Lycio intricati-Euphorbietum balsamiferae (F)
Euphorbietum balsamiferae (C)
Ceropegio fuscae-Euphorbietum balsamiferae (T) (Fig. 6.152)
Neochamaeleo pulverulentae-Euphorbietum balsamiferae (G)
Echio breviramis-Euphorbietum balsamiferae (P)
Rubio fruticosae-Euphorbietum balsamiferae (H)

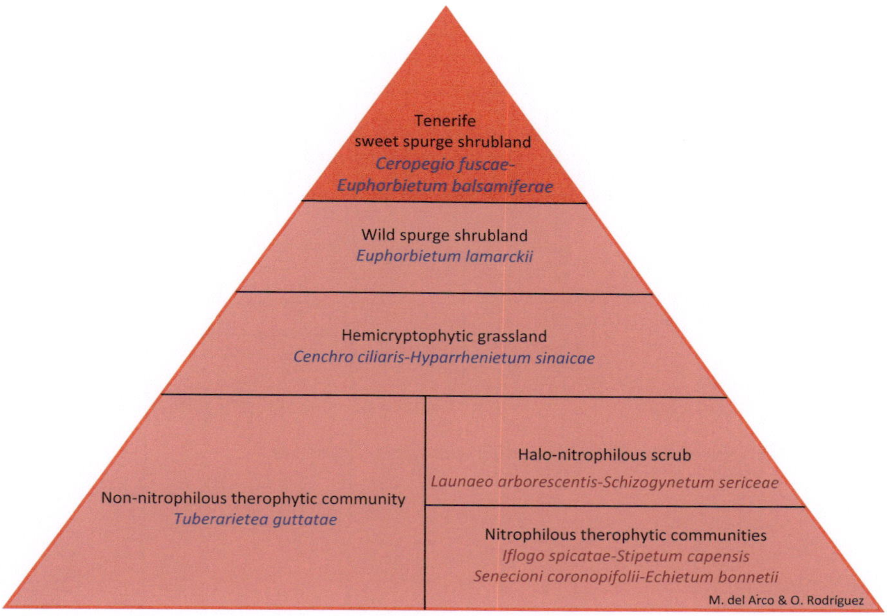

Fig. 6.152 *Ceropegio fuscae-Euphorbio balsamiferae Sigmetum*

Cardón shrublands

These are terminal communities of: (a) the upper-arid upper-Inframediterranean and lower-semiarid Inframediterranean climatophilous series; (b) the upper-semiarid Inframediterranean, semiarid and lower-dry Thermomediterranean (wc) edaphophilous rupicolous series; and (c) the lower-arid Inframediterranean edapho-hygrophilous series on lava flows, of the *Euphorbia canariensis* shrublands.

Asterisco intermedii-Euphorbietum balsamiferae subass. aeonietosum lacerotensis (L)
Kleinio neriifoliae-Euphorbietum canariensis (F)
Aeonio percarnei-Euphorbietum canariensis (C)
Periploco laevigatae-Euphorbietum canariensis (T) (Fig. 6.153)
Euphorbietum berthelotii-canariensis (G)
Echio breviramis-Euphorbietum canariensis (P)
Aeonio valverdensis-Euphorbietum canariensis (H)

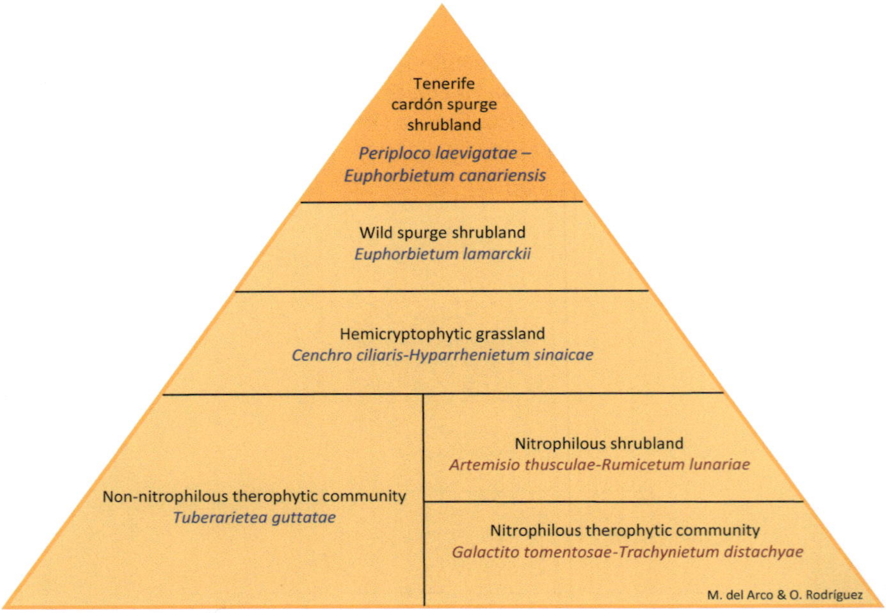

Fig. 6.153 *Periploco laevigatae –Euphorbio canariensis Sigmetum*

Thermo-sclerophyllous woodlands

These are terminal communities of the upper-semiarid to subhumid Inframediterranean (wc), semiarid Thermomediterranean and lower-dry Thermomediterranean (wc) climatophilous series of the juniper (Juniperus turbinata subsp. canariensis) and wild olive (Olea cerasiformis) woodlands.

Convolvulo lopezsocasi-Oleetum cerasiformis (L)
Micromerio rupestris-Oleetum cerasiformis (F)
Pistacio lentisci-Oleetum cerasiformis (C)
Junipero canariensis-Oleetum cerasiformis (T) (Fig. 6.154)
Brachypodio arbusculae-Juniperetum canariensis (G)
Rhamno crenulatae-Juniperetum canariensis (P)
Rubio fruticosae-Juniperetum canariensis (H)

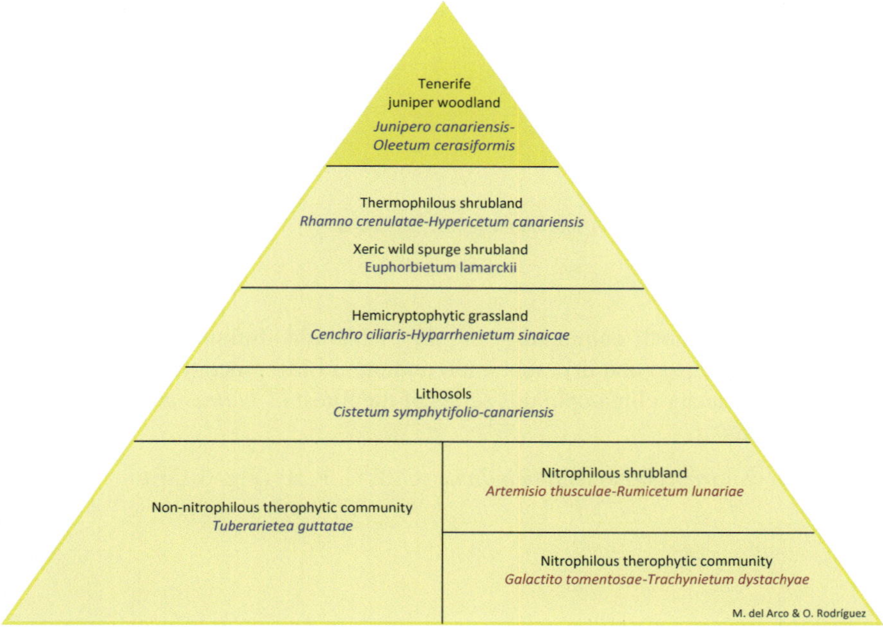

Fig. 6.154 *Junipero canariensis-Oleo cerasiformis Sigmetum*

Dry laurel forest

These are terminal communities of the dry Inframediterranean, upper-semiarid and dry Thermomediterranean (c), and subhumid Thermomediterranean (oc) climatophilous series of the subxeric forest of *Arbutus canariensis*, *Visnea mocanera*, *Apollonias barbujana* and *Picconia excelsa*.

Echio handiensis-Visneetum mocanerae (F)
Visneo mocanerae-Arbutetum canariensis (C, T, G, P, H) (Fig. 6.155)

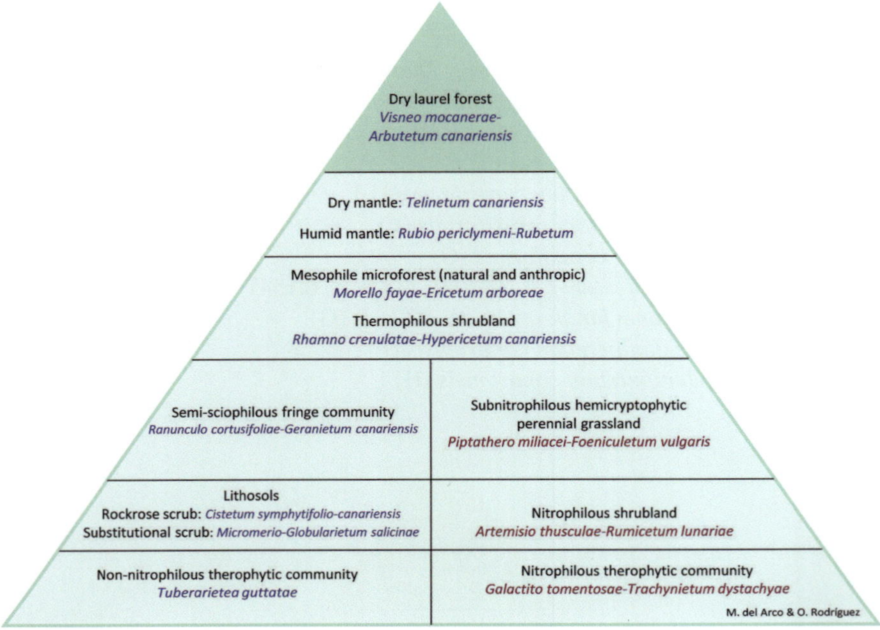

Fig. 6.155 *Visneo mocanerae-Arbuto canariensis Sigmetum*

Humid laurel forest

This is the terminal community of the subhumid Inframediterranean, dry Thermomediterranean and dry, subhumid and humid Mesomediterranean belts (c) with western Canary climatophilous series of the forest of *Laurus novocanariensis* and *Persea indica*.

Lauro novocanariensis-Perseetum indicae (C, T, G, P, H) (Fig. 6.156)

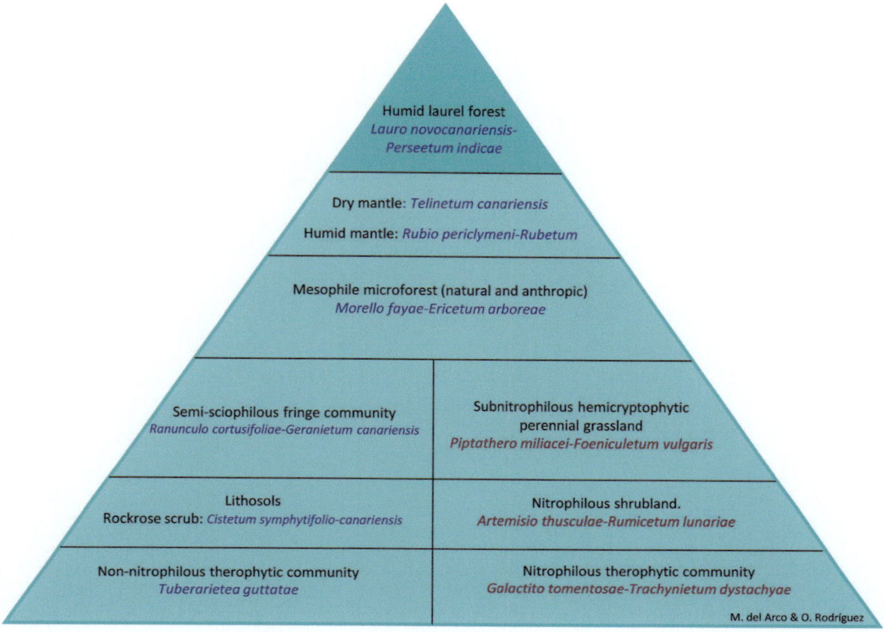

Fig. 6.156 *Lauro novocanariensis-Perseo indicae Sigmetum*

Summer-dry laurel forest

These are terminal communities of the dry to subhumid Mesomediterranean belts (ces) of the western Canary climatophilous series of the *Morella faya* forests.

Violo rivinianae-Morelletum fayae (G)
Pericallido murrayii-Morelleteum fayae (C, T, G, P, H)

Pine forests

These are the terminal communities of the upper-dry, subhumid and humid Thermomediterranean (wc) and upper-semiarid, subhumid and humid Mesomediterranean belts (wc) climatophilous series, and also the terminal community of salic upper-semiarid and lower-dry Thermomediterranean belts (wc) edaphoxerophilous series, all of the Canary pine (*Pinus canariensis*) forests.

Pinetum canariensis (C)
Sideritido solutae-Pinetum canariensis (T) (Fig. 6.157)
Cisto gomerae-Pinetum canariensis subass. *typicum* (G)
Loto hillebrandii-Pinetum canariensis (P)
Bystropogono ferrensis-Pinetum canariensis (H)

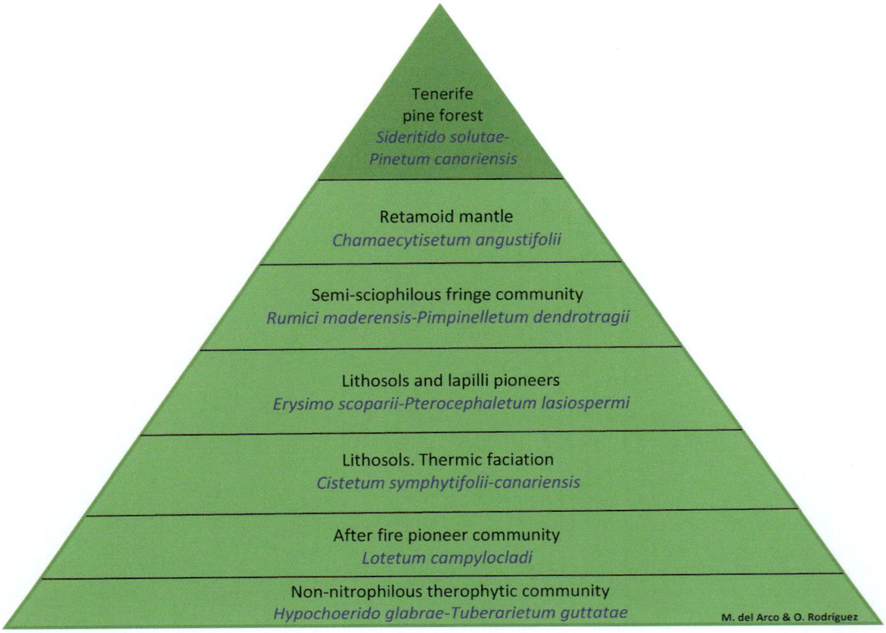

Fig. 6.157 *Sideritido solutae-Pino canariensis Sigmetum*

Summit Broom Shrublands

These are the terminal communities of the dry (Meso-) Supra- and Oromediterranean climatophilous series of the Teide broom (*Spartocytisetum supranubii*) shrublands and subhumid and humid Supramediterranean climatophilous series of the *G. benehoavensis* and *A. viscosus* subsp. *spartioides* shrublands.

Spartocytisetum supranubii (T) (Fig. 6.158)
Genisto benehoavensis-Adenocarpetum spartioidis (P)

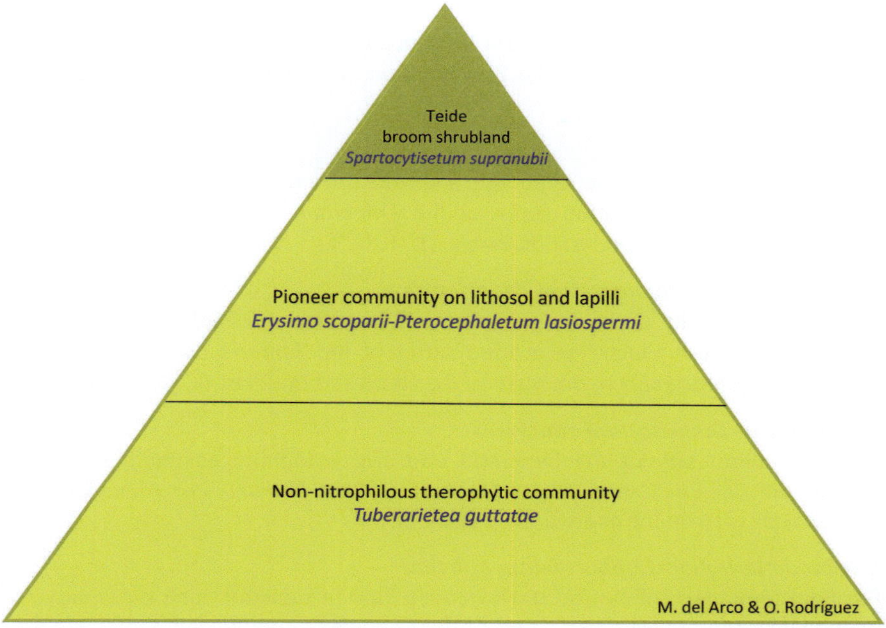

Fig. 6.158 *Spartocytiso supranubii Sigmetum*

6.3.2 Edaphophilous Communities

These are azonal plant communities whose development is not directly dependent on macroclimate but on particular climatic, edaphic, geologic or topographic conditions: extra or less water supply, soil type, rock type, heavy metals, particular climatic agents, drainage, or other factors. They can be structured into edaphophilous series or permaseries.

6.3.2.1 Terminal Communities of Edaphophilous Series

These are communities that can show a biomass development and dynamics similar to that of the climatophilous communities. Among these are the edaphohygrophilous series that are favored by a higher water supply and are found in the lower sector of the typical concatenation of series in the topo-sequence, crest – slope – valley: rivers, ravines, ramblas, lagoons, etc., and the edaphoxerophilous series that are subjected to dryness because of a much lower water supply. The latter series occurs on lithosols, leptosols, debris, mountain screes and other rocky environments, which differentiate them from the most humid adjacent potential vegetation. In an elevation sequence, the following can be distinguished.

Frankenio capitatae-Zygophylletum gaetuli
Terminal community of the hyperarid Inframediterranean psammophile edaphoxerophilous vegetation series of the *Zygophyllum gaetulum* scrub with *Frankenica capitata*, on microdunes. (F).

Traganetum moquinii
Terminal community of the hyperarid and arid Inframediterranean psammophile edaphoxerophilous vegetation series of the *Traganum moquinii* shrublands with *Atriplex glauca* subsp. *ifniensis* on dunes. (L, F, C, T)

Atriplici ifiniensis-Tamaricetum canariensis
Terminal community of the arid and semiarid brackish and hard water edaphohygrophilous Western Canary vegetation series of the *Tamarix canariensis* thicket with *Atriplex glauca* subsp. *ifniensis* on the shore and ramblas. (C, T, G and P)

Suaedo verae-Tamaricetum canariensis
Terminal community of the hyperarid and arid halophilic edaphohygrophilous Central and Eastern Canary vegetation series of the *Tamarix canariensis* thicket with *Suaeda vera* on the shore and ramblas. (F)

Astydamio latifoliae-Euphorbietum aphyllae
Terminal community of the arid and lower-semiarid Inframediterranean edaphoxero-anemogenic littoral vegetation series of the *Euphorbia aphylla* and *Astydamia latifolia* scrub. (C)

Ceropegio dichotomae-Euphorbietum aphyllae
Terminal community of the semiarid Inframediterranean edaphoxero-anemogenic littoral vegetation series of the *Euphorbia aphylla* and *Ceropegia dichotoma* scrub. (T)

Euphorbietum aphyllae
Terminal community of the semiarid Inframediterranean edaphoxero-anemogenic littoral vegetation series of the *Euphorbia aphylla* and *Ceropegia dichotoma* subsp. *krainzii* scrub. (G)

Schizogino sericeae-Euphorbietum wildpreti
Terminal community of the arid and lower-semiarid Inframediterranean edaphoxerophilous vegetation series of the *Schizogyne sericea* and *Euphorbia lamarckii* subsp. wildpretii shrubland, on volcanic ashes. (P, H)

Euphorbio berthelotii-Schizogynetum sericeae
Terminal community of the arid to semiarid Inframediterranean edaphoxerophyllous colluvial vegetation series of the *Euphorbia berthelotii* and *Schizogyne sericea* shrubland. (G)

Euphorbietum atropurpureae
Terminal community of the semiarid and lower-dry Thermomediterranean edaphoxerophilous vegetation series of the *Euphorbia atropurpurea* shrubland. (T)

Euphorbietum handiensis
Terminal community of the hyperarid and arid Inframediterranean edaphophilous colluvial vegetation series of the *Euphorbia handiensis* succulent scrubland. (F)

Echio decaisnei-Retametum rhodorhizoidis (C); ***Echio aculeati-Retametum rhodorhizoidis*** (T); ***Euphorbio berthelotii-Retametum rhodorhizoidis*** (G); ***Euphorbio wildpretii-Retametum rhodorhizoidis*** (P); ***Echio hierrensis-Retametum rhodorhizoidis*** (H)
These correspond to the terminal communities of the upper-semiarid and dry Inframediterranean, and semiarid and lower-dry Thermomediterranean (wc) edaphoxerophilous colluvial Western Canary vegetation series of the Retama rhodorhizoides shrublands. (C, T, G, P, H)

Periploco laevigatae-Phoenicetum canariensis
Terminal community of the arid to semiarid Inframediterranean and semiarid to dry Thermomediterranean edaphohygrophilous colluvial and riparian vegetation series of the *Phoenix canariensis* groves. (F, C, T, G, P)

Ilici canariensis-Ericetum platycodonis
Terminal community of the subhumid-humid Thermo- and Mesomemediterranean anemogenic and nebulous edaphophilous vegetation series of the ridge crest forest of *Ilex canariensis* and *Erica platycodon*. (T, G)

Micromerio lepidae-Ericetum arboreae
Terminal community of the subhumid-humid Thermo- and Mesomediterranean anemogenic and temporary-nebulous edaphophilous series of the ridge crest forest of *Erica arborea* (G).

Diplazio caudati-Ocoteetum foetentis
Terminal community of the subhumid-humid Infra- to Mesomediterranean edaphohygrophilous and nebulous vegetation series of the Western Canary forest of *Ocotea foetens* with *Ilex perado* subsp. *platyphylla* and *Diplazium caudatum*. (C, T, G, P, H)

Morello fayae-Pinetum canariensis
Terminal community of the dry Infra- and Thermomediterranean and subhumid Mesomediterranean subnebulous salic edaphophilous vegetation series of the Tenerife forest of *Pinus canariensis* and *Morella faya* (T).

Micromerio pineolentis-Pinetum canariensis
Terminal community of the upper-dry Thermo-, and Mesomediterranean subnebulous salic edaphophilous vegetation series of the Gran Canaria *Pinus canariensis* forest (C).

Cisto gomerae-Pinetum canariensis **subass.** ***juniperetosum canariensis***
Terminal community of upper-semiarid and lower-dry Thermomediterranean (wc), and upper-semiarid to subhumid Thermo- and Mesomediterranean (c) salic edaphophilous vegetation series of the La Gomera *Pinus canariensis* forest with *Juniperus turbinata* subsp. *canariensis* (G).

Telinetum spachianae
Terminal community of the dry Meso- and Supramediterranean edaphoxerophilous vegetation series of the *Teline stenopetala* subsp. *spachiana* broom shrubland (T).

Descurainio bourgeauanae-Spartocytisetum supranubii
Terminal community of the dry and subhumid Meso- and Supramediterranean edaphophilous vegetation series of the *Spartocytisus supranubius* and *Descurainia bourgeauana* shrubland. (T)

Adenocarpo spartioidis-Juniperetum cedri
Terminal community of the subhumid-humid Supramediterranean edaphophilous vegetation series on lithosols of the sparse woodland of *Juniperus cedrus* and A*denocarpus viscosus subsp. spartioides*, with *Echium wildpretii* subsp. *trichosiphon* and *Pterocephalus porphyranthus*. (P)

Adenocarpo viscosi-Juniperetum cedri
Terminal community of the subhumid-humid Supramediterranean edaphophilous vegetation series on lithosols of the sparse woodland of *Juniperus cedrus* and A*denocarpus viscosus* subsp. *viscosus*, with *Echium wildpretii* subsp. *wildpretii* and *Pterocephalus lasiospermus*. (T)

Plocametum pendulae
Terminal community of the alluvial arid and lower-semiarid Inframediterranean edaphohygrophilous Western Canary vegetation series in ramblas with the *Plocama pendula* shrubland.

Rubo-Salicetum canariensis
Terminal community of the xeric to pluviseasonal Infra- to Supramediterranean Western Canary edaphohygrophilous and riparian vegetation series of the forest of *Salix canariensis* with *Morella faya* and *Rubus ulmifolius.* (C, T, G, P) (Fig. 6.159)

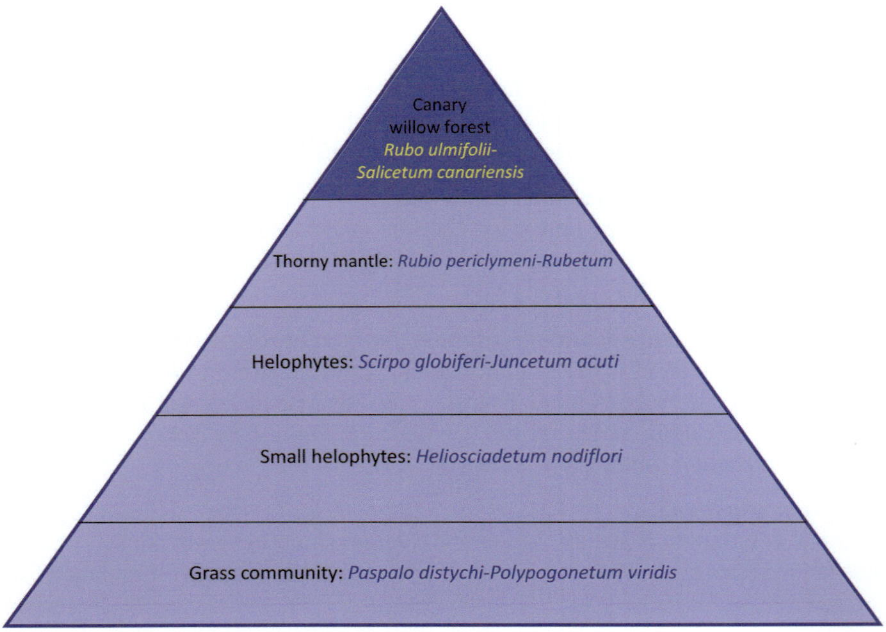

Fig. 6.159 *Rubo ulmifolii-Salici canariensis Sigmetum*

6.3.3 Permanent Communities

In addition to the edaphophilous series, there are permaseries characterized by **permanent communities** without vivacious substitutional stages. They are typical of territories subject to extreme climatic factors and also of other environments such as rocks, benthic areas, saline littoral rocks, saline tidal reaches, mobile littoral dunes, etc. They are the optimum vegetation of those environments. They can be concatenated in geopermaseries according to a gradient of their determining factor or be immersed in vegetation complexes. We also include here some pioneer communities of recent volcanic substrates whose evolution will be very slow and will take many human generations to develop. These often have dual behavior and can be substitutional communities of disturbed potential series.

Marine vegetation
Halophiletum decipientis (C, T)
Zosteretum noltii (L; conservation status critical).
Cymodoceetum nodosae (L, F, C, T, G, H, P)

Marshes
Halimiono portulacoides-Salicornietum perennis (L, F)
Zygophyllo fontanesii-Arthrocnemetum macrostachyi (L, F)

Brackish water
Enteromorpho intestinalis-Ruppietum maritimae (L, F, C, T, C)

Fresh water
Adianto-Hypericetum coadunati (C)
Camptolometum canariensis (C)
Eucladio-Adiantetum capilli-veneris (L, C, T, G, P)
Mentho-Caricetum calderae (T)

Coastal halophilous vegetation
Frankenio ericifoliae-Astydamietum latifoliae (C, T, G, P, H)
Frankenio ericifoliae-Zygophylletum fontanesii (L, F, C, T)

Sandy beaches
Euphorbio paraliae-Cyperetum capitati (L, F, C, T, G)
Salsolo kali-Cakiletum maritimae (L, F, C, T)

Lapilli, volcanic sand and aeolian sand
Cenchro ciliaris-Hyparrhenietum sinaicae subass. *tricholaenetosum teneriffae* (L, F, C, T, G, H)
Echietum auberiani (T)
Echio breviramis-Micromerietum herpyllomorphae (P)
Echio aculeati-Micromerietum hyssopifoliae (H)
Descurainio gilbae-Plantaginetum webbii (P)
Erysimo scoparii-Pterocephaletum lasiospermi (T)
Pelargonium capitatum-Launaea arborescens community (L)
Rumex lunaria community on lapilli (H)

Alluvium
Rumex lunaria alluvial community (P)

Screes
Periploca laevigata community (G, H)
Arrhenathero calderae-Plantaginetum webbii (T)
Violetum cheiranthifoliae (T)

Rocks
Adianto pusilli-Cheilanthetum pulchellae (C, T, G, P, H)
Aeonietum aizoi (T)
Aeonietum aureae (C, T, G, H, P)
Aeonietum aureo-caespitosi (C)
Aeonietum canariensis (T)
Aeonietum christii (H) (=*Aeonietum longithyrsii*)
Aeonietum cuneati (T)
Aeonietum diplocyclae (P)
Aeonietum lancerottensis (L)
Aeonietum lindleyi (T)
Aeonietum christi-davidbramwellii (P)
Aeonietum latifolii (G)
Aeonietum undulato-percarnei (C)
Aeonietum virginii (C)
Aeonietum viscati (G)
Aeonio davidbrramwellii-Ceropegietum hiantis (P)
Aeonio decoris-Atalanthetum pinnati (G)
Aichrysetum immaculati (T)
Aichrysetum pachycauli (F)
Aichrysetum punctati (C, T, G, P, H)
Aichrysetum tortuosii (L)
Cheilantho guanchicae-Aeonietum smithii (T)
Chrysoprenantho pendulae-Atalanthetum pinnatae (C)
Davallio canariensis-Aichrysetum laxi (C, T, G, P, H)
Lolio saxatile-Aichrysetum bethencourtiani (F)
Monanthetum anagensis (T)
Monanthetum pallentis (T)
Parietarietum filamentosae (T, G, P)
Pericallido lanatae-Sonchetum gummiferi (T)
Phyllido viscosae-Aeonietum sedifolii (T)
Reichardio famarae-Helichrysetum gossypini (L)
Soncho-Monanthetum laxiflorae (G)
Soncho congesti-Aeonietum holochrysi (T)
Soncho radicati-Aeonietum tabulaeformis (T)
Soncho hierrensis-Aeonietum diplocyclae (G, H)
Tolpidetum calderae (P)
Umbilico gaditani-Aeonietum urbici (T)
Vieraeo laevigatae-Polycarpaeetum carnosae. (T)

Epiphytic and comophytic
Davallio canariensis-Polypodietum macaronesici (C, T, G, P, H)
Hymenophylletum tunbrigensis (T, G)
Selaginello denticulatae-Anogrammetum leptophyllae (L, F, C, T, G, P, H)

6.3.4 Vegetation Complexes

These are landscape units in which several permanent communities grow together, with scarce or no dynamic interrelations and no association with an ecological gradient.

A good example of a vegetation complex is shown in rocky environments (Fig. 6.160) with several types of substrate (rock, earthy rock, crevices, ledges, etc.), changes in moisture, exposure, etc. resulting in different permanent communities, with a dominance of cryptogamic units.

They are abundant on our islands but only some of them are mentioned in the maps of Potential Natural Vegetation for cliffs and lava-flows.

There several lichen communities are found:

***Xanthorietum resendei* subass. *caloplacetosum gloriae* Llimona 1975**
Saxicolous community typical of dry sunny rocks and lava-flows of the Inframediterranean belt. *Caloplaca gomerana* (crustose thallus) and *Xanthoria resendei* (foliose thallus) are their most characteristic species which provide a strong identifying signal through their orange color.

Fig. 6.160 Vegetation complex developed on cliffs (Añavingo, Arafo, Tenerife). Apart from fragments of potential climatophilous communities (in this case pine forest), it comprises several communities established in different microteselar environments typical for permanent communities

Ramalinetum bourgeanae **Follmann 1976**
Saxicolous community typical of the Inframediterranean belt. It colonizes both pahoehoe lava-flows and volcanic scoria. It requires some moisture and also has greater saline and nitrogen requirements.

The fruticose lichen *Ramalina bourgeauana* is the best representative of the community.

Pertusarietum gallicae **Llimona & Egea 1987**
This is a heliophilous saxicolous community of crustose lichens typical of the lower Inframediterranean territories, with a dominant greenish color.

Pertusaria gallica, Rhizocarpon viridiatrum and *Lecanora sulphurella* are its most characteristic species.

Parmelietum pseudotinctorum **Klement 1965**
This is a coastal community on rocks with faint slopes characterized by *Parmelia tinctorum*, a lichen with a foliose, loosely adnate gray thallus.

Stereocauletum vesuviani **Klement 1965**
This is a lichenic-bryophytic pioneer community on recent aa lava flows and walls subject to the humid NE trade-wind, mainly in the dry, subhumid and humid Thermo- and Mesomediterranean bioclimatic belts.

The fruticose *Stereocaulon vesuvianum* is dominant, and is found together with diverse bryophytes, forming dense whitish carpets.

Roccelletum tinctoreae **Klement 1965** ***(Roccelletum canariensis)***
This is an anemophilous, aero-hygrophilous, saxicolous community typical of vertical walls facing the humid NE trade-wind.

It is characterized by the abundance of fruticose lichens of the genus *Roccella* (*R. canariensis, R. vicentina*, etc.).

Dimelaenion radiatae **Llimona 1965**
This is a phytosociological alliance that includes xerophilic and thermophilic communities of predominantly crustose lichens, typical of the lower parts of the islands. Among other associations it includes the *Xanthorietum resendei* and *Ramalinetum bourgeanae*.

6.3.5 Geosigmeta and Geopermaseries

The geosigmeta (=geoseries) and geopermaseries are communities concatenated along an ecological gradient (Fig. 6.161). Geosigmeta group series, and geopermaseries group permaseries (permanent communities). They are named after the most representative series or permanent community, respectively, modifying the association suffix "*etum*" with a connecting vowel and adding "geosigmetum" or "geopermaseries" at the end of the complete name (e.g. Western Canary edaphohygrophilous geosigmetum of the Canary willow forests: *Rubo ulmifolii-Salici canariensis geosigmetum*; or rocky, saline, coastal, semiarid to dry Inframediterranean geopermaseries of *Astydamia latyfolia* with *Frankenia ericifolia* scrub: *Frankenio-Astydamio latifoliae* geopermaseries.

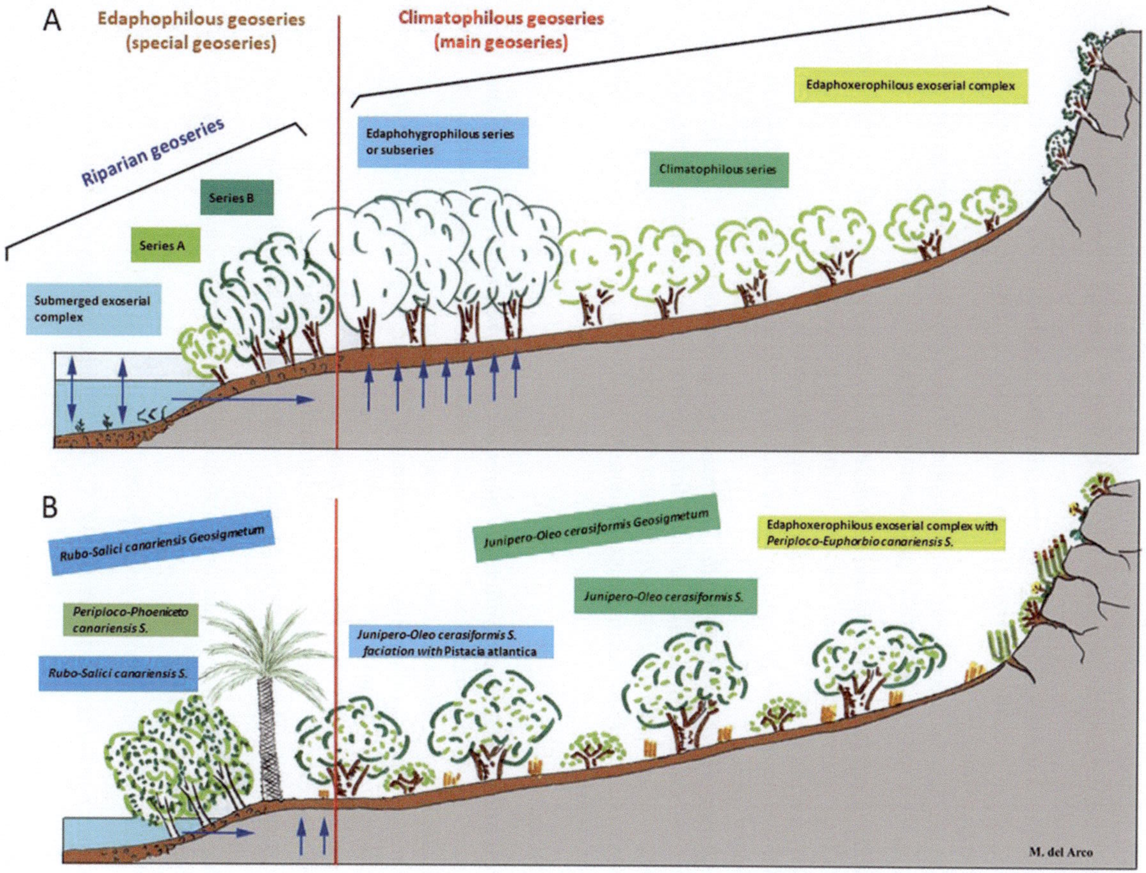

Fig. 6.161 A. General scheme for special and main geosigmeta. B. Examples for Canary Islands. *Rubo ulmifolii-Salici canariensis G.* and *Junipero canariensis-Oleo cerasiformis G.*

Geosigmeta can be climatophilous (main) or edaphophilous (special). Climatophilous geosigmeta are formed from a climatophilous series and adjacent series in the typical geomorphological topo-sequence, valley-slope-crest, within a precise bioclimatic and biogeographic context; thus they include that series, as well as the edaphohygrophilous and the edaphoxerophylous or exoserial edaphophilous complex. Edaphophilous geosigmeta are typical of special environments, such as river-banks.

We have mentioned some geopermaseries and edaphophilous geosigmeta in the legend of our potential vegetation maps for littoral communities, salt marshes, ramblas and ravines.

On saline coastal rocks, we have considered the following geopermaseries, mentioning their most characteristic permanent community:

Frankenio ericifoliae-Astydamio latifoliae geopermaseries (C, T, G, H, P). Rocky, saline, coastal, semiarid to dry Inframediterranean geopermaseries of the *Astydamia latyfolia* with *Frankenia ericifolia* scrub.
Most characteristic permanent community: *Frankenio ericifoliae-Astydamietum latifoliae.*

Frankenio ericifoliae-Zygophyllo fontanesii geopermaseries (T, C, F, L). Rocky, saline, coastal, hyperarid to arid Inframediterranean geopermaseries of the *Frankenia ericifolia* with *Zygophyllum fontanesii* scrub.
Most characteristic permanent community: *Frankenio ericifoliae-Zygophylletum fontanesii.*

In salt marshes:

Halimiono portulacoidis-Salicornio perennis geopermaseries (L, F). Saline, littoral, tidal, hyperarid and arid Inframediterranean geopermaseries of the *Halimione portulacoides* with *Salicornia perennis* scrub (Fig. 6.162).
Most characteristic permanent community: *Halimiono portulacoidis-Salicornietum perennis* scrub.

Zygophyllo fontanesii-Arthrocnemo macrostachyi geopermaserie (L, F, C). Saline, littoral, hyperarid and arid Inframediterranean geopermaseries of the *Zygophyllum fontanesii* with *Arthrocnemum macrostachyum* succulent scrub (Fig. 6.162).
Most characteristic permanent community: *Zygophyllo fontanesii-Arthrocnemetum macrostachyi.*

For ramblas we consider two geosigmeta:

Plocamo pendulae geosigmetum (C, T, G). Alluvial hyperarid to lower-semiarid Inframediterranean geosigmetum of ramblas.

1. *Atriplici ifniensis-Tamarici canariensis sigmetum* (arid and semiarid territories) or *Suaedo verae-Tamaricetum canariensis* (hyperarid or arid territories), on the most saline zone of the rambla.
2. *Plocamo pendulae sigmetum.*
3. Fragments of adjacent climatophilous vegetation (variable depending on the island).

Fig. 6.162 Salt marshses geopermaseries: *Halimiono portulacoidis-Salicornio perennis* and *Zygophyllo fontanesii-Arthrocnemo macrostachyi*

Gran Canaria
Hyperarid faciation:

1. *Suaedo verae-Tamarici canariensis sigmetum*
2. *Plocamo pendulae sigmetum*
3. *Euphorbietum balsamiferae plocametosum pendulae*

Arid faciation:

1. *Atriplici ifniensis-Tamarici canariensis sigmetum*
2. *Plocamo pendulae sigmetum*
3. *Euphorbietum balsamiferae plocametosum pendulae*

Tenerife
1. *Atriplici ifniensis-Tamarici canariensis sigmetum*
2. *Plocamo pendulae sigmetum*
3. *Ceropegio fuscae-Euphorbietum balsamiferae plocametosum pendulae*

La Gomera
Arid faciation:

1. *Atriplici ifniensis-Tamarici canariensis sigmetum*
2. *Plocamo pendulae sigmetum*
3. *Neochamaeleo pulverulentae-Euphorbietum canariensis* facies of *Plocama pendula*

Fig. 6.163 *Suaedo verae-Tamarici canariensis geosigmetum: Suaedo verae-Tamarici canariensis sigmetum*, tamarisk thicket, and *Periploco laevigatae-Phoenico canariensis sigmetum*, palm community

Semiarid faciation:

1. *Atriplici ifniensis-Tamarici canariensis sigmetum*
2. *Plocamo pendulae sigmetum*
3. *Euphorbio berthelotii-canariensis* facies of *Plocama pendula*

Suaedo verae-Tamarici canariensis geosigmetum (F). Saline, hyperarid and arid Inframediterranean geosigmetum of ramblas (Fig. 6.163).

1. *Suaedo verae-Tamarici canariensis sigmetum*. In the most saline zone.
2. *Periploco laevigatae-Phoenico canariensis sigmetum*. In wet area of the rambla with scarce or no salinity.

For ravines we have considered one geosigmetum, pointing out their two representative vegetation series:

Rubo-Salici canariensis geosigmetum (P, G, T, C). Semiarid to subhumid Infra-Thermomediterranean edaphohygrophilous geosigmetum in ravines:

Rubo-Salici canariensis sigmetum. Linked to flowing water or frequent flooding of the watercourse bed and edges. Only with summer water deficit.

Periploco laevigatae-Phoenico canariensis sigmetum. Linked to humid soils of the watercourse edge.

In the Northern half of Gran Canaria we have considered a desertic faciation for the lowest stretch of large ravines which flow from potential areas of laurel forest. This faciation includes *Euphorbietum balsamiferae plocametosum pendulae* as a differential.

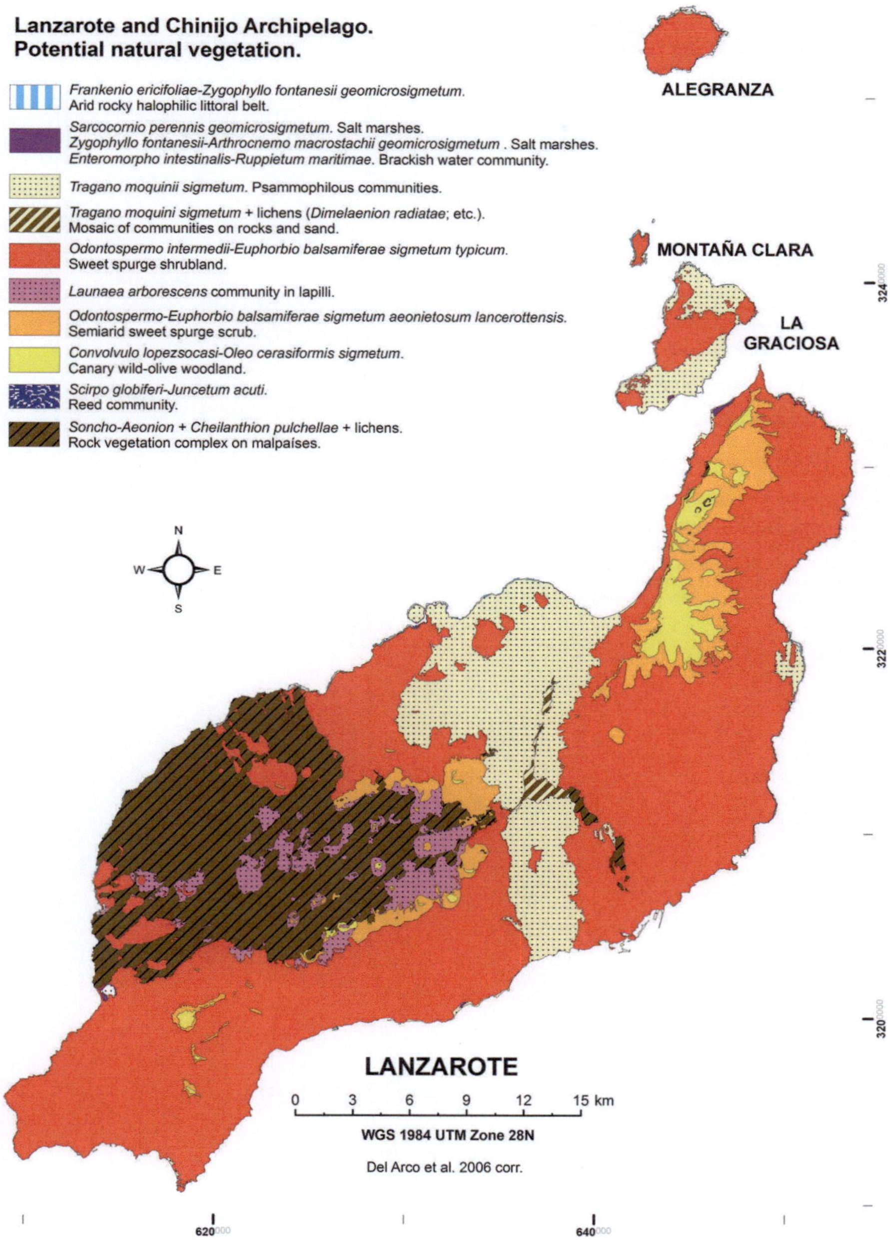

Fig. 6.164 Potential natural vegetation map of Lanzarote and Archipelago Chinijo

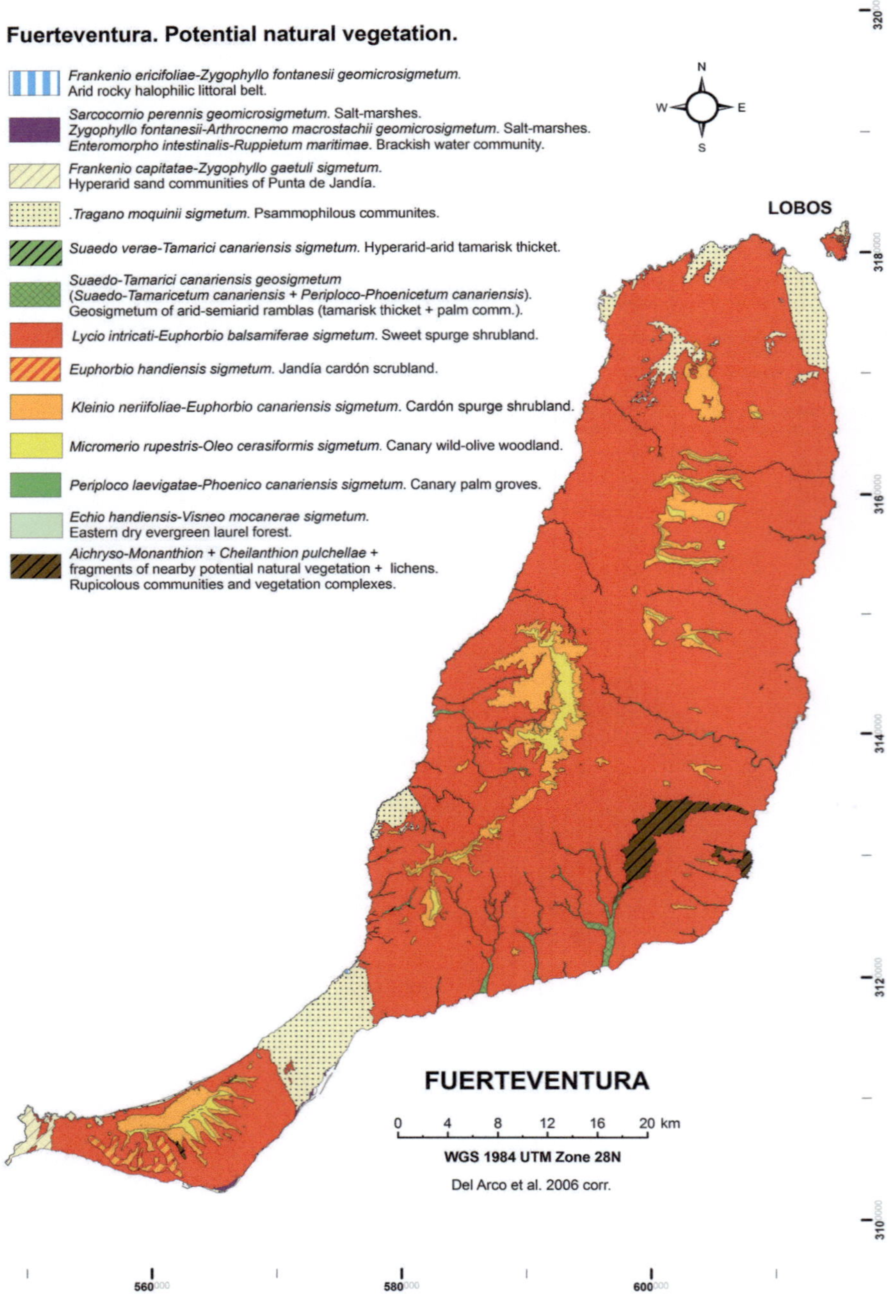

Fig. 6.165 Potential natural vegetation map of Fuerteventura

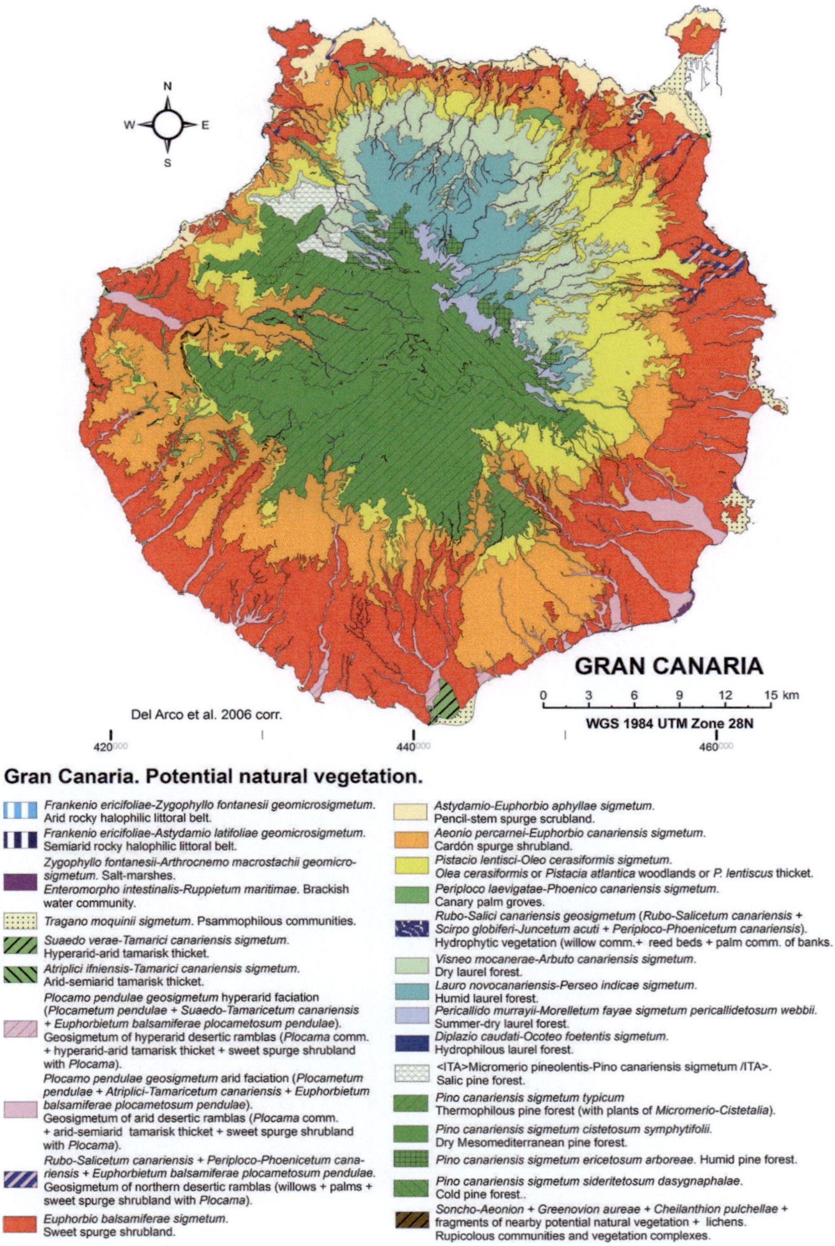

Fig. 6.166 Potential natural vegetation map of Gran Canaria

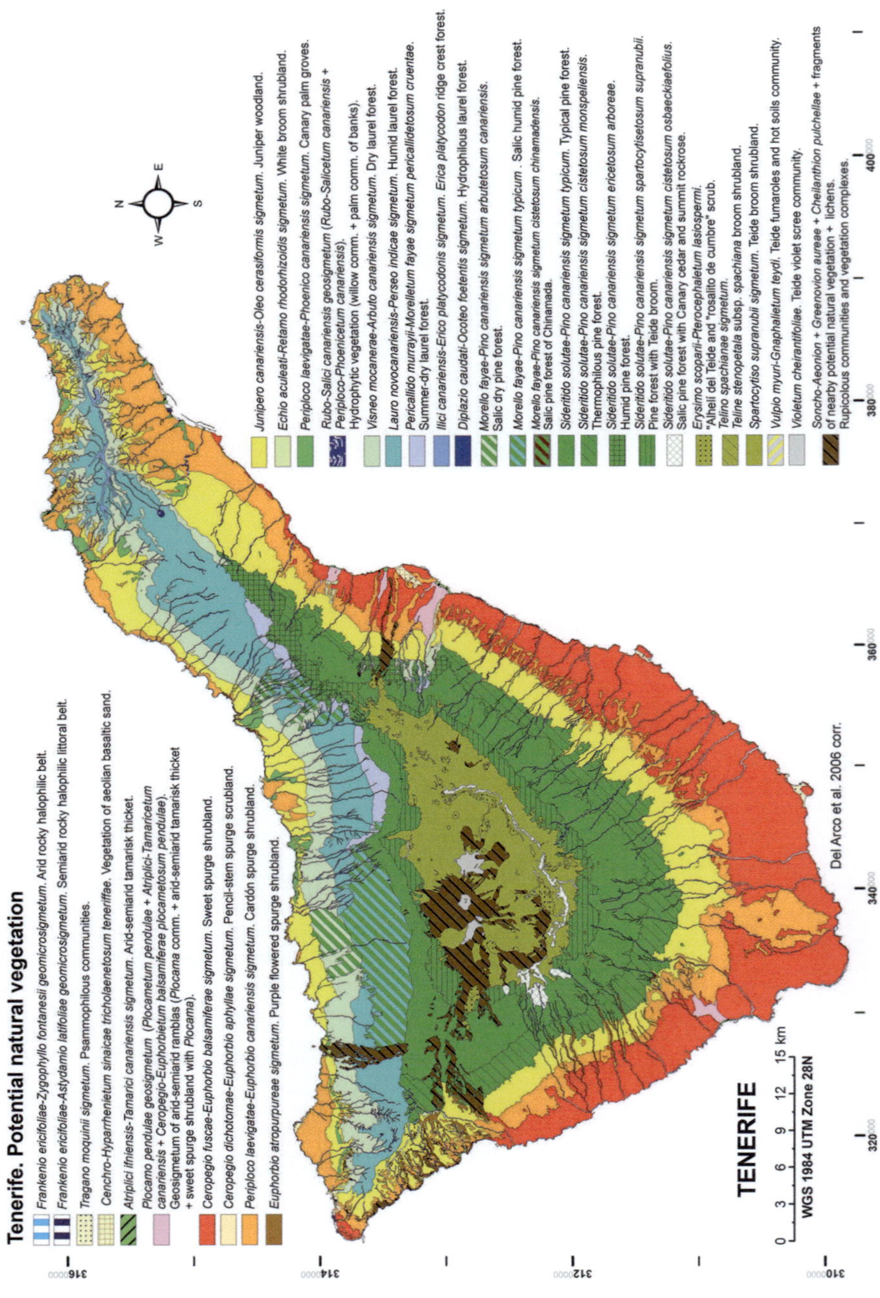

Fig. 6.167 Potential natural vegetation map of Tenerife

Fig. 6.168 Potential natural vegetation map of La Gomera

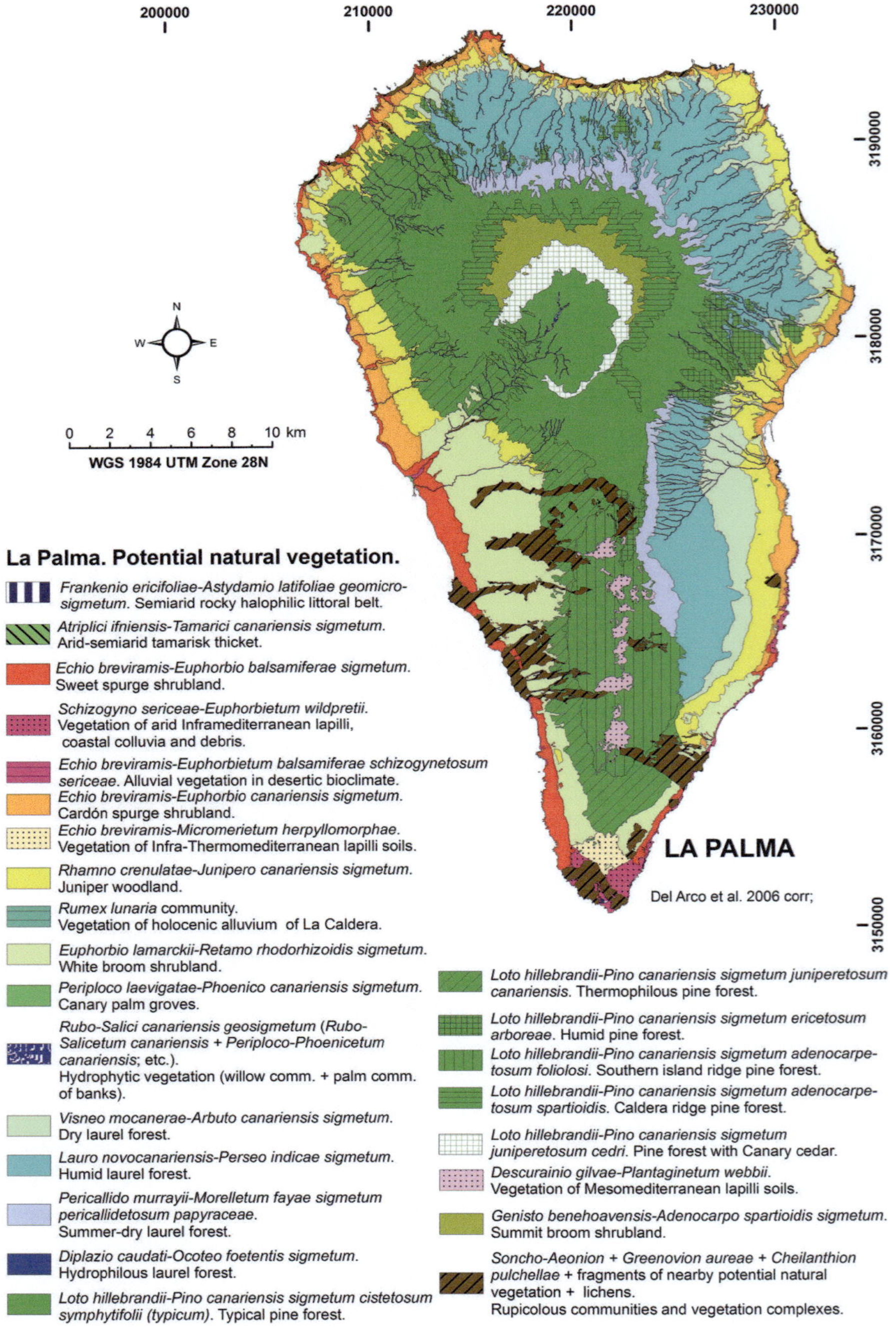

Fig. 6.169 Potential natural vegetation map of La Palma

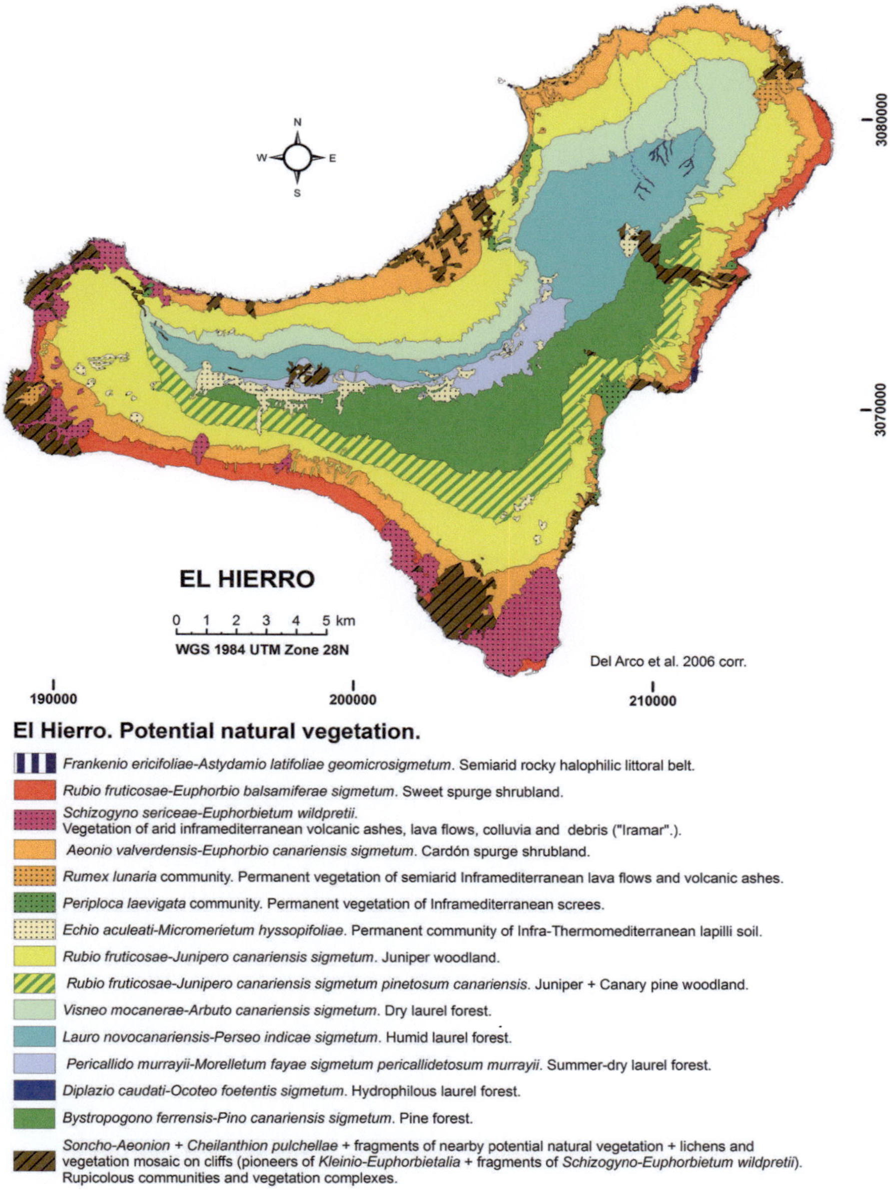

Fig. 6.170 Potential natural vegetation map of El Hierro

Chapter 7
Changes in the Natural Landscape Through Human Influence

The arrival of the human being in the islands caused a gradual alteration of the plant cover. Initially, changes started slowly, because the aboriginal population was relatively scarce and lacked metal tools for massive felling. But changes in the vegetation accelerated after the Castilian conquest, which brought a major population increase, better tools, a greater demand for land and wood, and new large livestock farm animals with their intensive plant consumption.

7.1 Uses and Degradation of the Plant Cover by the Aboriginal Population

The Canary Islands were colonized approximately at the tenth century BCE from North Africa by various Berber peoples. They were the only inhabitants until the Castilian conquest during the fifteenth century AD. Canary Islanders long had different names depending on the islands they lived on: Majes on Fuerteventura and Lanzarote, Canarios on Gran Canaria, Guanches on Tenerife, Gomeritas or Gomeros on La Gomera, Benahoaritas or Auaritas on La Palma, and Bimbaches on El Hierro. More recently, the demonym "Guanches" has become is use for all of them (Fig. 7.1).

These aboriginal islanders made varied use of the island flora, in some cases even of a ritual character. They indeed proclaimed their kings in the shade of a dragon tree (*Dracaena*). For a long time it was believed that the disturbance caused by their actions on the landscape was not very significant, but evidence from archeological sites indicates that they affected it largely negatively, both directly and indirectly.

Among the food resources used were the dates and sap obtained from the local palm tree *Phoenix canariensis*, the fruits of *Visnea mocanera* ("mocán"), *Morella faya* ("faya"), *Arbutus canariensis* (strawberry tree), *Canarina canariensis* ("bicácaro"), *Rubus spp.* (blackberries), pine nuts from *Pinus canariensis*, seeds

M. J. del Arco Aguilar, O. Rodríguez Delgado, *Vegetation of the Canary Islands*, Plant and Vegetation 16, https://doi.org/10.1007/978-3-319-77255-4_7

Fig. 7.1 Guanches kings. Candelaria. Tenerife

from *Cistus spp.* (rockrose), fern rhizomes mainly of *Davallia canariensis*, etc. In their folk medicine, they used the latex of various *Euphorbia spp.*, the protective resin or blood of *Dracaena draco* (dragon tree), the berry-like cones of *Juniperus turbinata* subsp. *canariensis* (Canary juniper), the bark and fruits of *Pistacia atlantica* (mastic tree), the "chercequén", an alcoholic drink made from the fruit of *Visnea mocanera* ("mocán"), the resin of the pine, infusions of *Labiatae*, etc. Wood was used for many purposes: from the palm tree or the drago for making shields and bucklers, drums and even coffins; *Olea cerasiformis* (wild olive tree) for rods of leadership ("añepas") and spears ("banots"); *J. turbinata* subsp. *canariensis* (Canary juniper) for tools, weapons, body ornaments, coffered ceilings in caves, funerary boards as biers ("chajascos"); *V. mocanera* for carved figurines and adornments; *P. canariensis* for weapons or toasted rods from its dark core wood "madera tea", rods of leadership, bowls, torches, funerary boards, pendants, etc.). And they made fabrics with reeds, bulrushes and palm leaves.

However, the most significant damage was due to the intense plant consumption as fodder (Fig. 7.2a). The aboriginals were mainly nomadic pastoral people that according to the seasons of the year moved looking for food for the livestock they had brought from Africa (goats, Pelibuey type sheep and pigs). These animals became very abundant and constituted the basis for their food and clothing. They ate the native species with highest nutritional value, the broom type legumes *Chamaecytisus proliferus* ("tagasaste" and "escobón"), *Genista benehoavensis, Teline spp., Spartocytisus spp.*, *Adenocarpus spp.* ("codesos"), and also *Rumex lunaria* ("vinagrera") and several other woody and herbaceous plants and many seedlings. This went along with deforestation for incipient agriculture, the growing of seeds brought from the neighboring continent (barley, oats, beans, peas, etc.). Other serious damage to the insular forest mass was due to their demand for wood consumption for fuel, especially from the noblest trees of the thermo-sclerophyllous forest, laurisilva or pine forest. In addition, they caused forest fires (Fig. 7.2b) to produce pasture for their flocks and open areas for crop growing. Besides the direct impact on the plants, all these uses also caused, to a greater or lesser extent, erosion of the soils.

Fig. 7.2 (**a**) Goats, that were introduced in the islands, still are a serious threat to autochthonous flora. (**b**) Fires to open up land for grazing were usual among the islands' shepherds. (**c**) Altarpiece. Adeje church. Tenerife. (**d**) Rest of a furnace to obtain pitch from pine wood

Due to these uses, the vegetation cover of the islands was clearly affected. It was more significant on the more arid islands such as Fuerteventura, where tree species of the laurel forest became extinct, such as the strawberry tree, laurel, etc., and the Canary pine and Canary willow of which there is evidence from charcoal found at several sites. The typical plant formations of those species are today absent from that island. The reduction of *Euphorbia balsamifera* scrubland by goats and sheep has also been confirmed in times of forage shortage, and the reduction in *E. canariensis* shrubland through fuel use. To this must be added the nitrification of the territory and the introduction of exotic species, whose diaspores came with people, animals or mixed with seeds for farming.

7.2 Intense Landscape Disturbance and Transformation After the Conquest

The incorporation of the islands to the Castilian Crown was very slow, lasting about a century. It started with Lanzarote in 1402 and finished with Tenerife in 1496. The causes were the lack of funding of the conquerors and the strong resistance offered by the aboriginals on the more populated islands.

After the conquest, the ancient ways of using the native Canarian vegetation continued, but with increased intensity due to the availability of metal tools. Thus, wood was widely used in construction, pine being highly appreciated, but also juniper, wild olive and diverse trees from the laurisilva like *Laurus novocanariensis* – "laurel", *Visnea mocanera* -"mocán", *Picconia excelsa* -"paloblanco", *Erica arborea* -heath, *Morella faya* -"faya", etc., especially for boats, roofs, churches (Fig. 7.2c) and sugar mills, water channels and piping, winepresses, troughs, carts, tillage implements, etc. Thousands of hives were also made with the trunks of palm and dragon trees. A large amount of firewood and charcoal was obtained from pine and from different laurel forest species, as well as from *Hypericum canariensis* ("granadillo"), *Olea cerasiformis* ("acebuche") and several shrubs. Enormous quantities of these were consumed in sugar-mill boilers, earthenware and lime kilns, as well as in household fire places and stoves. *Juniperus cedrus* (Canary cedar), *J. turbinata* subsp. canariensis (Canary juniper), *V. mocanera* ("mocán"), *Apollonias barbujana* ("barbuzano") (the so-called Canary ebony), *Olea cerasiformis* ("acebuche") and *Pistacia atlantica* (mastic tree) were greatly used in furniture work and folk handicrafts. The pine was massively cut down to burn its wood in furnaces to obtain pitch (Fig. 7.2d), of great importance in ship-building and to a lesser extent in construction. Pitch production caused a serious retreat of the pine forests until the end of the seventeenth century. Orchil for purple dyes stood out as a commercial product; it was extracted from *Roccella* lichens and also *Murex* molluscs provided a purple dye.

New ways of using forestry products were developed, such as brushwood and pine litter for livestock bedding and compost. Honey production grew and continues until today. The ornamental use of some plants like palms, dragon tree, *Echium spp.* ("tajinastes"), *Convolvulus floridus* ("guaidil"), etc., and festive decorations from palm leaves, branches of *Morella faya* ("faya") and *Erica*, etc. were promoted. Certain rituals were encouraged, such as bonfires which required large quantities of firewood, or the massive collection of palm leaves and branches of wild olive for the celebration of Palm Sunday. In handicrafts, the palm tree still had a prominent role. From its leaves, several household articles like brooms, mats, baskets, were obtained. Likewise, from thin branches of *Pistacia atlantica* (mastic tree), pipe-stems were made, and spinning tops were made out of its thick roots, for children's games. In addition, indigenous plants continued to be used for gastronomic, medicinal and forage purposes, as during the aboriginal era.

Recently phytochemical studies of the insular flora have been carried out, discovering products of great pharmaceutical interest. As major damage to the vegetation cover, intense forest clearings to convert the land to agricultural and grazing uses should be added to the list. This increased as the population grew, and new population nuclei appeared, initially at mid-elevations and later in the coastal zones. This has been increased because of tourist development in the second half of last century and continues at present. In addition, the frequency of largely intentional wild-fires, despite the adaptation of some species (e.g. Canary pine), has seriously damaged forest formations and favored erosion processes (Fig. 7.3).

Fig. 7.3 (**a**) Devastation after fire in the laurel forest of Garajonay, La Gomera, in 2012. (**b**) Wildfire in the North of Tenerife in 2007. (**c**) *Pinus canariensis* is able to resprout after fire. (**d**) The *Pinus canariensis* afforestation in North Tenerife in the 1940s favored the pine forest, but afforestation task also transformed potential areas of laurel forest extensively planted with *P. radiata*

Likewise, as a result of European colonization, the plant diversity of the archipelago has increased through the voluntary or involuntary introduction of new plants: crops (sugar-cane, chestnut, grapevine, potato, corn (maize), sisal, tobacco, banana, tomato, …); ornamental plants (geraniums, roses, carnations, cacti, oleander, *Ficus microcarpa* (Indian laurel or banyan), some of which have even become touristic symbols of the islands such *Strelitzia reginae* (bird of paradise), native to South Africa, and *Euphorbia pulcherrima* (Poinsettia or Christmas star), a native of Mexico; medicinal plants and spices (rosemary, chamomile, parsley, sage, mints, *Chenopodium ambrosiodes* (epazote). Along with the above herbs many weeds arrived, usually involuntarily, such us *Oxalis pes-caprae* (Bermuda buttercup), a native of South Africa, *Chenopodium spp.* (goosefoots) of worldwide distribution), thistles, etc., which quickly adapted to the new territory. Many plants have become wild, some with great aggressiveness, displacing the native flora, such as *Pennisetum setaceum* (crimson fountain grass), native of Africa and Asia, *Nicotiana glauca* (tree tobacco), native of South America, *Eschscholzia californica* (California poppy), native of the United States and Mexico, *Tropaeolum majus* (Indian cress or nasturtium), native of the Andes, and acacias of tropical and subtropical regions of the World.

In short, during the last six centuries all the Canarian plant formations have been disturbed by human activity (Figs. 7.4, 7.5, 7.6 and 7.7; Tables 7.1 and 7.2). The halophilic vegetation of the coast, settled on rocky or sandy soil, the salt marshes, the tamarisk thickets, as well as *Euphorbia* scrubland and shrubland, have been particularly subjected to strong human pressure. This has been caused particularly by the expansion of intensive crops, like sugar cane, banana and tomato. More recently, particularly during the last seventy years up to the present, this has been succeeded by a largely unbridled tourist and urban development, which also stimulates road building. This has led to a severe transformation of the coastal and lowland plant communities.

The thermo-sclerophyllous and laurel forests were drastically reduced after the conquest of the islands. Apart from timber harvesting for diverse uses, the mid-elevations of the islands were chosen by the new colonists for settlement as principal population nuclei, cultivated lands and grazing space. This was due to their well-watered and -structured soils, besides being away from coastal pirate raids. For such reasons, they constitute today the most degraded Canarian vegetation communities. Throughout history the pine forest has suffered repeated fires and uses (lumber, pitch production, etc.), which have transformed its natural composition and appearance. In addition, reforestations and forest plantations with Canary pine or other pine trees (*Pinus radiata*, *P. halepensis*, etc.) have been established in its potential area and beyond its limits. These have contributed to maintaining a large current wooded area, although sometimes at the expense of other indigenous formations of the islands of great biological interest, such as Canary juniper woodland, laurel forest and summit broom shrubland. The latter has been subject to intense grazing, frequent fires, firewood harvesting and plantations of conifers in its area. Also within its area several buildings have been constructed, notably the IAC (Canary Islands Institute of Astrophysics) observatories, still expanding, and some public facilities (antennas, meteorological stations, restaurants, cable car, etc.). Today this legume shrubland is subject to increasing tourist pressure.

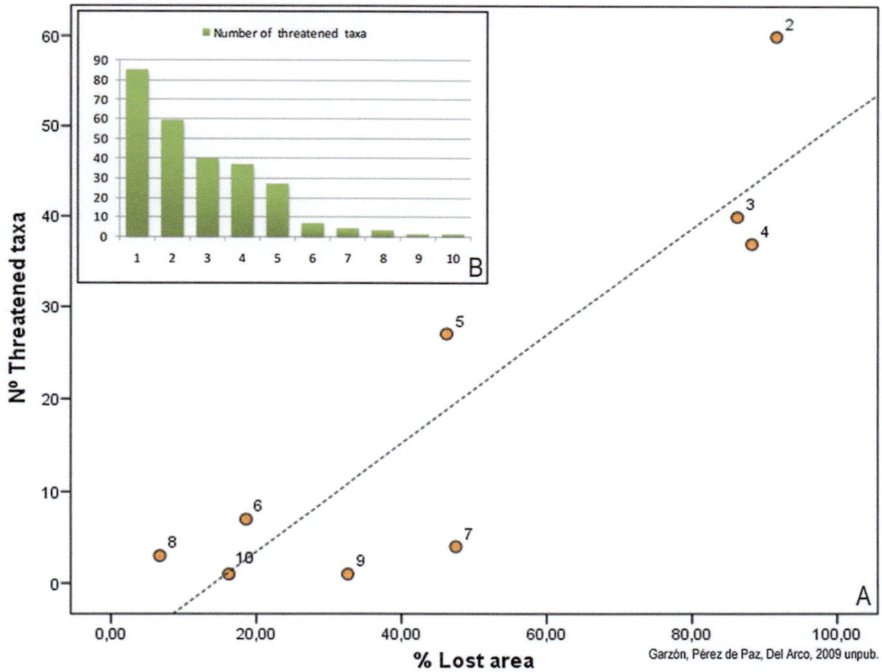

Fig. 7.4 (**a**) Dispersion graph between number of threatened taxa and % lost area of potential natural vegetation (r = 0.82). (**b**) Number of threatened taxa assigned to each of the main potential communities, based on the 127 most threatened plant species selected from several catalogs and red-lists (Communities are non-exclusive for species and therefore do not add up to 127). Communities: *1*. Rupicolous communities and vegetation complexes. *2*. Thermo-sclerophyllous forest. *3*. *Euphorbia* scrublands and shrublands. *4*. Laurel forest (laurisilva). *5*. Canary pine forest. *6*. Rocky halophilic belt. *7*. Sandy beaches and dune vegetation. *8*. Meso-Oromediterranean summit broom shrubland. *9*. Teide violet community. *10*. Coastal salt-marsh vegetation

Table 7.1 Potential cover, current cover and percentage of remaining area of main zonal communities in Canary Islands

Main zonal communities	% Potential cover[a]	% Current cover	% Remaining from potential cover	% Current cover adding main previous substitutional community	% Remaining adding main previous substitutional community
Euphorbia scrubland and shrubland	47,8	6,7	13,8	12,9	27[b]
Thermo-sclerophyllous forest	10,4	0.9	8,4		
Laurel forest	11,7	1.4	11,8	4,1	35[c]
Pine forest	15,2	8,2	53,8	8,4	67[d]
Summit broom shrubland	2,2	2,1	93,3		

Referred to archipelago surface area of 743.611 ha
[a]Potential cover of zonal communities is 87.3%. The remaining 12.7% corresponds to azonal communities
[b]Adding secondary *Euphorbia* shrubland
[c]Adding secondary *Morello fayae-Ericetum arboreae*
[d]Adding Canary pine reforestation areas

Table 7.2 Potential cover, current cover and percentage of remaining area of main zonal communities in the Canary Islands

	% Potential cover	% Current cover	% Remaining from potential cover
Forest	39	14	36
Scrubland and shrublands	61	61[a]	26[b]

Referred to archipelago surface area of 743.611 ha
[a]74% of them are secondary
[b]only potential scrubland and shrublands

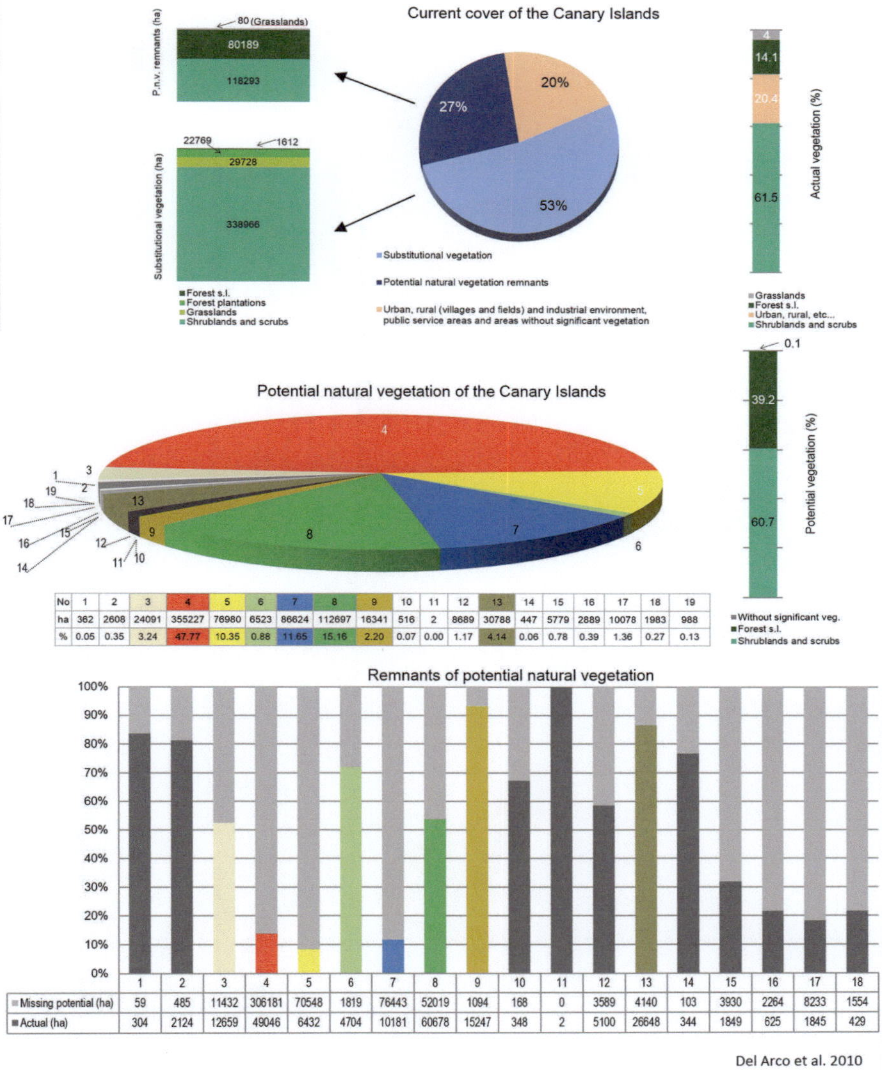

No	1	2	3	4	5	6	7	8	9	10	11	12	13	14	15	16	17	18	19
ha	362	2608	24091	355227	76980	6523	86624	112697	16341	516	2	8689	30788	447	5779	2889	10078	1983	988
%	0.05	0.35	3.24	47.77	10.35	0.88	11.65	15.16	2.20	0.07	0.00	1.17	4.14	0.06	0.78	0.39	1.36	0.27	0.13

	1	2	3	4	5	6	7	8	9	10	11	12	13	14	15	16	17	18
Missing potential (ha)	59	485	11432	306181	70548	1819	76443	52019	1094	168	0	3589	4140	103	3930	2264	8233	1554
Actual (ha)	304	2124	12659	49046	6432	4704	10181	60678	15247	348	2	5100	26648	344	1849	625	1845	429

Fig. 7.5 Vegetation cover of the Canary Islands. *1*: Brackish ponds and coastal salt-marsh vegetation; *2*: Rocky coastal halophilic belt. *3*: Sandy beaches and dune vegetation; *4*: *Euphorbia* scrub and shrublands; *5*: Thermo-sclerophyllous forest; *6*: White broom shrubland (*Retametum rhodorhizoidis*); 7: Evergreen laurel forest (laurisilva); *8*: Canary pine forest (*Pinus canariensis* communities); *9*: Meso-oromediterranean summit broom shrubland; *10*: Teide violet community (*Violetum cheiranthifolii*); *11*: Fumarole community (*Vulpio myuri-Gnaphalietum teydei*); *12*: Permanent vegetation of basaltic sands, lapilli, malpaíses, alluvial soils, screes and debris; *13*: Rupicolous communities and vegetation complexes; *14:* Mosaics; *15*: Alluvial inframediterranean shrubby vegetation (*Plocametum pendulae*); *16*: Edaphohaline tamarisk ticket (*Tamariceta canariensis*); *17*: Canary palm community (*Periploco laevigatae-Phoenicetum canariensis*; *18*: Canary willow community (*Rubo-Salicetum canariensis*); *19*: Areas without significant vegetation

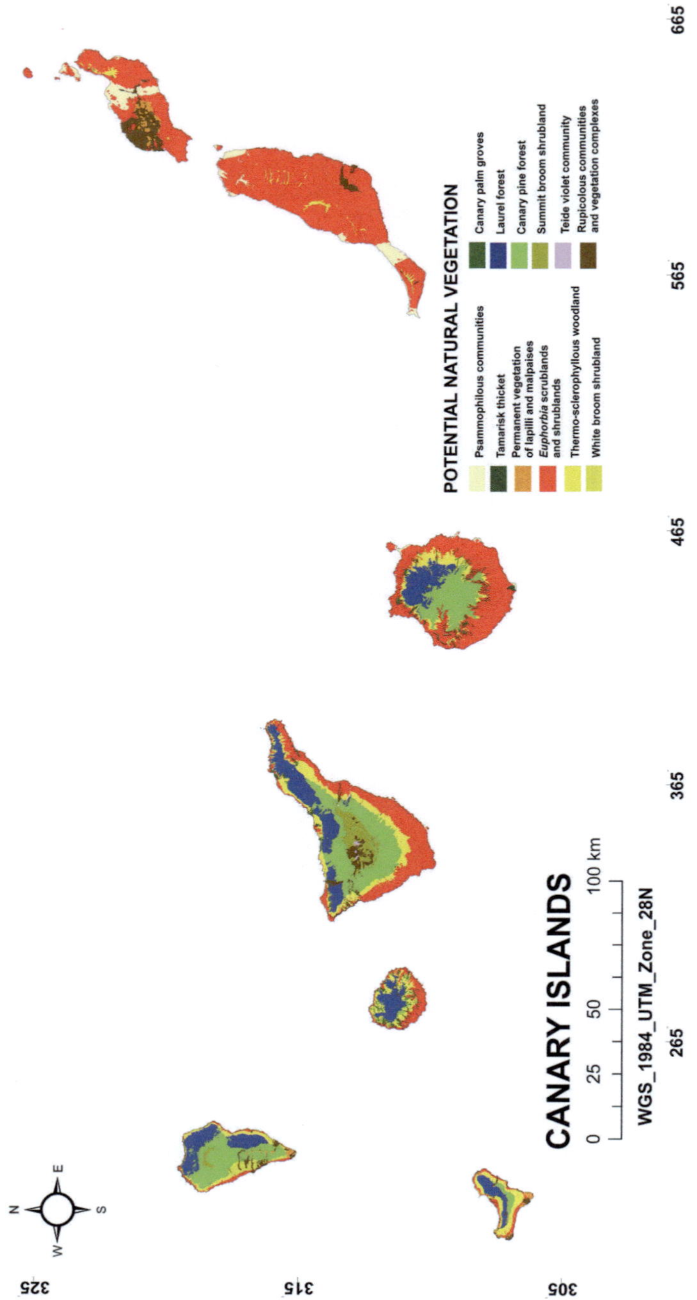

Fig. 7.6 Distribution map of the main potential natural communities of the Canaries

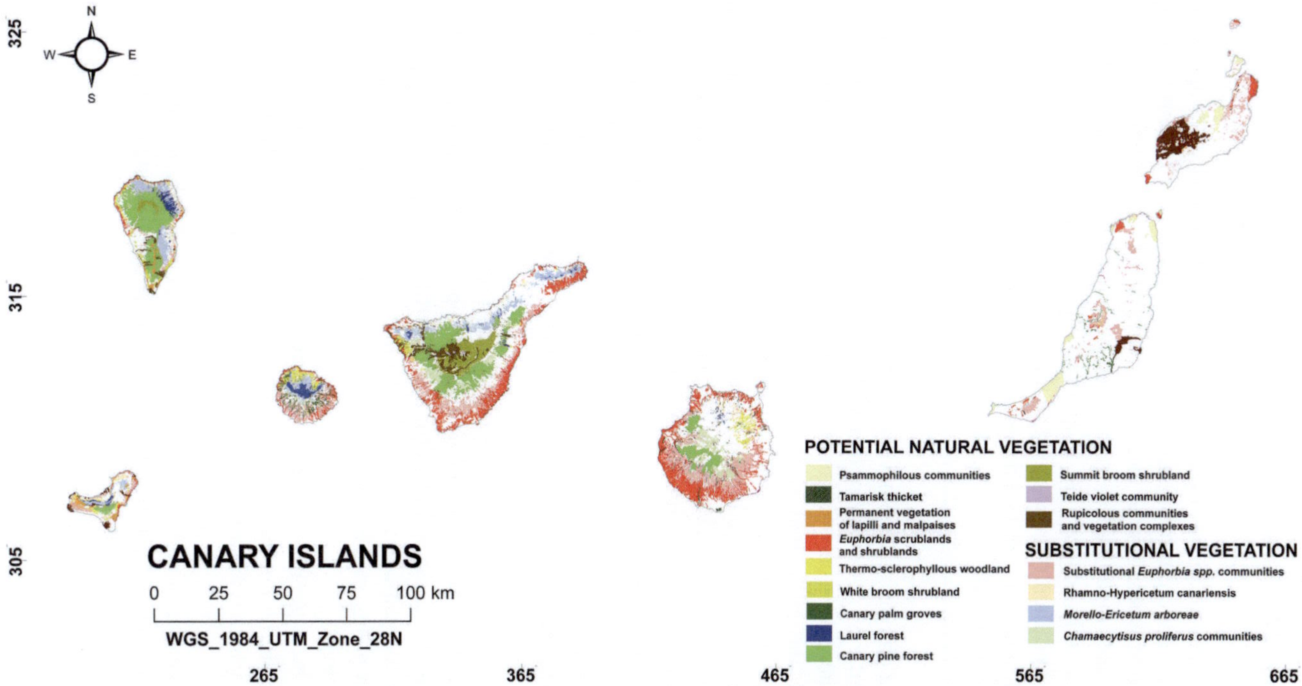

Fig. 7.7 Map showing the remnants of the main potential natural communities, adding the main substitutional communities of potential *Euphorbia* scrubland and shrubland -substitutional *Euphorbia spp.* communities, potential thermo-sclerophyllous woodland -*Rhamno-Hypericetum canariensis*, potential laurel forest -*Morello-Ericetum arboreae*, and potential pine forest –*Chamaecytisus proliferus* communities

7.3 Current Climate Change

7.3.1 General Comments

The current climate change is directly affecting plant biodiversity through increased temperature and reduced rainfall, and their consequences. The increase in temperature, which at first would lead to increasing plant productivity, implies a substantial change that causes modification in the species' phenological cycles. This will be followed by possibly adaptations, migrations, changes in species interactions, or extinction due to difficulty in adapting to the new conditions.

The rainfall reduction causes a decrease in productivity and will negatively affect reproduction and recruitment rates, also causing weakness. These effects may lead to an increased incidence of diseases and pests, and to the exclusion of less tolerant species.

This current climate change is very fast and therefore greatly affects long-lived species with little evolutionary potential. It is to be expected that many plants do not have the phenotypic capacity to resist such changes. The main response will be the latitudinal and altitudinal movements of vegetation. The variable speed of movement depends on the migratory capacity of the species, its degree of tolerance to change, longevity, production and type of diaspore, competitiveness, habitat loss, degree of habitat fragmentation, etc. The species with greater genetic diversity, widespread occurrence and with a greater range of habitats, will be favored. As said, it is to be expected that many species will not be able to withstand the change, as most of their migratory capacities will be exceeded. Under the present climate change, changes in interspecific interactions are occurring, highlighting changes in competitive relations, in phenological asynchronies, pests and invasions.

Soil changes due to a reduction of organic carbon in the soil are expected, as a result of a faster mineralization due to higher temperatures. Also, higher incidence of fires, increased erosion, salinization and desertification processes are foreseen. This will be accompanied by changes in the floristic composition, favorable to the development of plants and communities that are already adapted to the new range of climatic conditions.

There will be a greater incidence of extreme events, such as violent storms (Fig. 7.8) in territories not regularly subjected to them, dry periods at unusual times, or heat waves, which will accelerate the exclusion of species that are not tolerant to them.

The thermal expansion of the oceans and the melting of the ice will cause a sea level increase that will affect the coastal strip, its geomorphological processes and its associated biological systems.

Fig. 7.8 Damages caused by tropical storm Delta (November 28th–29th, 2005) in *Euphorbia* shrubland (**a**), laurel forest (**b**), and pine forest (**c**). (**a**) Malpaís de Güímar, Tenerife; (**b**) Vueltas de Taganana, Tenerife; (**c**) La Cumbrecita, La Palma

Foreseeable trends show a structural simplification of the vegetation and a predominance of local extinctions with the accompanying recolonizations by relatively widespread tolerant species.

7.3.2 Canary Islands

In the Canary Islands, it has been observed that between 1940 and 1970 there has been an increase in the average temperature of 0.09 ± 0.04 °C per decade, and that increased to 0.17 ± 0.09 °C per decade since then, demonstrating the acceleration of the process. These values are higher in the high mountain zone, especially regarding the minimum temperatures. A decrease in precipitation is detected, mainly affecting the northern slopes, and there is an increased frequency of the dry easterly winds and heat waves.

The possible effects of climate change on the Canary Islands flora and vegetation for the period 2070–2100 have been analyzed through bioclimatic studies. As a result of the predictable rise in temperatures and estimated decrease in precipitation, the bioclimatic belts already tend towards an altitudinal ascent of their thermotypes and an aridization of their ombrotypes. This would cause a displacement of the different types of potential climatophilous vegetation associated with those belts, with a reduction in their surface area except for the typical communities of the most arid belts and the pine forest (Fig. 7.9). The estimated rapid time lapse for these climatic changes does not seem long enough to be followed by a parallel altitudinal displacement of the climatophilous vegetation linked to the different bioclimatic belts or by the rest of the potential vegetation, because the vegetation has a slower response to change. This also applies to the flora that constitutes the vegetation, and thus a loss of floristic biodiversity is inevitable.

The territory previously occupied, the evolutionary potential of the species and their dispersive capability, as well as the new types of biological interactions, will be among the factors that will determine the dynamics of the response to the new situation. In any case, the species of these communities with easily dispersible diaspores, and the substitutional scrub and shrublands that are much more aggressive than the potential communities, will be the first to be favored in altitudinal displacement.

In general, the surface area of the Inframediterranean thermotype will increase considerably, although there will be a loss in area at its lower limit due to the rising sea level. A slight ombrotypic variation towards aridity is also predicted for the Inframediterranean. This will extend the potential area of *Euphorbia balsamifera* scrub of the hyper-arid and arid Inframediterranean belts, and especially that of *Euphorbia canariensis* shrubland, typical of the lower semiarid Inframediterranean and their edapho-xerophilous projections into the upper belts. Thus, the shrublands of *Euphorbia lamarckii*, *E. berthelotii, Launaea arborescens, Schizogyne sericea* and *Rumex lunaria*, among others, could be the main protagonists in the colonization of new territories. In addition, the grassland of *Hyparrhenia sinaica* and *Cenchrus ciliaris* can widely expand, but also some invaders such as *Pennisetum setaceum* will increase their distribution area.

The upper semiarid Inframediterranean territory of the North will enlarge slightly. Territory loss to the lower-semiarid Thermomediterranean in the South will be compensated by the persistence of the upper-semiarid and by the widening of the dry-Thermomediterranean. These territories are potential thermo-sclerophyllous woodlands, although a portion of the dry- Thermomediterranean is shared with pine forest. This will lead to a slight reduction in the potential area of *Juniperus turbinate* subsp. *canariensis* woodland, *Olea cerasiformis* woodland, *Pistacia atlantica* woodland or *Pistacia lentiscus* thicket. All these communities are currently poorly represented and have very scarce recovery capacities due to deforestation, fragmentation, urban expansion, and consolidated agricultural use of their territory.

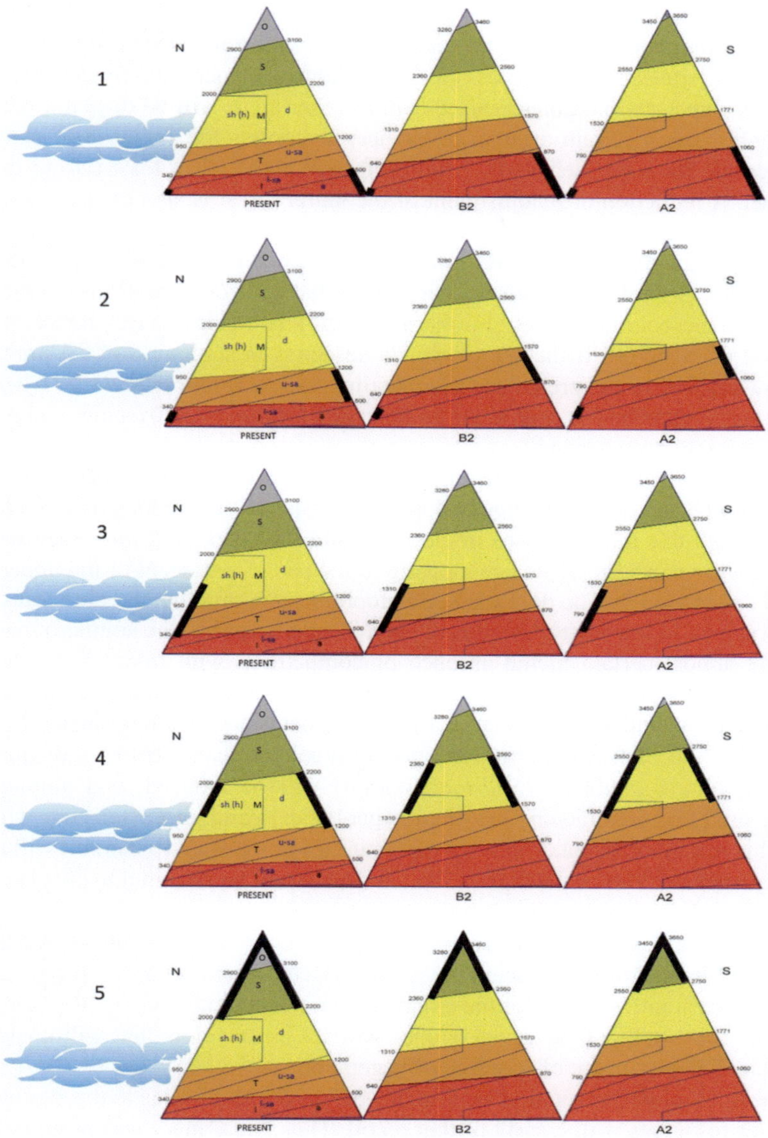

Fig. 7.9 Shift in the main climatophilous communities because of climatic change in three different scenarios in the ECCE project (De Castro et al. 2005): Present, B2 ([CO_2] = 760 ppm), and A2 ([CO_2] = 850; supposed in 2100). B_2 predicts an average annual temperature increase of 1.5 °C at the northernmost parts of the archipelago (Lanzarote) and 2 °C at the southernmost (Gran Canaria), while A_2 predicts an increase of 2.3 °C and 2.9 °C, respectively. Climatophilous communities (black lines): *1*. *Euphorbia* scrub and shrubland; *2*. Thermo-sclerophyllous woodland; 3. Laurel forest; *4*. Pine forest; *5*. Summit broom shrubland. Thermotypes (*I* Inframediterranean, *T* Thermomediterranean, *M* Mesomediterranean, *S* Supramediterranean, *O*: Oromediterranean). Ombrotypes (*a* arid, *l-sa* lower semiarid, *u-sa* upper semiarid, *d* dry, *sh* subhumid, *h* humid)

The expected drift in the Azores high pressure nucleus and other climatic data suggest a reduction in the cloudy area on the windward slopes under Infra- to Mesomediterranean thermotypes. This will lead to a reduction in the potential area of laurel forest. Its possible displacement to other areas will be difficult, because, although it has a certain drought resistance, it requires the presence of clouds to compensate its water deficit, by limiting the evaporation, at least during the dry summer. A reduction or readjustment of the laurel forest is also predictable, if the expected shrinkage of the sea of clouds is combined with a rising temperature. It will favor the thermophilic laurel forest (of warmer and dry places), and reduce the mesophile types (of cooler and moister places) that will continue to grow at the most favorable mesoclimatic sites. Furthermore, the crest-ridge laurel forest will be reduced or displaced if the sea of clouds descends in summer and the prevailing trade-winds weaken as predicted. The substitutional *Rhamnus crenulata-Hypericum canariense* shrubland at mid-altitudes and the *Morello faya-Ericetum canariensis* above it might both expand.

The Mesomediterranean territory of the North outside the influence of trade-wind clouds, and that of the south slope of the islands may reduce in surface area, even though this territory rises in elevation and its lower limit may even become lower on the northern slope, again owing to the possible descent of the upper limit of the sea of clouds. The dominant climatophilous community, the pine forest, can easily spread in the territories gained; even its persistence in the southern lowlands will be almost certain in the absence of competition with trees of the thermo-sclerophyllous woodlands. The *Chamaecytisus proliferus* subsp. *angustifolius* community can expand, especially in rocky environments and at sunny spots.

Finally, the Supramediterranean and Oromediterranean territories of potential summit broom shrubland (*Spartocytisetalia*) will be reduced, and although its ascent might be possible on Tenerife, the geological youth of the territory will prevent this to a great extent. The scree communities of the island's summit will not be greatly affected. Part of the high mountain flora seems endangered by displacement limitations due to the limited altitude of the islands.

The rupicolous vegetation, adapted to the drought and particular stress conditions of its habitat, does not seem especially affected by this climate change, except for the communities typical of the cloud zone in the potential area of the laurel forest. The rock walls of the ravine slopes will act as corridors that will enable the altitudinal displacement of this type of vegetation.

The hydrophytic vegetation, now in decline on the islands due to the exploitation of water resources, will clearly further recoil. Thus, the Canary willow community and Canary palm groves will be seriously damaged.

The coastal vegetation of dunes, beaches, cliffs and salt marshes will undergo major changes, due to the expected sea level rise, which will lead to a great remodeling of the coast line. The current rise of the sea around the islands is almost one cm per year. The coastal vegetation will be even more restrictedly conditioned due to human impact. At present the coasts of the islands are already largely transformed, mainly by tourist and residential development.

Chapter 8
Conservation Status of the Canarian Flora and Vegetation

After the conquest, the increasing disturbance of the island ecosystems, and their fragility, forced the authorities to implement conservation measures to protect them. Thus, since the sixteenth century, provisions were enacted and sometimes implemented to control the destruction of forests and water sources. In later centuries, reforestation was regulated in the most sensitive areas, to avoid damage that could have a negative impact on the island's economy and development. However, it would not be until the mid-twentieth century when conservation was taken seriously from an ecological point of view.

8.1 Protection of Natural Reserves

Effective regulations for the conservation of nature in the Canaries began in 1954, with the legal constitution of the Teide National Park (NP) (Tenerife) (Fig. 8.1) and the Caldera de Taburiente NP (La Palma). They were followed by the Timanfaya NP (Lanzarote) in 1974 and the Garajonay NP (La Gomera) in 1981 (Fig. 8.2). Nowadays the Canary Islands is the Spanish Autonomous Community with the greatest number of NPs. They host representative samples of the high mountain vegetation of Tenerife, broom shrubland and pine forest of La Palma, pioneer lichen communities of recent lava flows and *Euphorbia balsamifera* scrub of Lanzarote, and laurel forest of La Gomera. Since the enactment of its constitution laws, there have been several redrawings of their bounderies, in some cases with enlargement of their area, as well as approval of use and management plans.

M. J. del Arco Aguilar, O. Rodríguez Delgado, *Vegetation of the Canary Islands*,
Plant and Vegetation 16, https://doi.org/10.1007/978-3-319-77255-4_8

Fig. 8.1 Teide National Park. Tenerife. Together with Caldera de Taburiente were the first National Parks of the Canaries. Teide NP contains the most complete collection of communities of the high mountain vegetation of the islands, as well as numerous endemic species

Fig. 8.2 Garajonay National Park, on La Gomera, hosts the most important reserve of laurisilva of the Macaronesia

Without forgetting the importance of the national legislation concerning natural heritage (laws governing land usage, forests, coasts, etc.), a peak event in the protection of the islands was the approval of the Declaration of Natural Reserves of the Canary Islands in 1987.[1] This allowed the creation of 34 natural parks and 70

[1] LEY 12/1987, de 19 de junio, de declaración de Espacios Naturales de Canarias. BOC 085 (1987/07/01). http://www.gobiernodecanarias.org/boc/1987/085/001.html

natural sites, which were added to the 4 existing NPs. These included the best representations of the original plant formations that characterize the landscape of the Canaries. Later, after the enactment of the new law of Natural Reserves of the Canaries in 1994,[2] the mentioned reserves were integrated into the Canarian Network of Protected Natural Reserves (Fig. 8.3).This network includes the protection of 145 natural areas, which occupy 40.4% of the surface area of the territory: 4 NPs, 11 natural parks, 7 rural parks (Fig. 8.4), 11 integral natural reserves, 15 special natural reserves, 19 sites of scientific interest, 51 natural monuments and 27 protected landscapes. All of them were incorporated into the new, reformulated text of the laws of Ordinance of the Territory and Natural Areas of the Canary Islands, in 2000.[3]

Subsequently, at the end of 2009,[4] 177 Sites of Community Importance (SCIs) proposed by the Autonomous Community of the Canaries were declared as Special Areas of Conservation (SAC) (under the European Union's Habitat Directive, by the Spanish State. To these were added the Special Protection Areas for birds (SPAs). As stated in the text, it should be noted that the land-based SACs are 89% in agreement with the Protected Natural Areas provided for in the above "reformulated text". Thus, they now have the protection measures specified in the planning regulations of these areas, in addition to those established in the plans for the recovery or conservation of species. For those Sites of Community Importance (SCIs) not coinciding geographically with the Protected Natural Areas of the Canary Network, this decree provides specific conservation regulations to complement their protection measures.

Other protected entities of relevant international prestige granted by UNESCO exist. The following seven Biosphere Reserves have been established: La Palma (1983), Lanzarote (1993), El Hierro (2000), Gran Canaria – 46% of the island- (2005), Fuerteventura (2009), La Gomera (2012) and the Anaga Massif on Tenerife (2015). In addition, two Natural World Heritage sites, Garajonay National Park (1986) and Teide National Park (2007), have been declared.

[2] LEY 12/1994, de 19 de diciembre, de Espacios Naturales de Canarias. BOC 157 (1994/12/24). http://www.gobiernodecanarias.org/boc/1994/157/001.html

[3] Decreto 1/2000, de 8 de mayo, por el que se aprueba el Texto Refundido de las Leyes de Ordenación del Territorio de Canarias y de Espacios Naturales de Canarias. BOC 60 (2000/05/15). https://www.boe.es/buscar/pdf/2000/BOC-j-2000-90006-consolidado.pdf

[4] DECRETO 174/2009, de 29 de diciembre, por el que se declaran Zonas Especiales de Conservación integrantes de la Red Natura 2000 en Canarias y medidas para el mantenimiento en un estado de conservación favorable de estos espacios naturales. BOC 007 (2010/01/13). http://www.gobcan.es/boc/2010/007/001.html

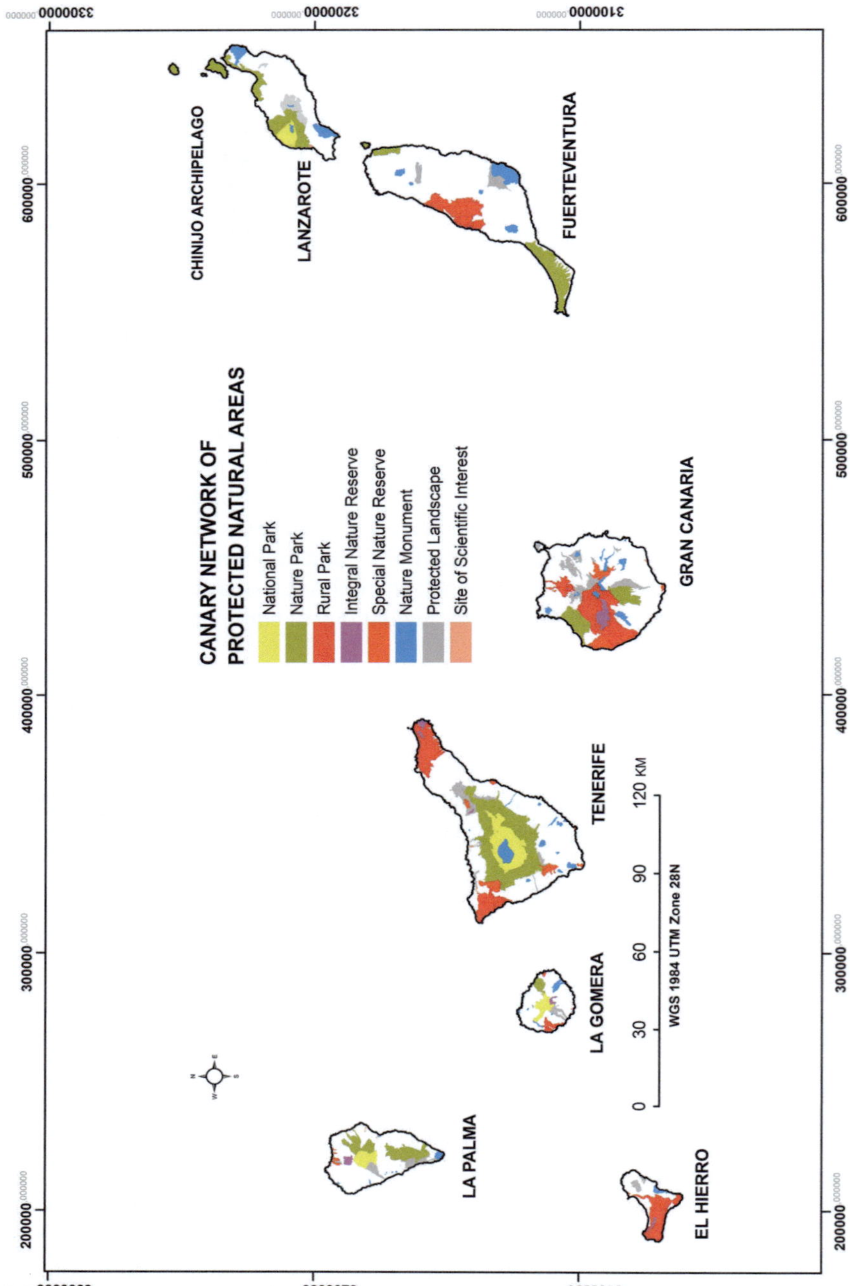

Fig. 8.3 Canary network of protected natural areas

Fig. 8.4 In the Rural Parks the agricultural, livestock or fishing activities coexist, with others of special natural and ecological interest. Anaga Rural Park, in Tenerife, is a large reserve that includes the old Massif of Anaga. It host good samples of *Euphorbia* scrub and shrublands, laurel forest, relics of thermo-sclerophyllous forest, and rupicolous communities, among other botanical and geological treasures

8.2 Protection of Species and Habitats

With the focus on vegetation, the main protection measures derive from the EE Council Directive 92/43/EEC (21 May 1992) on the conservation of natural habitats, wild fauna and flora (known as the Habitats Directive). At that time, a detailed study and mapping of the natural habitats of the archipelago was carried out, mostly characterized by plant communities delimited according to phytosociological methodology. The annexes to the Directive included 24 natural habitats of Community interest (4 marine and 20 terrestrial), 7 of which were priority areas. A high percentage of them were included in the Natura 2000 network. This allowed for effective protection, since it provided a significant financial investment in the Canaries for the management plans of these areas, contributing greatly to the conservation of their natural habitats and species. Also, the annexes included 66 Canary Islands species of Community interest (60 spermatophytes, 4 ferns and 2 mosses), among which 35 were given priority.

The Canary flora knows other protection measures. In addition to inclusion in several red books or red lists, many native plants are considered in international conventions for conservation and protection of species. Annex I of the Berne Convention on the Conservation of European Wildlife and Natural Habitats includes 104 native vascular plants (98 endemic and 6 non-endemic), plus one alga, and four bryophytes. The Convention on International Trade in Endangered Species of Wild Fauna and Flora (also known as the Washington Convention or CITES) includes 22 species of *Orchidaceae* and succulent species of *Ceropegia* and *Euphorbia*.

Also, there is a legal regulation with a regional scope. The annexes to the Order of the Territorial Policy Council of 1991[5] about protection of wild species of vascular flora in the Autonomous Community of the Canaries included those endemic or autochthonous taxa considered threatened to a greater or lesser extent. 116 taxa in Annex I, 271 taxa in Annex II, apart from all ferns (except *Pteridium aquilinum*) and all endemic taxa of the genus *Aeonium, Greenovia, Monanthes* and *Euphorbia* (except for *E. obtusifolia*), and 17 in Annex III. In addition, 300 vascular plants were recorded in a red book[6] in 1996, mostly insular endemics, applying threat categories established by IUCN.

However, the most important protection legislation began with the national 1989[7] Law of Conservation of Natural Reserves and Wild Flora and Fauna. This created a national catalog of threatened species that later incorporated 1 fern and 69 flowering plants in the category of "endangered" for the Canaries, along with another fern in the category "sensitive to habitat disturbance". Then, through a Canary decree in 2001,[8] a catalog of threatened species of the Canary Islands was approved, under the 1989 law, including 2 ferns and 72 spermatophytes "in danger of extinction"; 3 algae, 6 ferns and 125 spermatophytes "sensitive to habitat disturbance"; 7 algae and 12 spermatophytes "vulnerable"; and 5 algae, 6 ferns and 24 spermatophytes of "special interest".

The following year, the Government of the Canary Islands launched the Program for the Monitoring of Threatened Species ("SEGA"), in order to obtain updated accurate information on the conservation status of the endangered species of the archipelago. Since then, numerous terrestrial or marine species or island populations from all over the islands have been studied. Later, through a 2005 legal order the criteria were determined to govern the evaluation of endangered species of wild flora and fauna. Since then, plans for the recovery of endangered species, and for conservation of "vulnerable" species, as well as some for conservation of the habitat of species "susceptible to habitat disturbance ", have being developed and carried out.

The Law of Natural Heritage and Biodiversity of 2007[9] repealed the previous Law and created the List of Wild Species under Special Protection Regime and the Spanish Catalog of Threatened Species, which were detailed in a decree in 2011.[10] It included 133 plant taxa for the Canaries: 70 "in danger of extinction" (2 ferns and 68

---

[5] ORDEN de 20 de febrero de 1991, de la Consejería de Política Territorial, sobre protección de especies de la flora vascular silvestre de la Comunidad Autónoma de Canarias. BOC 35 (1991/03/18). http://www.gobiernodecanarias.org/libroazul/pdf/6359.pdf

[6] GÓMEZ CAMPO, C. & col. 1996. Libro Rojo de Especies Vegetales Amenazadas de las Islas Canarias. Viceconsejería de Medio Ambiente, Consejería de Política Territorial, Gobierno de Canarias. 672 pp.

[7] Ley 4/1989, de 27 de marzo, de Conservación de los Espacios Naturales y de la Flora y Fauna Silvestres. BOE 74 (1989/03/28). https://www.boe.es/boe/dias/1989/03/28/pdfs/A08262-08269.pdf

[8] DECRETO 151/2001, de 23 de julio, por el que se crea el Catálogo de Especies Amenazadas de Canarias. BOC 97 (2001/08/01). http://www.gobcan.es/boc/2001/097/boc-2001-097-004.pdf

[9] Ley 42/2007, de 13 de diciembre, del Patrimonio Natural y de la Biodiversidad. BOE 299 (2007/12/14). https://www.boe.es/buscar/pdf/2007/BOE-A-2007-21490-consolidado.pdf

[10] Real Decreto 139/2011, de 4 de febrero, para el desarrollo del Listado de Especies Silvestres en Régimen de Protección Especial y del Catálogo Español de Especies Amenazadas. BOE 46 (2011/02/23). https://www.boe.es/buscar/pdf/2011/BOE-A-2011-3582-consolidado.pdf

Fig. 8.5 Endangered species. (**a**) *Anagyris latifolia*; (**b**) *Atractylis preauxiana*; (**c**) *Bencomia exstipulata*; (**d**) *Caralluma burchardii*; (**e**) *Lotus maculatus*; (**f**) *Teline pallida* subsp. *silensis*

spermatophytes) (Fig. 8.5), 13 "vulnerable" (2 ferns and 11 spermatophytes) and 50 under "special protection regime" (2 bryophytes, 5 ferns and 43 spermatophytes).

Three years later, in 2010,[11] the Law of the Canary Catalog of Protected Species was approved, repealing the 2001 Decree and updating the list, including: 62 plants "in danger of extinction" (all of them spermatophytes), 38 "vulnerable" (4 algae, 1 fern, and 33 spermatophytes), 99 of "interest for Canary ecosystems" (8 algae, 9 ferns and 82 spermatophytes), 13 under "special protection" (1 fern and 12 sper-

[11] 3310 LEY 4/2010, de 4 de junio, del Catálogo Canario de Especies Protegidas. BOC 112 (2010/06/09). http://www.gobiernodecanarias.org/boc/2010/112/001.html

matophytes), and 28 in a "supplementary category" (2 ferns and 26 spermatophytes) (Tables 8.1, 8.2 and 8.3). Subsequently, by a Decree of 2014,[12] numerous species of the Habitats Directive that had not been included in it were added to an annex of the

Table 8.1 Plants included in Annex I (Species in danger of extinction). Law 4/2010, 4 June, Canary Catalog of Protected Species. BOC 112 (2010/06/09)

Scientific taxon name	Canary common name	Islands
Adenocarpus ombriosus	Codeso herreño	H
Atractylis arbuscula	Piñamar mayor	C,L
Atractylis preauxiana	Piñamar	T,C
Bencomia brachystachya	Bencomia de Tirajana	C,L
Bencomia exstipulata	Bencomia de cumbre	P,T
Bencomia sphaerocarpa	Bencomia herreña	H
Caralluma burchardii	Cuernúa	F,L
Cheirolophus dariasi	Cabezón del Cabrito	G
Cheirolophus ghomerythus	Cabezón gomero	G
Cistus chinamadensis subsp. *ombriosus*	Amagante herreño	H
Crambe scoparia	Colderrisco de La Aldea	C
Crambe sventenii	Colino majorero	F
Crambe wildpretii	Colderrisco de Arure	G
Dorycnium broussonetii	Trébol de risco de Broussonet	T,C
Dracaena tamaranae	Drago de Gran Canaria	C
Echium acanthocarpum	Tajinaste gomero	G
Echium handiense	Tajinaste de Jandía	F
Euphorbia mellifera	Tabaiba de monteverde	P,G,T
Globularia ascanii	Mosquera de Tamadaba	C
Helianthemum aganae	Jarilla de Anaga	T
Helianthemum bramwelliorum	Jarilla de Guinate	L
Helianthemum bystropogophyllum	Jarilla peluda	C
Helianthemum gonzalezferreri	Jarilla de Famara	L
Helianthemum inaguae	Jarilla de Inagua	C
Helianthemum juliae	Jarilla de Las Cañadas	T
Helianthemum teneriffae	Jarilla de Agache	T
Helianthemum tholiforme	Jarilla de Gran Canaria	C
Helichrysum alucense	Yesquera de Aluce	G
Ilex perado subsp. *lopezlilloi*	Naranjero salvaje gomero	G
Kunkeliella canariensis	Escobilla de Guayadeque	C
Kunkeliella psilotoclada	Escobilla de Masca	T
Kunkeliella subsucculenta	Escobilla carnosa	T

(continued)

[12] DECRETO 20/2014, de 20 de marzo, por el que se modifican los anexos de la Ley 4/2010, de 4 de junio, del Catálogo Canario de Especies Protegidas. BOC 63 (2014/03/31). http://www.gobiernodecanarias.org/boc/2014/063/index.html

Table 8.1 (continued)

Scientific taxon name	Canary common name	Islands
Lavatera acerifolia var. *hariensis*	Malvarrisco rosada	L
Limonium bourgeaui	Siempreviva de Bourgeau	F,L
Limonium dendroides	Siempreviva gigante	G
Limonium relicticum	Siempreviva de Tejeleche	G
Lotus berthelotii	Picopaloma	T
Lotus eremiticus	Picocernícalo	P
Lotus kunkelii	Yerbamuda de Jinámar	C
Lotus maculatus	Pico de El Sauzal	T
Lotus pyranthus	Pico de fuego	P
Morella rivas-martinezii	Faya herreña	H,P,G
Normania nava	Tomatillo de Nava	T
Onopordon carduelium	Cardo de Tenteniguada	C
Onopordon nogalesii	Cardo de Jandía	F
Parolinia glabriuscula	Dama de Bandama	C
Pericallis appendiculata var. *preauxiana*	Alamillo de Doramas	C
Pericallis hadrosoma	Flor de mayo leñosa	C
Pulicaria burchardii	Pulicaria majorera	F
Salvia herbanica	Conservilla majorera	F
Sideritis amagroi	Salviablanca de Amagro	C
Sideritis discolor	Salviablanca de Doramas	C
Sideritis marmorea	Chajorra de Aluce	G
Silene sabinosae	Canutillo de Sabinosa	H
Solanum lidii	Pimentero de Temisas	C
Solanum vespertilio subsp. *doramae*	Rejalgadera de Doramas	C
Stemmacantha cynaroides	Cardo de plata	T
Tanacetum oshanahanii	Magarza de Guayedra	C
Teline pallida subsp. *silensis*	Gildana de Los Silos	T
Teline salsoloides	Retamón de El Fraile	T
Teucrium heterophyllum subsp. *hierrense*	Jocama	H
Zostera noltii	Seba fina	L

Islands: *H* El Hierro, *P* La Palma, *G* La Gomera, *T* Tenerife, *C* Gran Canaria, *F* Fuerteventura, *L* Lanzarote

Catalog within the category of "special protection", together with other species previously considered of "interest for the Canary Islands ecosystems".

It should be noted that from the 100 taxa under priority management on the islands, 79 are flowering plants (2 marine and 77 terrestrial). The most threatened plants in the Canary Islands, defined as having less than 50 individuals, in their total populations are: *Gonospermum oshanahanii, Helianthemum aganae, Helianthemum inaguae, Kunkeliella canariensis, Lotus eremiticus, Lotus kunkelii, Lotus maculatus, Lotus pyranthus* and *Solanum vespertilio* subsp. *doramae*.

Table 8.2 Plants included in Annex II (Vulnerable species). Law 4/2010, 4 June, Canary Catalog of Protected Species. BOC 112 (2010/06/09)

Scientific taxon name	Canary common name	Islands
Aichryson tortuosum (=*A. bethencourtianum*)	Gongarillo majorero	F (L)
Argyranthemum adauctum subsp. *jacobaeifolium*	Magarza de Doramas	C
Argyranthemum winteri	Margarita de Jandía	F
Asparagus nesiotes subsp. *purpuriensis*	Esparraguera majorera	F
Bupleurum handiense	Anís de Jandía	F
Bystropogon odoratissimus	Poleo de cardonal	T
Carduus bourgeaui	Cardo majorero	F
Cheirolophus arboreus	Cabezón arbóreo	P
Cheirolophus arbutifolius	Cabezón de Gran Canaria	C
Cistus chinamadensis subsp. *gomerae*	Amagante gomero	G
Convolvulus caput-medusae	Chaparro canario	C,F
Convolvulus lopezsocasi	Corregüelón de Famara	L
Dactylis metlesicsii	Jopillo de cumbre	T
Dendriopoterium pulidoi	Algafitón de La Aldea	C
Gonospermum ptarmiciflorum (=*Tanacetum ptarmiciflorum*)	Magarza plateada	C
Hymenophyllum wilsonii	Helechilla de Wilson	G,T,C
Hypericum coadunatum	Cruzadilla de naciente	C
Limonium sventenii	Siempreviva azul	C
Limonium tuberculatum	Siempreviva espinocha	C,F
Limonium vigaroense	Siempreviva de Inagua	C
Lolium saxatile	Ballico de risco	F,L
Ononis christii	Taboire de Jandía	F
Plantago famarae	Pinillo de Famara	L
Pleiomeris canariensis	Delfino	P,G,T,C
Polygonum maritimum	Treintanudos de mar	P,T,C,F,L
Salix canariensis	Sauce canario	H,P,G,T,C
Scrophularia calliantha	Fistulera de Gran Canaria	C
Sideritis guayedrae	Salvia blanca	C
Silene nocteolens	Canutillo del Teide	T
Solanum vespertilio subsp. *vespertilio*	Rejalgadera	T
Sonchus gandogeri	Cerrajón de El Golfo	H
Teline splendens	Herdanera	P
Traganum moquinii	Balancón	G,T,C,F,L
Viola palmensis	Violeta de La Palma	P

Islands: *H* El Hierro, *P* La Palma, *G* La Gomera, *T* Tenerife, *C* Gran Canaria, *F* Fuerteventura, *L* Lanzarote

Table 8.3 Plants included in Annex V (Temporary category), Law 4/2010, 4 June, Canary Catalog of Protected Species. BOC 112 (2010/06/09), their categories in the former Catalog of threatened species of the Canaries – (Decree 151/2001), and category in the current Spanish Catalog (R/D 139/2011)

Scientific name	Common name	Canary catalogue (2001)	Canary catalogue (2010)	Spanish catalogue (2011)	Islands
Anagyris latifolia	Oro de risco	EPE	V	EPE	P,G,T,C
Androcymbium hierrense	Cebollín estrellado herreño	EPE	IEC	EPE	H,P,G
Argyranthemum lidii	Magarza de Lid	EPE	V	EPE	C
Argyranthemum sundingii	Margarita de Sunding	EPE	IEC	EPE	T
Asparagus fallax	Esparraguera de monteverde	EPE	IEC	EPE	G,T
Cheirolophus duranii	Cabezón herreño	EPE	V	EPE	H
Cheirolophus falcisectus	Cabezón de Güí-Güí	EPE	IEC	EPE	C
Cheirolophus metlesicsii	Cabezón de Añavingo	EPE	V	EPE	T
Cheirolophus santos-abreui	Cabezón de Las Nieves	EPE	V	EPE	P
Cheirolophus sventenii subsp. *gracilis*	Cabezón de Tijarafe	EPE	V	EPE	P
Convolvulus subauriculatus	Corregüelón gomero	EPE	IEC	EPE	G
Diplazium caudatum	Helecho de monte	SAH	V	V	P,G,T,C
Dorycnium spectabile	Trébol de risco rosado	EPE	V	EPE	T
Euphorbia bourgeauana	Tabaiba amarilla de Tenerife	EPE	V	EPE	T
Globularia sarcophylla	Mosquera de Tirajana	EPE	V	EPE	C
Himantoglossum metlesicsianum	Orquídea de Tenerife	EPE	IEC	EPE	T
Hypochoeris oligocephala	Lechuguilla de El Fraile	EPE	IEC	EPE	T
Isoplexis chalcantha	Crestagallo de Doramas	EPE	V	EPE	C
Isoplexis isabelliana	Crestagallo de pinar	EPE	V	EPE	C
Limonium spectabile	Siempreviva de Guelgue	EPE	IEC	EPE	T
Micromeria glomerata	Tomillo de Taganana	EPE	IEC	EPE	T

(continued)

Table 8.3 (continued)

Scientific name	Common name	Canary catalogue (2001)	Canary catalogue (2010)	Spanish catalogue (2011)	Islands
Pteris incompleta	Rasguño dentado	EPE	IEC	V	P,G,T,C
Ruta microcarpa	Ruda gomera	EPE	IEC	EPE	G
Sambucus nigra subsp. palmensis (=*S. palmensis*)	Saúco canario	EPE	V	EPE	P,G,T,C
Sideritis cystosiphon	Chajorra de Tamaimo	EPE	IEC	EPE	T
Teline nervosa	Gildana peluda	EPE	V	EPE	C
Teline rosmarinifolia subsp. *eurifolia*	Gildana de Faneque	EPE	IEC	EPE	C
Tolpis glabrescens	Lechuguilla de Chinobre	EPE	IEC	EPE	T

Categories: *EPE* in danger of extiction, *SAH* sensitive to habitat disturbance, *V* vulnerable, *IEC* of interest fot Canary ecosystems. Islands: *P* La Palma, *G* La Gomera, *T* Tenerife, *C* Gran Canaria

The Canary Islands Biodiversity Data Bank was created in 2010.[13] It includes the list and known distribution of all wild species of plants, algae, fungi, animals and other living organisms that regularly inhabit or reproduce in the archipelago and in its waters without direct intervention of man. It specifies the chorology of the taxa, the category in which they are included in the Canary Catalog of Protected Species, and gathers documents related to them. It is a very useful tool, both for research, scientific outreach, and environmental management.

As a curiosity, in order to raise awareness among the Canary islanders about the native species as part of their identity, on April 17, 1991[14] the Canary Parliament unanimously approved a law determining the official "Symbols of Nature for the Canary Islands", assigning a vascular plant and an animal to each of the islands. The plant symbols were *Euphorbia balsamifera* (Lanzarote), *E. handiensis* (Fuerteventura), *E. canariensis* (Gran Canaria), *Dracaena draco* (Tenerife), *Persea indica* (La Gomera), *Pinus canariensis* (La Palma) and *Juniperus turbinata* subsp. *canariensis* (El Hierro). In addition, the palm *Phoenix canariensis* was designated the archipelago's symbol. This law sought to improve environmental education, bringing the natural heritage of the archipelago closer to its inhabitants, to encourage greater respect and affection for their environment.

[13] Banco de datos de Biodiversidad de Canarias. http://www.biodiversidadcanarias.es/atlantis/common/index.jsf

[14] Ley 7/1991, de 30 de abril, de símbolos de la naturaleza para las Islas Canarias. BOE 151 (1991/06/25). https://www.boe.es/diario_boe/txt.php?id=BOE-A-1991-16276

8.3 Botanical Gardens, Seed and Germplasm Banks and Nurseries

Fig. 8.6 Nursery for recovering threatened plants at El Portillo, Teide National Park

The Canary Islands also have two official botanical gardens: Puerto de la Cruz in Tenerife, and Jardín Canario Viera y Clavijo in Tafira, Gran Canaria. They are of great value in the conservation of the plant heritage of the islands, particularly by housing seed banks. The one in Tafira also has an established DNA bank, as well as a magnificent live collection of plants endemic to the islands, having adopted cutting-edge "in vitro" technologies to save plants threatened with extinction. In the same line, the island councils (Cabildos) that now manage all natural reserves including NPs maintain specialized nurseries to then the most threatened species, reinforcing their populations and organizing re-planting operations at the most degraded sites within the protected natural areas (Fig. 8.6).

8.4 Invasive Plants

At present, in addition to those species that continue to be cultivated (agricultural, medicinal, ornamental, etc.), it is estimated that more than 700 of the wild species of the Canaries are exotic, having been introduced intentionally or accidentally by man, the great majority during or since the nineteenth century. After a certain period, some of them manage to adapt to the new environment and produce changes in the natural ecosystems, threatening the biological diversity by becoming invasive. One fern and about 80 spermatophytes are considered invasive (Fig. 8.7).

Although the damage caused to plants by introduced herbivores is well known, the negative impact of invasive species on the native Canary flora is not so obvious, despite the fact that they alter ecosystems and compete with native species for habitat, water resources, nutrients and light. For this reason, a national decree regulating the Spanish list and catalog of invasive alien species, was passed in 2011.[15] Its Annex I included 1 fungus, 7 algae and 45 spermatophytes for the Canaries. In addi-

[15] Real Decreto 1628/2011, de 14 de noviembre, por el que se regula el listado y catálogo español de especies exóticas invasoras. BOE 298 (2011/12/12). https://www.boe.es/boe/dias/2011/12/12/pdfs/BOE-A-2011-19398.pdf

tion, another 3 algae and 126 "potentially invasive" spermatophytes were added in Annex II.

Among the most widespread invasive plant species on the Canaries are: *Ageratina adenophora*, *Agave americana*, *Arundo donax*, *Cytisus scoparius*, *Eschscholzia californica*, *Nicotiana glauca*, *Opuntia maxima*, *Oxalis pes-caprae*, *Pennisetum setaceum*, *Spartium junceum*, *Tropaelum majus* and *Ulex europaeus*.

Fig. 8.7 Invasive plants. (**a**) *Agave americana*; (**b**) *Opuntia maxima* and *Eschscholzia califormica*; (**c**) Slopes of Bajamar, La Palma, with *Pennisetum setaceum*; (**d**) *Opuntia dillenii*; (**e**) *Pennisetum setaceum*

Appendices

Appendix 1: Syntaxonomic Scheme[1]

ADIANTETEA Br.-Bl. *in* Br.-Bl., Roussine & Nègre 1952

ADIANTETALIA CAPILLI-VENERIS Br.-Bl. *ex* Horvatic 1934

Adiantion capilli-veneris Br.-Bl. *ex* Horvatic 1934

-*Adianto-Hypericetum coadunati* Deil *ex* Quintana, Salas & E. Fernández 2006	C
-*Camptolometum canariensis* Sunding 1972 [*nom. mut. propos.*]	C
-*Eucladio-Adiantetum capilli-veneris* Br.-Bl. *ex* Horvatic 1934	L,C,T,G,P

[1] In order to facilitate reading, the following authorships have been abbreviated because of its excessive length:

- Capelo *et al.* 2000 [= Capelo, J.C. Costa, Lousã, Fontinha, Jardim, Sequeira & Rivas-Martínez 2000].
- Del Arco *et al.* 2006 [= Del Arco, Wildpret, Pérez de Paz, O. Rodríguez, Acebes, García Gallo, Martín, Reyes, Salas, M.A. Diaz, Bermejo, R. González, M.V. Cabrera & S. García 2006].
- Del Arco *et al.* 2009 [= Del Arco, O. Rodríguez, Acebes, García Gallo, Pérez de Paz, J.M. González, R. González & Garzón 2009].
- Lohmeyer *et al.* 1962 [=Lohmeyer, A. Matuszkiewics, W. Matuszkiewics, Merker, Moore, Müller, Oberdorfer, Poli, Seibert, Sukopp, Trautmann, J. Tüxen, Tüxen & Westhoff 1962].
- Martín *et al.* 2007 [= Martín, Wildpret & Rivas-Martínez *in* Martín, Wildpret, Del Arco, Pérez de Paz, B. Hernández, O. Rodríguez, Acebes & García Gallo 2007].
- Rivas-Martínez *et al.* 1993 [=Rivas-Martínez, Wildpret, Del Arco, O. Rodríguez, Pérez de Paz, García Gallo, Acebes, T.E. Díaz & Fernández-González 1993].
- Rivas-Martínez *et al.* 2002 [=Rivas-Martínez, T.E. Díaz, Fernández-González, Izco, Loidi, Lousã & Penas 2002].

Clarifications to nomenclatural changes proposed in this work are shown between brackets after the new proposed name

M. J. del Arco Aguilar, O. Rodríguez Delgado, *Vegetation of the Canary Islands*, Plant and Vegetation 16, https://doi.org/10.1007/978-3-319-77255-4

AEONIETEA A. Santos 1976 [*nom. mut. propos.*] (*Greenovio-Aeonietea* Santos 1976, *Anales Inst. Bot. Cavanilles* 33: 358. Art. 45-ICPN -Weber et al. 2000)

SONCHO ACAULIS-AEONIETALIA Rivas Goday & Esteve *ex* Sunding 1972 [*nom. mut. propos.*]

Aichryso laxi-Monanthion laxiflorae A. Santos & Reyes 2009

-*Aichrysetum inmaculati* A. Santos & Reyes 2009	T
-*Aichrysetum pachycauli* Martín, Scholz & Wildpret 2013	F
-*Aichrysetum punctati* A. Santos & Reyes 2009	C,T,G,P,H
-*Aichrysetum tortuosi* Reyes, Wildpret & León 2001	L
-*Davallio canariensis-Aichrysetum laxi* Wildpret, García Gallo & Carqué *in* Rivas-Martínez *et al.* 1993	C,T,G,P,H
-*Lolio saxatile-Aichrysetum bethencourtiani* Martín, Scholz & Wildpret 2013	F
-*Monanthetum anagensis* A. Santos & Reyes 2009	T
-*Monanthetum pallentis* J. García, Wildpret & O. Rodríguez 2002	T
-*Soncho-Monanthetum laxiflorae* M. Fernández 1983 *nom. prov.*	G

Aeonion aureae A. Santos *ex* Rivas-Martínez *et al.* 1993 [*nom. mut. propos.*] (*Greenovion aureae* A. Santos ex Rivas-Martínez et al. 1993, *Itinera Geobotanica* 7: 313. Art. 45-ICPN)

-*Aeonietum aizoi* Rivas-Martínez *et al.* 1993 [*nom. mut. propos.*] (*Greenvietum aizoii* Rivas-Martínez et al. 1993, *Itinera Geobotanica* 7: 316. Art. 45-ICPN)	T
-*Aeonietum aureae* Rivas-Martínez *et al.* 1993 [*nom. mut. propos.*] (*Greenovietum aureae* Rivas-Martínez et al. 1993, *Itinera Geobotanica* 7: 314. Art. 45-ICPN)	T
-*Aeonietum aureo-caespitosi* Sunding 1972 [*nom. mut. propos.*] (*Greenovio-Aeonietum caespitosi* Sunding 1972, *Skr. Vidensk.-Akad Oslo, Mat-Naturvidensk Kl*, N.S. 29: 188. Art. 45-ICPN)	C
-*Aeonietum diplocyclae* A. Santos 1983 [*nom. mut. propos.*] (*Greenovietum diplocyclae* A. Santos 1983, *Vegetación y Flora de La Palma*: 88, Art. 45-ICPN)	P
-*Cheilantho guanchicae-Aeonietum smithii* Rivas-Martínez *et al.* 1993	T
-*Soncho hierrensis-Aeonietum diplocyclae* A. Santos 1976 [*nom. mut. propos.*] (*Soncho-Greenovietum diplocyclae* A. Santos 1976, *Anales Inst. Bot. Cavanilles* 33: 260. Art. 45-ICPN)	G,H
-*Tolpidetum calderae* A. Santos 1983	P

Soncho acaulis-Aeonion Sunding 1972 [*nom. mut. propos.*] (*Soncho-Sempervivion* Sunding 1972, *Skr. Vidensk.-Akad Oslo, Mat-Naturvidensk Kl, N.S.* 29: 94. Art. 45-ICPN)	
-*Aeonietum canariensis* Rivas-Martínez *et al.* 1993	T
-*Aeonietum christii* A. Santos 1976 [*nom. mut. propos.*] (*Aeonietum longithyrsii* A. Santos 1976, *Anales Inst. Bot. Cavanilles* 33: 259. Art. 45-ICPN)	H
-*Aeonietum christii-davidbramwellii* A. Santos ex Del Arco & O. Rodríguez [*nom. novum prov.*] (*Aeonietum palmensis* A. Santos 1976, *Vegetación y Flora de La Palma*: 92, Art. 45, 39-ICPN)	P
subass. *typicum*	P
subass. *aeonietosum goochiae* A. Santos 1983 [*nom. inval.*]	P
subass. *aeonietosum nobile* A. Santos 1983 [*nom. inval.*]	P
-*Aeonietum cuneati* Voggenreiter *ex* Rivas-Martínez *et al.* 1993	T
-*Aeonietum lancerottensis* Reyes, Wildpret & León 2001	L
-*Aeonietum latifolii* M. Fernández ex Del Arco & O. Rodríguez *nom. prov.*	P
-*Aeonietum lindleyi* Voggenreiter *ex* Rivas-Martínez *et al.* 1993	T
subass. *aeonietosum lindleyi* [*typicum*]	T
subass. *aeonietosum canariensis* Rivas-Martínez *et al.* 1993	T
-*Aeonietum undulato-percarnei* Carqué, Wildpret & García Gallo *in* Wildpret, Carqué & García Gallo 1996	C
-*Aeonietum valverdensis* A. Santos ex M. Gaisberg 2005	H
-*Aeonietum virginii* Suárez & Pérez de Paz 1993	C
-*Aeonietum viscati* Del Arco & O. Rodríguez *nom. prov.*	G
-*Aeonio decoris-Atalanthetum pinnati* M. Fernández 1983 [*nom. mut. propos.*] (*Aeonio decoris-Sonchetum leptocephali* M. Fernández 1983, *II Congreso Pro Flora Macaronesica. Funchal* p. 289, Art. 45, 39-ICPN)	G
-*Chrysoprenantho-Atalanthetum pinnati* Sunding 1972 [*nom. mut. propos.*] (*Prenantho-Taeckholmietum pinnate* Sunding 1972, *Skr. Vidensk.-Akad Oslo, Mat-Naturvidensk Kl*, N.S. 29: 97. Art. 45-ICPN)	C
-*Parietarietum filamentosae* Rivas-Martínez *et al.* 1993	T,G,P
-*Pericallido lanatae-Sonchetum gummiferi* Rivas-Martínez *et al.* 1993	T
-*Phyllido viscosae-Aeonietum sedifolii* A. Santos & M. Fernández 1983	T
-*Polycarpaea smithii* community Stiertorfer 2005	P,H

-*Reichardio famarae-Helichrysetum gossypini* Reyes, L
Wildpret & León 2001
-*Soncho congesti-Aeonietum holochrysi* O. Rodríguez, García T
Gallo & Wildpret 2006
-*Soncho hierrensis-Aeonietum diplocyclae* A. Santos 1976 H
[*nom. mut. propos.*]
subass. *typicum* H
subass. *phyllidetosum noblae* Stiertorfer 2005 H
-*Soncho radicati-Aeonietum tabulaeformis* A. Santos & T
M. Fernández 1983
subass. *aeonietosum tabulaeformis* [*typicum*] T
subass. *hypochaeridetosum oligocephalae* T
Rivas-Martínez *et al.* 1993 [*nom.mut. propos.*]
-*Umbilico gaditani-Aeonietum urbici* García Gallo & T
Wildpret *in* Rivas-Martínez *et al.* 1993 *corr.*
Rivas-Martínez *et al.* 2002
-*Vieraeo laevigatae-Polycarpaeetum carnosae* T
Rivas-Martínez *et al.* 1993
ANOMODONTO VITICULOSI-POLYPODIETEA CAMBRICI
Rivas-Martínez 1975
ANOMODONTO VITICULOSI-POLYPODIETALIA CAMBRICI O. Bolòs &
Vives *in* O. Bolòs 1957
Hymenophyllion tunbrigensis Tüxen *in* Tüxen & Oberdorfer 1958
-*Hymenophylletum tunbrigensis* Br.-Bl. *in* Br.-Bl. & Tüxen 1952 T,G
Polypodion cambrici Br.-Bl. in Br.-Bl., Roussine & Nègre 1952
[*nom. mut. propos.*]
Bartramio strictae-Polypodienion cambrici
(O. Bolòs & Vives 1957) Rivas-Martínez 2002
-*Davallio canariensis-Polypodietum macaronesici* C,T,G,P,H
Rivas-Martínez, Wildpret, Del Arco, O. Rodríguez,
Pérez de Paz, García Gallo, Acebes, T.E. Díaz &
Fernández-González *ex* Capelo *et al.* 2000
Selaginello denticulatae-Anogrammion leptophyllae
Rivas-Martínez, Fernández-González & Loidi 1999
-*Selaginello denticulatae-Anogrammetum leptophyllae* T,H
R.Molinier 1937
ARTEMISIETEA VULGARIS Lohmeyer, Preising & Tüxen *ex*
Von Rochow 1951
ONOPORDENEA ACANTHI Rivas-Martínez, Báscones,
T.E. Díaz, Fernández-González & Loidi 2002
CARTHAMETALIA LANATI Brullo *in* Brullo & Marcenò 1985
Bromo madritensis-Piptatherion miliacei O. Bolòs 1970
[*nom. mut. propos.*]
-*Dittrichia viscosa* community Suárez 1994 C,T,H

-*Pipthathero miliacei-Foeniculetum vulgaris* Rivas-Martínez *et al.* 1993	L,F,C,T,G,P
-*Pericallido murrayi-Bituminarietum bituminosae* Stiertorfer 2005	H
Urtico piluliferae-Silybion mariani Sissingh *ex* Br.-Bl. & O. Bolòs 1958 [*nom. inv. propos.*]	
-*Conio maculati-Silybetum mariani* Lohmeyer & Trautmann 1970	C,T,G,P,H
-*Scolymo maculati-Cynaretum ferocissimae* Wildpret, Del Arco & García Gallo 1989	L,F,C,T,G,P,H
ASPLENIETEA TRICHOMANIS (Br.-Bl. *in* Meier & Br.-Bl. 1934) Oberdorfer 1977	
NOTHOLAENO MARANTAE-CHEILANTHETALIA MADERENSIS (Sáenz & Rivas-Martínez 1979) Rivas-Martínez & col. 2011	
Cheilanthion pulchellae Sáenz & Rivas-Martínez 1979	
-*Adianto pusilli-Cheilanthetum pulchellae* Sáenz & Rivas-Martínez 1979	T,P
-*Cheilanthes guanchica* community Stiertorfer 2005	H
-*Notolaena (Cheilanthes) marantae* community A. Santos 1983	P
CAKILETEA MARITIMAE Tüxen & Preising *ex* Br.-Bl. & Tüxen 1952	
CAKILETALIA INTEGRIFOLIAE Tüxen *ex* Oberdorfer 1949 *corr.* Rivas-Martínez, Costa & Loidi 1992	
Cakilion maritimae Pignatti 1953	
-*Salsolo kali-Cakiletum maritimae* M.Costa & Mansanet 1981 [*nom. mut. propos.*]	L,F,T
CARDAMINO HIRSUTAE-GERANIETEA PURPUREI (Rivas-Martínez, Femández-González & Loidi 1999) Rivas-Martínez, Fernández-González & Loidi 2002	
CARDAMINO HIRSUTAE-GERANIETALIA PURPUREI Brullo *in* Brullo & Marcenò 1985 [*nom. inv. propos.*]	
Geranio purpurei-Torilidion neglectae Lohmeyer & Trautmann 1970 *corr.* Lohmeyer 1975	
-*Carduo clavulati-Urticetum stachyoidis* Rivas-Martínez *et al.* 1993	T
-*Drusa glandulosa-Parietaria debilis* community Rivas-Martínez *et al.* 1993	T,H
-*Galio aparines-Torilidetum neglectae* Lohmeyer & Trautmann 1970	T,G,P
Geranio pusilli-Anthriscion caucalidis Rivas-Martínez 1978	
-*Arabidopsis thaliana-Myosotis gracillima* community Rivas-Martínez *et al.* 1993	T
-*Myosotidetum canariensis* A. Santos 1980 *nom. prov.*	H

CHAMAECYTISO-PINETEA CANARIENSIS Rivas Goday & Esteve *ex* Esteve 1969 [*nom. mut. propos.*]

CHAMAECYTISO-PINETALIA CANARIENSIS Rivas Goday & Esteve *ex* Esteve 1969 [*nom. mut. propos.*]

Cisto symphytifolii-Pinion canariensis Rivas Goday & Esteve *ex* Esteve 1969

-*Bystropogono ferrensis-Pinetum canariensis* Del Arco, Acebes & Pérez de Paz 1996	H
subass. *typicum*	H
subass. *ericetosum arboreae* Del Arco, Acebes & Pérez de Paz 1996	H
-*Cisto gomerae-Pinetum canariensis* Del Arco *et al.* 2009	G
subass. *typicum*	G
subass. *juniperetosum canariensis* Del Arco *et al.* 2009	G
-*Loto hillebrandii-Pinetum canariensis* A. Santos 1983	P
subass. *cistetosum symphytifolii* [*typicum*]	P
subass. *adenocarpetosum foliolosi* A. Santos 1983 [*nom. inval.*]	P
subass. *adenocarpetosum spartioidis* A. Santos 1983 [*nom. inval.*]	P
subass. *ericetosum arboreae* A. Santos 1983 [*nom. inval.*]	P
subass. *juniperetosum canariensis* Del Arco & O. Rodríguez *in* Del Arco *et al.* 2006 *nom. prov.*	P
subass. *juniperetosum cedri* Del Arco & O. Rodríguez *in* Del Arco *et al.* 2006 *nom. prov.*	P
subass. *spartocytisetosum filipis* Garzón, Del Arco & Pérez de Paz 2011 *nom. prov.*	P
-*Micromerio pineolentis-Pinetum canariensis* Esteve 1969	C
-*Morello fayae-Pinetum canariensis* Del Arco, O. Rodríguez & Pérez de Paz 2013	T
subass. *typicum*	T
subass. *arbutetosum canariensis* Del Arco, O. Rodríguez & Pérez de Paz 2013	T
subass. *cistetosum chinamadensis* Del Arco, O. Rodríguez & Pérez de Paz 2013	T
-*Pinetum canariensis* Ceballos & Ortuño *ex* Sunding 1972	C
subas. *typicum*	C
subass. *cistetosum horrentis* Del Arco & O. Rodríguez *in* Del Arco *et al.* 2006 *nom. prov. corr. hoc loco*	C
subass. *ericetosum arboreae* Pérez de Paz, Acebes, Del Arco & Salas 1993	C
subass. *juniperetosum canariensis* Salas, Del Arco & Pérez de Paz 1998	C
subass. *sideritidetosum dasygnaphalae* Del Arco & O. Rodríguez *in* Del Arco *et al.* 2006 *nom. prov.*	C

-*Sideritido solutae-Pinetum canariensis* Esteve 1973 T
subass. *pinetosum canariensis* [*typicum*] T
subass. *cistetosum canariensis* (Del Arco, Pérez de Paz & Wildpret 1987) Rivas-Martínez *et al.* 1993 [*nom. mut. propos.*] T
subass. *cistetosum osbaeckiaefolii* Rivas-Martínez *et al.* 1993 T
subass. *ericetosum arboreae* (Del Arco, Pérez de Paz & Wildpret 1987) Rivas-Martínez *et al.* 1993 T
subass. *juniperetosum canariensis* Del Arco & O. Rodríguez *in* Del Arco *et al.* 2006 *nom. prov.* T
subass. *spartocytisetosum supranubii* Rivas-Martínez *et al.* 1993 T

Juniperion cedri Martín, Wildpret & Rivas Martínez *in* Martín *et al.* 2007

-*Adenocarpo spartioidis-Juniperetum cedri* Martín, Wildpret & B. Hernández *in* Martín *et al.* 2007 P
-*Adenocarpo viscosi-Juniperetum cedri* Martín, Wildpret & B. Hernández *in* Martín *et al.* 2007 T

SPARTOCYTISETALIA SUPRANUBII Schönfelder & Voggenreiter 1994

Plantaginion webbii Martín, Wildpret & Rivas Martínez *in* Martín *et al.* 2007

Plantaginenion webbii Martín, Wildpret & Rivas Martínez *in* Martín *et al.* 2007
-*Descurainio gilvae-Plantaginetum webbii* A. Santos 1983 P

Pterocephalenion lasiospermi Martín, Wildpret & Rivas Martínez *in* Martín *et al.* 2007
-*Arrhenathero calderae-Plantaginetum webbii* Martín & Wildpret *in* Martín & B. Hernández 2003 T
-*Echietum auberiani* P.Schönfelder *in* P.Schönfelder & Voggenreiter 1994 T
-*Erysimo scoparii-Pterocephaletum lasiospermi* Rivas-Martínez *et al.* 1993 T

Spartocytision supranubii Oberdorfer *ex* Esteve 1973 [*nom. mut. propos.*]

-*Descurainio bourgeauanae-Spartocytisetum supranubii* (Esteve 1973) Martín, Wildpret & B. Hernández *in* Martín *et al.* 2007
subass. *typicum*
subass. *erigerontetosum calderae* Martín & Wildpret *in* Martín *et al.* 2007
-*Genisto benehoavensis-Adenocarpetum spartioidis* A. Santos 1983 [*nom. mut. propos.*] P
-*Spartocytisetum supranubii* Oberdorfer *ex* Esteve 1973 [*nom. mut. propos.*] T
subass. *typicum* T
subass. *aeonietosum* Esteve 1973 T

subass. *descurainietosum gonzalezii* Martín & Wildpret *in* Martín *et al.* 2007	T
subass. *descurainietosum lemsii* Martín & Wildpret *in* Martín *et al.* 2007	T
subass. *echietosum wildpretii* Schönfelder *in* Schönfelder & Voggenreiter 1994	T
-*Telinetum spachianae* Del Arco & Wildpret 1983	T
-*Violo cheiranthifoliae-Adenocarpetum viscosi* Wildpret & Martín 2005	
Violion cheiranthifoliae Voggenreiter *ex* Rivas-Martínez, Wildpret & Martín *in* Martín *et al.* 2007	T
-*Violetum cheiranthifoliae* Rivas-Martínez *et al.* 1993	T
CHARETEA FRAGILIS *Fukarek ex Krausch 1964*	
CHARETALIA HISPIDAE *Sauer ex Krausch 1964*	
Charion fragilis *Krausch 1964*	
-*Charetum conniventis* Velayos, Carrasco & Cirujano 1989	T
CRITHMO-LIMONIETEA Br.-Bl. *in* Br.-Bl., Roussine & Nègre 1952 [*nom. mut. propos.*]	
CRITHMO-LIMONIETALIA Moliner 1934 [*nom. mut. propos.*]	
Frankenio-Astydamion latifoliae A. Santos 1976	
-*Frankenio ericifoliae-Astydamietum latifoliae* Lohmeyer & Trautmann *ex* A. Santos 1976	C,T,G,P,H
subass. *typicum*	C,T,G,P,H
subass. *limonietosum imbricati* A. Santos 1983 [*nom. inval.*]	P
-*Frankenio ericifoliae-Zygophylletum fontanesii* Rivas-Martínez *et al.* 1993 *corr.* A. Santos 2002	L,F,C,T
subass. *zygophylletosum fontanesii* [*typicum*]	L,F,C,T
subass. *suaedetosum verae* (Biondi, Alegrezza, Taffetani & Wildpret 1994) O. Rodríguez, García Gallo & Reyes 2000	F
HALODULO WRIGHTII-THALASSIETEA TESTUDINUM Den Hartog *ex* Rivas-Martínez, Fernández-González & Loidi 1999	
THALASSIO-SYRINGODIETALIA FILIFORMIS Borhidi, Muñiz & Del Risco *in* Borhidi 1996	
Syringodio-Thalassion testudinum Borhidi 1996	
-*Cymodoceetum nodosae* Feldmann 1937	L,F,C,T,G,P,H
-*Halophiletum decipientis* Wildpret & M.C. Gil *in* Rivas-Martínez *et al.* 2002	C,T
ISOETO-NANOJUNCETEA Br.-Bl. & Tüxen *ex* Westhoff, Dijk & Passchier 1946	
NANOCYPERETALIA Klika 1935	
Nanocyperion Koch *ex* Libbert 1933	
-*Juncus bufonius-Laphangium luteoalbum* community Rivas-Martínez *et al.* 1993	T,P,H
Verbenion supinae Slavnic 1951	
-*Verbenetum supinae* Sunding 1972	C,T

KLEINIO NERIIFOLIAE-EUPHORBIETEA CANARIENSIS (Rivas Goday & Esteve 1965) A. Santos 1976	
KLEINIO NERIIFOLIAE-EUPHORBIETALIA CANARIENSIS (Rivas Goday & Esteve 1965) A. Santos 1976	
Euphorbion regisjubo-lamarckii Rivas-Martínez, Wildpret, O. Rodríguez & Del Arco *in* Rivas-Martínez & col. 2011	
Euphorbienion regisjubo-lamarckii Rivas-Martínez, Wildpret, O. Rodríguez & Del Arco *in* Rivas-Martínez & col. 2011	
-*Euphorbietum berthelotii* M. Fernández 1983	G
-*Euphorbietum lamarckii* Del Arco & O. Rodríguez *in* Del Arco *et al.* 2006 *nom. prov.*	P,H,T,C
-*Euphorbietum regis-jubae* Arco & O. Rodríguez *in* Del Arco *et al.* 2006 *nom. prov.*	C,L
-*Euphorbio berthelotii-Schizogynetum sericeae* Del Arco & O. Rodríguez *in* Del Arco *et al.* 2006 *nom. prov.*	H
-*Kleinio neriifoliae-Asparagetum pastoriani* Lems 1968 *corr.* A. Santos 1983	F
-*Schizogyno sericeae-Euphorbietum wildpretii* Pérez de Paz, Del Arco & Wildpret 1990 *corr.* Rivas-Martínez & Gaisberg *in* Rivas-Martínez & col. 2011	P,H
subass. *schizogynetosum sericeae* [*typicum*]	H
subass. *echietosum breviramis* Pérez de Paz, Del Arco & Wildpret 1991	P
subass. *ceballosietosum fruticosae* M. Gaisberg 2005	H
Periplocenion laevigatae Del Arco & O. Rodríguez *suball. nova prov.*	
-*Periploca laevigata* community Del Arco & O. Rodríguez *in* Del Arco *et al.* 2006	H,G
Plocamenion pendulae Rivas-Martínez, Wildpret, O. Rodríguez & Del Arco *in* Rivas-Martínez & col. 2011	
-*Plocametum pendulae* M.C. Marrero, O. Rodríguez & Wildpret 2003	G,T,C
subass. *euphorbietosum lamarckii* M.C. Marrero, O. Rodríguez & Wildpret 2003 [*typicum*]	T
subass. *euphorbietosum berthelotii* M.C. Marrero, O. Rodríguez & Wildpret 2003	G
subass. *euphorbietosum regis-jubae* M.C. Marrero, O. Rodríguez & Wildpret 2003	C
Kleinio neriifoliae-Euphorbion canariensis (Rivas Goday & Esteve 1965) A. Santos 1976	
Aeonio-Euphorbienion canariensis (Sunding 1972) A. Santos & Rivas-Martínez *in* Rivas-Martínez & col. 2011	
-*Aeonio percarnei-Euphorbietum canariensis* (Rivas Goday & Esteve 1965) Sunding 1972	C

subass. *typicum*	C
subass. *pistacietosum* Sunding 1972	C
-*Aeonio valverdensis-Euphorbietum canariensis* Del Arco, Acebes & Pérez de Paz 1996.	H
subass. *typicum*	H
subass. *euphorbietosum balsamiferae* Del Arco, Acebes & Pérez de Paz 1996	H
-*Echio breviramis-Euphorbietum canariensis* A. Santos 1983	P
-*Euphorbietum atropurpureae* Lems 1968	T
subass. *euphorbietosum atropurpureae* [*typicum*]	T
subass. *juniperetosum canariensis* O. Rodríguez, Wildpret, Del Arco & Pérez de Paz 1990 *corr.* Rivas-Martínez *et al.* 1993	T
-*Euphorbietum berthelotii-canariensis* Rivas-Martínez *et al.* 1993	G
-*Kleinio neriifoliae-Euphorbietum canariensis* O. Rodríguez, García Gallo & Reyes 2000	F
-*Periploco laevigatae-Euphorbietum canariensis* Rivas-Martínez *et al.* 1993	T
subass. *euphorbietosum canariensis* [*typicum*]	T
subass. *euphorbietosum balsamiferae* Rivas-Martínez *et al.* 1993	T
subass. *jasminetosum odoratissimi* Rivas-Martínez *et al.* 1993	T
subass. *salsoletosum divaricatae* Rivas-Martínez *et al.* 1993	T
Euphorbienion aphyllae A. Santos 1983	
-*Astydamio latifoliae-Euphorbietum aphyllae* (Rivas Goday & Esteve 1965) A. Santos 1983	C
-*Ceropegio dichotomae-Euphorbietum aphyllae* Rivas-Martínez *et al.* 1993	T
-*Euphorbietum aphyllae* (Rivas-Martínez *et al.* 1993) Del Arco *et al.* 2009	G
Helianthemo canariensis-Euphorbienion balsamiferae (Sunding 1972) A. Santos & Rivas-Martínez *in* Rivas-Martínez & col. 2011	
-*Asterisco intermedii-Euphorbietum balsamiferae* Esteve & Socorro 1977 [*nom. mut. propos.*]	L
subass. *euphorbietosum balsamiferae* [*typicum*]	L
subass. *aeonietosum lancerotensis Reyes*, Wildpret & León 2001.	L
-*Ceropegio fuscae-Euphorbietum balsamiferae* Rivas-Martínez *et al.* 1993	T
subass. *euphorbietosum balsamiferae* [*typicum*]	T
subass. *salsoletosum divaricatae* Rivas-Martínez *et al.* 1993	T
-*Echio breviramis-Euphorbietum balsamiferae* A. Santos 1983	P
subass. *typicum*	P
subass. *schizogynetosum sericeae* Del Arco & O. Rodríguez *in* Del Arco *et al.* 2006 *nom. prov.*	P

-*Euphorbietum balsamiferae* Sunding 1972	C
subass. *euphorbietosum balsamiferae* [*typicum*]	C
subass. *plocametosum pendulae* Sunding 1972	C
subass. *salsoletosum divaricatae* Del Arco & O. Rodríguez *in* Del Arco *et al.* 2006 *nom. prov.*	C
-*Euphorbietum handiensis* O. Rodríguez, García Gallo & Reyes 2000	F
-*Lycio intricati-Euphorbietum balsamiferae* O. Rodríguez, García Gallo & Reyes 2000	F
-*Neochamaeleo pulverulentae-Euphorbietum balsamiferae* Del Arco *et al.* 2009	G
-*Rubio fruticosae-Euphorbietum balsamiferae* Del Arco, Acebes & Pérez de Paz 1996	H
LEMNETEA MINORIS Tüxen *ex* O. Bolòs & Masclans 1955	
LEMNETALIA MINORIS Tüxen *ex* O. Bolòs & Masclans 1955	
Lemnion minoris Tüxen *ex* O. Bolòs & Masclans 1955	
-*Lemnetum gibbae* Miyawaki & J.Tüxen 1960	C,T,G,P,H
-*Lemnetum minoris* Oberdorfer *ex* Müller & Görs 1960	L,F,C,T,G
-*Pistia stratiotes* stands Pérez de Paz, Del Arco & Wildpret 1987	P
-*Salvinia molesta* stands Salas & Quintana 2016	C
LYGEO SPARTII-STIPETEA TENACISSIMAE Rivas-Martínez 1978 [*nom. conserv. propos.*]	
HYPARRHENIETALIA HIRTAE Rivas-Martínez 1978	
Hyparrhenion sinaicae Br.-Bl., P.Silva & Rozeira 1956 corr. J.C. Costa, Capelo, Espírito-Santo & Lousã 2001	
-*Cenchro ciliaris-Hyparrhenietum sinaicae* Wildpret & O. Rodríguez *in* Rivas-Martínez *et al.* 1993 *corr.* Díez-Garretas & Asensi 1999	L,F,C,T,G,P,H
subass. *hyparrhenietosum sinaicae* [*typicum*]	L,F,C,T,G,P,H
subass. *eremopogonetosum foveolati* García-Casanova, O. Rodríguez & Wildpret 1996	T
subass. *tricholaenetosum teneriffae* Rivas-Martínez *et al.* 1993	L,F,C,T,G,H
MAGNOCARICI ELATAE-PHRAGMITETEA AUSTRALIS Klika *in* Klika & V. Novák 1941*nom. inv.*	
NASTURTIO OFFICINALIS-GLYCERIETALIA FLUITANTIS Pignatti 1953	
Rorippion nasturtii-aquatici Géhu & Géhu-Franck 1987	
-*Helosciadietum nodiflori* Maire 1924	C,T,P
MAGNOCARICETALIA ELATAE Pignatti 1953	
Magnocaricion elatae Koch 1926	
-*Mentho-Caricetum calderae* Wildpret, Socorro & León 1987	T
PHRAGMITETALIA AUSTRALIS Koch 1926	
Phragmition australis Koch 1926 [*nom. mut. propos.*]	
Bolboschoenenion maritimi Rivas-Martínez *in* Rivas-Martínez, Costa, Castroviejo & E. Valdés 1980 [*nom. mut. propos.*]	

-*Cypero alopecuroidis-Bolboschoenetum maritimi* Rivas-Martínez *et al.* 1993 [*nom. mut. propos.*] T

Phragmitenion australis (Koch 1926) Rivas-Martínez *in* Rivas-Martínez, Costa, Castroviejo & E. Valdés 1980 [*nom. mut. propos.*]

-*Arundo donax* community Pérez de Paz, Del Arco & Wildpret 1987 C,T,G,P,H

-*Colocasia esculenta* community Pérez de Paz, Del Arco & Wildpret 1987 C,T,G,P

-*Cyperus eragrostis* community Pérez de Paz, Del Arco & Wildpret 1987 C,T,G,P

-*Cyperus involucratus* community Pérez de Paz, Del Arco & Wildpret 1987 C,T,G,P

-*Typha domingensis* community Pérez de Paz, Del Arco & Wildpret 1987 C,T,G

MOLINIO CAERULEAE-ARRHENATHERETEA ELATIORIS Tüxen 1937

Arrhenatheretalia elatioris Tüxen 1931

Arrhenatherion elatioris Koch 1926

-*Arrhenatherum elatius* subsp. *bulbosum* community Del Arco & O. Rodríguez *in* Del Arco *et al.* 2006 P

Holoschoenetalia vulgaris Br.-Bl. *ex* Tchou 1948

Molinio arundinaceae-Holoschoenion vulgaris Br.-Bl. *ex* Tchou 1948

-*Scirpo globiferi-Juncetum acuti* Rivas-Martínez *et al.* 1993 [*nom. mut. propos.*] L,C,T,G,P

Plantaginetalia majoris Tüxen & Preising *in* Tüxen 1950

Paspalo-Polypogonion viridis Br.-Bl. *in* Br.-Bl., Roussine & Nègre 1952 [*nom. mut. propos.*]

-*Aster squamatus* community Reyes, Wildpret & León 2001 L

-*Cyperetum laevigati* Sunding 1972 C

-*Cyperus laevigatus-Apium graveolens* community Reyes, Wildpret & León 2001 L, F?

-*Cyperus rotundus-Cynodon dactylon* community Reyes, Wildpret & León 2001 L

-*Paspalo distichi-Polypogonetum viridis* Br.-Bl. *in* Br.-Bl., Gajewski, Wraber & Walas 1936 [*nom. mut. propos.*] T

Potentillion anserinae Tuxen 1947

-*Juncus effusus-Epilobium parviflorum* community Rivas-Martínez *et al.* 1993 T

NERIO OLEANDRI-TAMARICETEA Br.-Bl. & O. Bolòs 1958

Tamaricetalia Br.-Bl. & O. Bolòs 1958 *em.* Izco, Fernández-González & A. Molina 1984

Tamaricion africanae Br.-Bl. & O. Bolòs 1958	
-*Atriplici ifniensis-Tamaricetum canariensis* Rivas-Martínez *et al.* 1993	C,T,G,P
Tamaricion boveano-canariensis Izco, Fernández-González & A. Molina 1984	
-*Suaedo verae-Tamaricetum canariensis* O. Rodríguez, García Gallo & Reyes 2000	F,C
PARIETARIETEA JUDAICAE Rivas-Martínez *in* Rivas Goday 1964	
PARIETARIETALIA JUDAICAE (Rivas-Martínez 1960) Rivas Goday 1964	
Cymbalario muralis-Asplenion quadrivalentis Segal 1969	
-*Adianto capilli-veneris-Parietarietum judaicae* Segal 1969	C,T
Parietario judaicae-Centhrantion rubri Rivas-Martínez 1960	
-*Umbilico gaditani-Parietarietum judaicae* Rivas-Martínez *et al.* 1993 *corr.* Rivas-Martínez *et al.* 2002	L,C,T
PEGANO HARMALAE-SALSOLETEA VERMICULATAE Br.-Bl. & O. Bolòs 1958	
CHENOLEOIDETALIA TOMENTOSAE Sunding 1972 [*nom. mut. propos.*]	
Chenoleoidion tomentosae Sunding 1972 [*nom. mut. propos.*]	
-*Chenoleoideo tomentosae-Salsoletum vermiculatae* Reyes, Wildpret & León 2001	L,F?
-*Chenoleoideo tomentosae-Suaedetum mollis* Sunding 1972 *corr.* Reyes, Wildpret & León 2001 [*nom. mut. propos.*]	L,F,C,T
subass. *typicum*	L,F,C,T
subass. *atractyletosum preauxianae* Sunding 1972	C
subass. *suaedetosum verae* Biondi 1988	F
FORSSKAOLEO ANGUSTIFOLIAE-RUMICETALIA LUNARIAE Rivas-Martínez *et al.* 1993	
Artemisio thusculae-Rumicion lunariae Rivas-Martínez *et al.* 1993	
-*Artemisio thusculae-Plantaginetum arborescentis* Rivas-Martínez *et al.* 1993	T
-*Artemisio thusculae-Rumicetum lunariae* Rivas-Martínez *et al.* 1993	C,T,P,H
-*Bidenti pilosae-Ageratinetum adenophorae* Rivas-Martínez *et al.* 1993 *nom. prov.*	T
-*Ceballosio fruticosae-Withanietum aristatae* Rivas-Martínez *et al.* 1993 *nom. prov.* [*nom. mut. propos.*]	T
-*Gonospermo fruticosi-Lavateretum acerifoliae* Rivas-Martínez *et al.* 1993 *nom. prov.*	T
-*Rumex lunaria* alluvial community Del Arco & O. Rodríguez *in* Del Arco *et al.* 2006	P
-*Rumex lunaria* community on lapilli Del Arco & O. Rodríguez *in* Del Arco *et al.* 2006	H

-*Salvio canariensis-Pterocephaletum dumetori* Salas, E. Fernández & Quintana 2009	C
-*Tricholaeno teneriffae-Rumicetum lunariae* Sunding 1972	C
Asteriscion sericei Rivas-Martínez, Wildpret, Martín & Scholz *in* Martín, Wildpret & Scholz 2011	
-*Andryalo variae-Asteriscetum sericei* O. Rodríguez, García Gallo & Reyes 2000	F
-*Lavandulo pinnatae-Asteriscetum intermedii* Reyes, Wildpret & León 2001	L
-*Asterisco stenophylli-Ononidetum ulicinae* Sunding 1972 [*nom. mut. propos.*]	C
Launaeo arborescentis-Schizogynion sericeae Rivas-Martínez *et al.* 1993	
-*Artemisietum ramosae* Del Arco, Acebes, A. Rodríguez, Padrón, O. Rodríguez, Pérez de Paz & Wildpret 1997	T,C
-*Cenchro ciliaris-Launaeetum arborescentis* Reyes, Wildpret & León 2001	L
-*Herniarietum canariensis* Rivas-Martínez *et al.* 1993 *nom. prov.*	T
-*Launaeo arborescentis-Schizogynetum sericeae* Rivas-Martínez *et al.* 1993	C,T,G
subass. *schizogynetosum sericeae* [*typicum*]	C,T,G
subass. *artemisietosum ramosae* Rivas-Martínez *et al.* 1993	T
subass. *reichardietosum crystallinae* Rivas-Martínez *et al.* 1993	T
subass. *salsoletosum divaricatae* Del Arco & O. Rodríguez *in* Del Arco *et al.* 2006 *nom. prov.*	T
subass. *zygophylletosum fontanesii* Rivas-Martínez *et al.* 1993	T
-*Pelargonium capitatum-Launaea arborescens* community Wildpret, Beltrán, J.M. González & Centellas 1995	L
NICOTIANO GLAUCAE-RICINETALIA COMMUNIS Rivas-Martínez, Fernández-González & Loidi 1999	
Nicotiano glaucae-Ricinion communis Rivas-Martínez, Fernández-González & Loidi 1999	
-*Polycarpo tetraphylli-Nicotianetum glaucae* Sunding 1972	L,F,C,T,G,P,H
-*Tropaeolo majoris-Ricinetum communis* Rivas-Martínez *et al.* 1993	L,F,C,T,G,P
POETEA BULBOSAE Rivas Goday & Rivas-Martínez *in* Rivas-Martínez 1978	
POETALIA BULBOSAE Rivas Goday & Rivas-Martínez *in* Rivas Goday & Ladero 1970	
Molineriello minutae-Trifolion subterranei Rivas Goday 1964 [*nom. inv. et nom. mut. propos.*]	
-*Trifolio subterranei-Poetum bulbosae* Rivas Goday 1964 [*nom. inv. propos.*]	C,T,G,H

POLYCARPAEO NIVEAE-TRAGANETEA MOQUINI A. Santos *ex* Rivas-Martínez & Wildpret 2002

ZYGOPHYLLO FONTANESII-POLYCARPAETALIA NIVEAE A. Santos *ex* Géhu, Biondi, Géhu-Franck, Hendoux & Mossa 1996

Polycarpaeo niveae-Euphorbion paraliae Rivas-Martínez & Wildpret 2002

-*Euphorbio paraliae-Cyperetum capitati* Sunding 1972 [*nom. mut. propos.*]	L,F,C,T,G
subass. *typicum*	L,F,C,T,G
subass. *atriplicetosum glaucae* Biondi, Allegrezza, Taffetani & Wildpret 1994	L
subass. *zygophylletosum fontanesii* Biondi, Allegrezza, Taffetani & Wildpret 1994 [*nom. inval.*]	L,F,T
-*Polycarpaeo niveae-Lotetum lancerottensis* Esteve 1968	L,F,C,T,G?
subass. *typicum*	L,F,C,T,G?
subass. *ononidetosum hesperiae* (Biondi, Allegrezza, Taffetani & Wildpret 1994) Reyes, Wildpret & León 2001	L

Traganion moquini Sunding 1972

-*Frankenio capitatae-Zygophylletum gaetuli* Del Arco & Wildpret 1991	F
-*Polycarpaeo niveae-Pulicarietum burchardii* Scholz, Wildpret & Martín 2003	F
-*Traganetum moquini* Sunding 1972	L,F,C,T

POLYGONO-POETEA ANNUAE Rivas-Martínez 1975

POLYGONO ARENASTRI-POETALIA ANNUAE Tüxen *in* Géhu, Richard & Tüxen 1972 *corr.* Rivas-Martínez, Báscones, T.E. Díaz, Fernández-González & Loidi 1991

Euphorbion prostratae Rivas-Martínez 1976

-*Polycarpo-Alternantheretum* Oberdorfer *ex* Lohmeyer & Trautmann 1970	C,T,G,P

Polycarpion tetraphylli Rivas-Martínez 1975

-*Polycarpo tetraphylli-Cotuletum australis* Wildpret, Pérez de Paz, Del Arco & García Gallo 1988	C,T,P
subass. *cotuletosum australis* [*typicum*]	C,T,P
subass. *solivetosum stoloniferae* Rivas-Martínez *et al.* 1993	T

Saginion procumbentis Tüxen & Ohba *in* Géhu, Richard & Tüxen 1972

-*Poo infirmae-Saginetum stoloniferi* Wildpret & Martín 2005	T

Sclerochloo durae-Coronopodion squamati Rivas-Martínez 1975

-*Polycarpo tetraphylli-Coronopodetum squamati* Rivas-Martínez *et al.* 1993	T

POTAMETEA Klika *in* Klika & V. Novák 1941

POTAMETALIA PECTINATI Koch 1926

Potamion pectinati (Koch 1926) Görs 1977

-*Elodea canadensis* stands Pérez de Paz, Del Arco & Wildpret 1987 — T,P

-*Myriophyllum spicatum* community Pérez de Paz, Del Arco & Wildpret 1987 — T,P

-*Potamogeton pusillus* community Sunding 1972 — C,T

Ranunculion fluitantis Neuhäusl 1959

-*Callitriche stagnalis-Potamogetum nodosus* community Rivas-Martínez *et al.* 1993 — C,T,G

-*Ranunculus penicillatus* community Del Arco & O. Rodríguez — T

Zannichellion pedicellatae Schaminée, Lanjouw & Schipper 1990

-*Zannichellietum pedicellatae* Nordhagen 1954 *em.* Pott 1992 — T

UTRICULARIETALIA VULGARIS Den Hartog & Segal 1964

Ceratophyllion demersi Den Hartog & Segal *ex* Passarge 1996

-*Potamo-Ceratophylletum demersi* (Hild & Rehnelt 1965) Passarge 1996 — T

PRUNO HIXAE-LAURETEA NOVOCANARIENSIS
Oberdorfer 1965 *corr.* Rivas-Martínez *et al.* 2002

ANDRYALO PINNATIFIDAE-ERICETALIA ARBOREAE Oberdorfer 1965

Morello fayae-Ericion arboreae Oberdorfer 1965 [*nom. mut. propos.*] (*Fayo-Ericion arboreae* Oberdorfer 1965, *Beitr. Naturk Forsch. Südwestdeutschland* 24(1): 78, Art. 41b, 45)

-*Morello fayae-Ericetum arboreae* Oberdorfer 1965 [*nom. mut. propos.*] (*Fayo-Ericetum arboreae* Oberdorfer 1965, *Beitr. Naturk Forsch. Südwestdeutschland* 24(1): 78, Art. 41b, 45) — C,T,G,P,H

subass. *ericetosum arboreae* [*typicum*] — C,T,G,P,H

subass. *telinetosum canariensis* Del Arco & Wildpret 1983 — T

Telino canariensis-Adenocarpion foliolosi
Rivas-Martínez *et al.* 1993

-*Adenocarpo foliolosi–Chamaecytisetum angustifolii* Del Arco *et al.* 2009 — G

-*Adenocarpus foliolosus* community Oberdorfer 1965 — T

-*Chamaecytisetum angustifolii* Del Arco, Pérez de Paz, O. Rodríguez, Salas & Wildpret 1992 *nom. prov.* — T

-*Chamaecytiso canariae-Adenocarpetum villosi* (Sunding 1972) Rivas-Martínez & Wildpret 2002 — C

-*Lotetum campylocladi* Del Arco, Pérez de Paz & Wildpret 1987 — T,P,H,P,H

-*Micromerio benthami-Telinetum microphyllae* Sunding 1972 [*nom. mut. propos.*] — C

subass. *typicum* — *C*

subass. *euphorbietosum regis-jubae* Naranjo, Salas & Almeida 2001 — C

subass. *sideritidetosum dasygnaphalae* (Sunding 1972) Naranjo, Salas & Almeida 2001 — C

-*Telinetum canariensis* Del Arco & Wildpret 1983	T
subass. *telinetosum canariensis* [*typicum*]	T
subass. *ericetosum platycodonis* Rivas-Martínez *et al.* 1993	T
subass. *telinetosum pallidae* Rivas-Martínez *et al.* 1993	T
Pruno hixae-Lauretalia novocanariensis Oberdorfer *ex* Rivas-Martínez, Arnáiz, Barreno & A.Crespo 1977 *corr.* Rivas-Martínez *et al.* 2002 [*nom. mut. propos.*]	
Ixantho viscosae-Laurion novocanariensis Oberdorfer *ex* A. Santos *in* Rivas-Martínez, Arnáiz, Barreno & A. Crespo 1977 *corr.* Rivas-Martínez *et al.* 2002	
-*Diplazio caudati-Ocoteetum foetentis* Rivas-Martínez *et al.* 1993	C,T,G,P
-*Ilici canariensis-Ericetum platycodonis* Rivas-Martínez *et al.* 1993	T,G
-*Lauro novocanariensis-Perseetum indicae* Oberdorfer *ex* Rivas-Martínez, Arnáiz, Barreno & A. Crespo 1977 *corr.* Rivas-Martínez *et al.* 2002	C,T,G,P,H
subass. *typicum*	C,T,G,P,H
subass. *ocoteetosum foetentis* Del Arco, Pérez de Paz, Acebes, A.B. Fernández & Wildpret 2009	G
-*Micromerio lepidae-Ericetum arboreae* Del Arco, Pérez de Paz, Acebes, A.B. Fernández & Wildpret 2009	G
-*Pericallido murrayii-Morelletum fayae* A. Santos 1976 [*nom. mut, propos.*] (*Senecio murrayii-Myricetum fayae* A. Santos 1976, *Anales Inst. Bot. Cavanilles* 33: 257. Art. 45-ICPN)	T,P,H
subass. *pericallidetosum murrayii* [*typicum*]	H
subass. *pericallidetosum cruenta* Del Arco & O. Rodríguez *in* Del Arco *et al.* 2006 *nom. prov.*	T
subass. *pericallidetosum papyracea* Del Arco & O. Rodríguez *in* Del Arco *et al.* 2006 *nom. prov.*	P
subass. *pericallidetosum webbii* Del Arco & O. Rodríguez *in* Del Arco *et al.* 2006 *nom. prov.*	C
-*Violo rivinianae-Morelletum fayae* M. Fernández *ex* Del Arco *et al.* 2009 [*nom. mut. propos.*] (*Violo rivinianae-Myricetum fayae* M . Fernández ex Del Arco et al. 2009, *Ann. Bot. Fennici* 46: 182, Art. 45-ICPN)	G
Visneo mocanerae-Apollonion barbujanae Rivas-Martínez *in* Capelo *et al.* 2000	
-*Echio handiensis-Visneetum mocanerae* Martín, Wildpret & Scholz 2011	F
-*Visneo mocanerae-Arbutetum canariensis* Rivas-Martínez *et al.* 1993	C,T,G,P,H

Rubo bollei-Salicetalia canariensis Rivas-Martínez *in* Capelo *et al.* 2000

Rubio periclymeni-Rubion ulmifolii (Oberdorfer 1965) Rivas-Martínez *et al.* 1993

-*Rubio periclymeni-Rubetum ulmifolii* Oberdorfer 1965 C,T,G,P,H

Salicion canariensis Rivas-Martínez, Wildpret, Del Arco, O. Rodríguez, Pérez de Paz, García Gallo, Acebes, T.E. Díaz & Fernández-González *ex* Rivas-Martínez, Fernández-González & Loidi 1999

-*Rubo ulmifolii-Salicetum canariensis* J.C. Rodríguez, Del Arco & Wildpret 1986 [*nom. mut. propos.*] C,T,G,P

subass. *salicetosum canariensis* [*typicum*] C,T,G,P

subass. *ericetosum arboreae* J.C. Rodríguez, Del Arco & Wildpret 1986 T,G

subass. *polygonetosum salicifolii* J.C. Rodríguez, Del Arco & Wildpret 1986 C,T,G

subass. *woodwardietosum radicantis* Himstedt 2010 T

RHAMNO CRENULATAE-OLEETEA CERASIFORMIS A. Santos *ex* Rivas-Martínez 1987 [*nom. inv. propos.*]

Micromerio hyssopifoliae-Cistetalia canariensis Pérez de Paz, Del Arco & Wildpret 1990 *corr.* Rivas-Martínez, Martín & Wildpret *in* Rivas-Martínez & col. 2011

Micromerio hyssopifoliae-Cistion canariensis Pérez de Paz, Del Arco & Wildpret 1990 *corr.* Rivas-Martínez, Martín & Wildpret *in* Rivas-Martínez & col. 2011

-*Carlino salicifoliae-Echietum famarae* Martín, Wildpret & Scholz 2011 F

-*Cistetum symphytifolio-canariensis* Rivas-Martínez *et al.* 1993 *corr.* Rivas-Martínez *in* Rivas-Martínez & col. 2011 T

-*Echio aculeati-Micromerietum hyssopifoliae* Pérez de Paz, Del Arco & Wildpret 1990 H

subass. *typicum* H

subass. *rubietosum fruticosae* Stiertorfer 2005 H

-*Echio breviramis-Micromerietum herpyllomorphae* Pérez de Paz, Del Arco & Wildpret 1990 P

-*Euphorbio regis-jubae-Cistetum canariensis* Esteve *ex* Salas, Del Arco & Pérez de Paz 1998 *corr.* Rivas-Martínez, Martín & Wildpret *in* Rivas-Martínez & col. 2011 C

-*Micromerio gomerensis-Cistetum canariensis* Del Arco *et al.* 2009 *corr.* Rivas-Martínez, Martín & Wildpret *in* Rivas-Martínez & col. 2011 G

-*Micromerio herpyllomorphae-Cistetum canariensis* Del Arco & O. Rodríguez *in* Del Arco *et al.* 2006 *nom. prov. corr. hoc loco* P

-*Micromerio hyssopifoliae-Cistetum canariensis* A. Santos *in* Pérez de Paz, Del Arco & Wildpret 1990 *corr.* Rivas-Martínez, Martín & Wildpret *in* Rivas-Martínez & col. 2011	H
subass. *cistetosum canariensis* [*typicum*]	H
subass. *hypericetosum canariensis* A. Santos 1980 [*nom. inval.*]	H
-*Micromerio variae-Globularietum salicinae* Rivas-Martínez *et al.* 1993	T
-*Spergulario fimbriatae-Helianthemetum canariensis* Reyes, Wildpret & León 2001	L,F?
RHAMNO CRENULATAE-OLEETALIA CERASIFORMIS A. Santos 1983 [*nom. inv. propos.*]	
Mayteno canariensis-Juniperion canariensis A. Santos & M. Fernández *ex* A. Santos 1983 *corr.* Rivas-Martínez *et al.* 1993	
-*Brachypodio arbusculae-Juniperetum canariensis* M. Fernández 1983 *corr.* Rivas-Martínez *et al.* 1993	G
subass. *typicum*	G
subass. *ericetosum arboreae* Del Arco & O. Rodríguez *in* Del Arco *et al.* 2006 *nom. prov.*	G
-*Convolvulo lopezsocasii-Oleetum cerasiformis* Reyes, Wildpret & León 2001.	L
subass. *oleetosum cerasiformis* [*typicum*].	L
subass. *euphorbietosum balsamiferae* Reyes, Wildpret & León 2001	L
-*Junipero canariensis-Oleetum cerasiformis* O. Rodríguez, Wildpret, Del Arco & Pérez de Paz 1990 *corr.* Rivas-Martínez *et al.* 1993	T
subass. *typicum*	T
subass. *ericetosum arboreae* O. Rodríguez, Wildpret, Del Arco & Pérez de Paz 1990	T
-*Micromerio rupestris-Oleetum cerasiformis* O. Rodríguez, García Gallo & Reyes 2000	F
-*Pistacio lentisci-Oleetum cerasiformis* Del Arco, Salas, Acebes, M.C. Marrero, Reyes & Pérez de Paz 2002	C
-*Rhamno crenulatae-Hypericetum canariensis* Rivas-Martínez *et al.* 1993	C,T,P,H
-*Rhamno crenulatae-Juniperetum canariensis* A. Santos 1983 *corr.* O. Rodríguez, Del Arco, García Gallo, Acebes, Pérez de Paz & Wildpret 1998 [*nom.inv. propos.*]	P
-*Rubio fruticosae-Juniperetum canariensis* A. Santos *in* Rivas-Martínez *et al.* 1993	H
subass. *typicum*	H
subass. *ericetosum arboreae* Del Arco, Acebes & Pérez de Paz 1996	H

subass. *pinetosum canariensis* (Del Arco, Pérez de Paz, Wildpret, Lucía & Salas 1990) Del Arco, Acebes & Pérez 1996 — H

Phoenicion canariensis Rivas-Martínez & Del Arco 2011

Periploco laevigatae-Phoenicetum canariensis Rivas-Martínez *et al.* 1993 — F,C,T,G,P

Retamion rhodorhizoidis Del Arco, O. Rodríguez, Acebes, Salas & Garzón 2009

-*Echio aculeati-Retametum rhodorhizoidis* Rivas-Martínez *et al.* 1993 — T

-*Echio decaisnei-Retametum rhodorhizoidis* Salas, Del Arco, Acebes, Naranjo & R. González 2003 — C

-*Echio hierrensis-Retametum rhodorhizoidis* Del Arco, Acebes & Pérez de Paz 1996 — H H

-*Euphorbio berthelotii-Retametum rhodorhizoidis* Del Arco *et al.* 2009 — G

-*Euphorbio wildpretii-Retametum rhodorhizoidis* A. Santos 1983 *corr.* Rivas-Martínez & Gaisberg *in* Rivas-Martínez & col. 2011 — P

RUPPIETEA J.Tüxen 1960

RUPPIETALIA MARITIMAE J.Tüxen 1960

Ruppion maritimae Br.-Bl. *ex* Westhoff *in* Bennema, Sissingh & Westhoff 1943

-*Enteromorpho intestinalis-Ruppietum maritimae* Westhoff *ex* Tüxen & Böckelmann 1957 — L,F,C,T,G

SARCOCORNIETEA FRUTICOSAE Br.-Bl. & Tüxen *ex* A. & O. Bolòs 1950 [*nom. mut. propos.*]

SARCOCORNIETALIA FRUTICOSAE Br.-Bl. 1933 [*nom. mut. propos.*]

Arthrocnemion macrostachyi Rivas-Martínez & Costa 1984 [*nom. mut. propos.*]

Arthrocnemenion macrostachyi (Rivas-Martínez & Costa 1984) Rivas-Martínez & Costa 2011

-*Zygophyllo fontanesii-Arthrocnemetum macrostachyi* M. Fernández & A. Santos 1984 — L,F

Sarcocornion fruticosae Br.-Bl. 1933 [*nom. mut. propos.*]

Sarcocornienion perennis Rivas-Martínez *in* Rivas-Martínez, Costa, Castroviejo & Valdés 1980 [*nom. mut. propos.*]

-*Halimiono portulacoides-Salicornietum perennis* (M. Fernández & A. Santos 1984) Del Arco, Izco & O. Rodríguez *nom. novum* (in press) — L,F

subass *typicum* — L,F

subass. *limonietosum canariensis* M. Fernández & A. Santos 1984 — F(Lobos)

Suaedion verae (Rivas-Martínez, Lousã, T.E. Díaz, Fernández-González & J.C. Costa 1990) Rivas-Martínez, Fernández-González & Loidi 1999	
-*Frankenio capitatae-Suaedetum verae* Reyes, Rivas-Martínez & Wildpret 2002	L,F,C
-*Zygophyllym fontanesii-Suaeda mollis* community Del Arco & O. Rodríguez *in* Del Arco *et al.* 2006	L,C
STELLARIETEA MEDIAE Tüxen, Lohmeyer & Preising *ex* von Rochow 1951	
CHENOPODIO-STELLARIENEA Rivas Goday 1956	
Chenopodietalia muralis Br.-Bl. *in* Br-Bl., Gajewski, Wraber & Walas 1936	
Chenopodion muralis Br.-Bl. *in* Br.-Bl., Gajewski, Wraber & Walas 1936	
Malvenion parviflorae Rivas-Martínez 1978	
-*Chenopodio muralis-Malvetum parviflorae* Lohmeyer & Trautmann 1970	L,F,C,T,G,P,H
subass. *malvetosum parviflorae* [*typicum*]	L,F,C,T,G,P,H
subass. *patellifolietosum patellaris* Lohmeyer 1975 [*nom. mut. propos.*]	C,P
Mesembryanthemion crystallini Rivas-Martínez *et al.* 1993	
-*Mesembryanthemetum crystallini* Sunding 1972	L,F,C,T,G,P,H
Sisymbrietalia officinalis J. Tüxen *in* Lohmeyer *et al.* 1962 *em.* Rivas-Martínez, Báscones, T.E. Díaz, Fernández-González & Loidi 1991	
Hordeion leporini Br.-Bl.*in* Br-Bl., Gajewski, Wraber & Walas 1936 *corr.* O. Bolòs 1962	
-*Bromo-Hirschfeldietum incanae* Oberdorfer *ex* Lohmeyer 1975	C,T,G,P,H
-*Euphorbio terracinae-Anacycletum coronati* Reyes, Wildpret & León 2001	C,T,G,P,H
Thero-Brometalia (Rivas Goday & Rivas-Martínez *ex* Esteve 1973) O. Bolòs 1975	
Echio plantaginei-Galactition tomentosae O. Bolòs & R.Molinier 1969	
-*Galactito tomentosae-Trachynietum distachyae Rivas*-Martínez *et al.* 1993 [*nom. mut. propos.*] (*Galactito tomentosae-Brachypodietum distachyae* Rivas-Martínez et al. 1993, *Itinera Geobotanica* 7: 264. Art. 45-ICPN)	C,T,G,P
-*Raphanus raphanistrum-Bromus diandrus community* Stiertorfer 2005	H
Resedo lanceolatae-Moricandion F. Casas & M.E. Sánchez 1972	
-*Bupleuro semicompositi-Mairetetum microspermae* Reyes, Wildpret & León 2001	L,F?

-*Iflogo spicatae-Stipetum capensis* (Esteve & Socorro 1977) Rivas-Martínez *et al.* 1993	L,F,C,T
-*Launaeo nudicaulis-Resedetum lancerotae* O. Rodríguez, García Gallo & Reyes 2000	L,F
-*Senecionetum incrassatii* Pérez de Paz, Del Arco & Wildpret 1985	H
-*Senecioni coronopifolii-Echietum bonnetii* Rivas-Martínez *et al.* 1993	T
-*Volutaria tubuliflora* community Reyes, Wildpret & León 2001	L
STELLARIENEA MEDIAE	
APERETALIA SPICAE-VENTI J. Tüxen & Tüxen *in* Malato-Beliz, J. Tüxen & Tüxen 1960	
Scleranthion annui (Kruseman & Vlieger 1939) Sissingh *in* Westhoff, Dijk & Passchier 1946	
Scleranthenion annui Kruseman & Vlieger 1939	
-*Raphanus raphanistrum-Scandix pecten-veneris* community Rivas-Martínez *et al.* 1993	C,T
SOLANO NIGRI-POLYGONETALIA CONVOLVULI (Sissingh *in* Westhoff, Dijk & Passchier 1946) O. Bolòs 1962	
Fumarion wirtgenii-agrariae Brullo *in* Brullo & Marcenò 1985	
-*Convolvulus arvensis* community Reyes, Wildpret & León 2001	L
-*Sinapis arvensis-Fumaria muralis* community Rivas-Martínez *et al.* 1993	C,T
Polygono-Chenopodion polyspermi Koch 1926	
Digitario ischaemi-Setarienion viridis (Sissingh *in* Westhoff, Dijk & Passchier 1946) Oberdorfer 1957	
-*Setario verticillatae-Echinochloetum cruris-galli* Peinado, Bartolomé & Martínez-Parras 1985	T
THERO-SALICORNIETEA Tüxen in Tüxen & Oberdorfer *ex* Gèhu & Gêhu-Frank 1984	
THERO-SUAEDETALIA Br.-Bl. & O. Bolòs 1958	
Thero-Suaedion Br.-Bl. *in* Br-.Bl., Roussine & Nègre 1952	
-*Cressetum villosae* Rothmaler 1943 *corr.* Rivas-Martínez *et al.* 2002	C
-*Frankenio boissieri-Suaedetum spicatae* Reyes, M.A. Padrón, Guma, A. Santos & J. Navarro 2009	F,C
TRIFOLIO MEDII-GERANIETEA SANGUINEI Müller 1962	
ORIGANETALIA VULGARIS Müller 1962	
Ranunculo cortusifolii-Geranion canariensis Rivas-Martínez *et al.* 1993	
-*Ranunculo cortusifolii-Geranietum canariensis* Rivas-Martínez *et al.* 1993	T,G,P
-*Rumici maderensis-Pimpinelletum dendrotragii* Rivas-Martínez *et al.* 1993	T

TUBERARIETEA GUTTATAE (Br.-Bl. *in* Br.-Bl., Rosussine & Nègre 1952) Rivas Goday & Rivas-Martínez 1963 [*nom. mut. propos.*]

MALCOLMIETALIA Rivas Goday 1958

Ononidion tournefortii Géhu, Biondi, Géhu-Franck, Hendoux & Mossa 1996

-*Ononido tournefortii-Cyperetum capitati* Wildpret, Del Arco & Acebes *in* Del Arco, Acebes & Wildpret 1983 — T

TUBERARIETALIA GUTTATAE Br.-Bl. *in* Br.-Bl., Molinier & Wagner 1940 [*nom. mut. propos.*]

Tuberarion guttatae Br.-Bl. *in* Br.-Bl., Molinier & Wagner 1940 [*nom. mut. propos.*]

-*Hypochaerido glabrae-Tuberarietum guttatae* Rivas-Martínez *et al.* 1993 — C,T,G,P,H

-*Tuberario guttatae-Ophioglossetum lusitanici* Br.-Bl. in Br.-Bl., Molinier & Wagner 1940 [*nom. mut. propos.*] — H

-*Vulpio myuri-Laphangietum teydei* Wildpret, O. Rodríguez & M.C.León *in* Rivas-Martínez *et al.*1993 *nom. mut propos.* (*Vulpio myuri-Gnaphalietum teydei Wildpret*, O. Rodríguez & M.C.León in Rivas-Martínez et al. 1993, *Itinera Geobotanica* 7: 247. Art. 45-ICPN) — T

ZOSTERETEA MARINAE Pignatti 1953

ZOSTERETALIA MARINAE Béguinot 1941

Zosterion marinae Christiansen 1934

-*Zosteretum noltii* Harmsen 1936 — L

Appendix 2: Brief History of the Botanical Studies and Explorations in the Islands

Many scientists have come to the Canary Islands attracted by its variety, to study the peculiarities of its flora. The insular nature of the territory allows to test out theories and assumptions difficult to cover in larger land units. Although botanical research in the Canaries did not begin in depth until the eighteenth century, there are references to the islands' terrestrial flora and vegetation since their discovery. Its history goes back to accounts by Phoenicians, Carthaginians, Greeks and Romans. At the beginning of the first century C.E., based on the writings of King Juba of Mauritania, the Roman Pliny alludes to certain plants observed on some islands. Commercial search for plants of alimentary, medicinal or industrial interest stimulated the interest of some peoples of antiquity in certain Canary plants. In this way, the purple dye obtained from orchil (the lichen *Roccella*) and the supposed magical and medicinal properties of "dragon blood" extracted from *Dracaena draco* were among the plant products that attracted their attention.

From the chronicles of the Conquest a series of sources of information such as island council ("cabildo") agreements, land assignments records, local bylaws and ordinances, notarial and official records by scribes, begin to appear. They provide information about the state of and changes in the natural vegetation of the archipelago and, of course, about the introduction of new crops. Several authors of the sixteenth and seventeenth centuries also provide botanical references of great interest.

The Canary Islands quickly became an excellent transit station on the routes between the Old and New World. The growing interest awakened in Europe by the knowledge of plant species from the newly discovered lands in the course of these centuries determined the collection and export of Canary plants by unknown collectors, who brought this plant heritage to the countries that had most commercial relations with the islands. Soon, commercial interest led to scientific interest, and so the material that served for the first botanical description of the *dragon tree*, published by Charles de L'Ecluse in Antwerp in 1576, came to European gardens. Similarly, throughout the seventeenth century, other European botanists, *e.g.*, Plukenet, Commelin, Morison, Petiver, Ray, etc. began to describe Canarian plants cultivated in European botanical gardens for their novelty and medicinal uses. The English doctor and traveler James Cunningham is the first person known to collect plants on the islands, after arriving in Santa Cruz de La Palma in 1697. Soon later, the prestigious French naturalist and clergyman Feuillée visited the archipelago in 1724. He was the first in describing and classifying in situ some thirty endemic plant species of the Canary Islands. For that, he has been considered as the father of the Canary Islands' Botany.

Throughout the eighteenth century, without visiting the archipelago, other prestigious European botanists continued to describe Canary plants: Linnaeus, Willdenow, Linnaeus filius, Solander, Banks, L'Heritier, Lamarck, Poiret, Desfontaines, De Candolle, etc. Almost all of them were sent plants from Canaries by the British gardener Francis Masson, who in 1777 by order of William Aiton, then senior gardener of the Royal Botanic Gardens of Kew, carried out the first botanical collection of importance.

Fig. A2.1 José de Viera y Clavijo

The history of Canarian Botany has in its beginnings an important local name: José de Viera y Clavijo, born at Realejo Alto (Tenerife) (Fig. A2.1). He was a historian and the first Canarian naturalist, who in the last stage of his life worked on plant uses and created a Natural History Cabinet. Among his scientific works, his Dictionary of Natural History of the Canaries deserves to be highlighted. We also should not forget Alonso de Nava Grimón, the 6th Marquis of Villanueva del Prado, a resident of La Laguna, the then capital of Tenerife, founder and first head of the Acclimatization Garden of La Orotava (now the Botanical Garden of Puerto de la Cruz). He was a friend of the French naturalist and historian André-Pierre Ledrú who stayed for several months on Tenerife in 1796, compiling a herbarium of the island and writing a catalog of plants for the Garden. He was proposed to the Madrid Court as head of the Garden but was refused. Bory de Saint-Vincent arrived at the islands in 1798 and stayed for four years. He was the author of the interesting book "Essais sur les Isles Fortunées et l'antique Atlantide, ou Précis de l'Histoire générale de l'Archipel des Canaries" published In Paris in 1803. It included an interesting catalog of 467 plant species, which for the first time included cryptogams (algae, fungi, lichens, bryophytes). In 1799 the famous German naturalist and traveler Alexander von Humboldt, accompanied by the French botanist Aimé Bonpland, also visited Tenerife on their way to the Americas. During the days they were on the island they climbed the Pico del Teide. During the ascent Humboldt described for the first time the elevational zonation of the vegetation, accompanied by other diverse botanical observations. Auguste Broussonet, who was consul of France in Tenerife for four years, arrived in the Canary Islands in the same year. During his stay, he began to prepare a catalog of plants of the archipelago, which was never published. However, he provided interesting examples of the Canary flora to famous European scientists, among whom was the Spanish botanist Cavanilles, who described several plants new to science.

Throughout the nineteenth century, attracted by the richness of the Canary flora and above all by the ascent of the Pico del Teide, scientific explorations proliferated in the Canaries. In 1815 the naturalists Christen Smith and Leopold von Buch toured the archipelago in successive explorations. After the death of the first, Buch published the results of their research under the title "Physicalische Beschreibung der Canarischen Inseln" in 1825. It was the first comprehensive geography of the islands and the most important contribution on their flora published until then. Its catalog amounted to 578 species. This period preceded what has been considered the "Golden Age" of Canary Islands Botany, the preparation and subsequent publication in Paris (between 1835 and 1850) of the "Histoire Naturelle des Iles Canaries" by the English botanist Philip Barker Webb and the French naturalist Sabin Berthelot (Fig. A2.2). This work still is indispensable for consultation about historical, floristic and taxonomic issues. The botanical section, Phytographia Canariensis, consists of three volumes and constitutes the main treatise written until then on Canary Island flora. Webb wrote it with the collaboration of the most prestigious specialists on systematic botany of that time (Decaisne, CH Schultz, Parlatore, Reichenbach, Moquin-Tandon, etc.). 1116 species of phanerogams were catalogued, many of them new to science. The "cellular" plants (cryptogams) were described by Montagne in a masterful and exhaustive way (141 algae, 53 fungi, 88 lichens and 108 bryophytes), including numerous endemics. The study was complemented with the Géographie Botanique by Berthelot. Eugène Bourgeau, a French traveler and botanist, visited the archipelago in 1846 sent by Webb to collect Canary plants. In 1855 Bourgeau returned to the islands to continue his collections, but this time by order of the Museum of Paris.

Fig. A2.2 (**a**) Philip Barker Webb. (**b**) Sabin Berthelot

During the second half of the nineteenth century, more European botanists came to the islands, no doubt attracted by the work of Webb and Berthelot. Between 1855 and 1893, Carl Bolle explored the poorer known regions of the eastern islands and published several studies, particularly on their flora. In 1859, the German Schacht published a study of the vegetation of Madeira and Tenerife. At that time, Joseph Dalton Hooker, director of the Royal Botanic Gardens of Kew, visited the Canary Islands. In 1880, Sauer's catalog of "Plantes des Canaries" already compiled 1296 taxa. Between 1881 and 1883, the Spanish military doctor and botanist Ramón Masferrer y Arquimbau published his "Recuerdos botánicos de Tenerife". In 1859 the Swiss Hermann Wildpret came to Tenerife, hired as senior Gardener of the Botanical Garden of La Orotava, and was its soul and support during the second half of the nineteenth century. The botanist Wilhelm Hillebrand gathered important plant collections on the islands around 1882. Almost simultaneously the Swiss botanist Hermann Christ explored the islands and later published two interesting works: "Vegetation und Flora der Canarischen Inseln" (1885) and "Spicilegium Canariense" (1888).

Within the 19th and 20th centuries several Canarian naturalists became interested in botany. The politician and writer Manuel de Ossuna y Saviñón was director of the Botanical Garden of La Orotava and author of several studies, among them the "Catálogo de las plantas más curiosas que nacen en la Isla de Tenerife" (1834). Botany was the favorite science of the lawyer, politician and painter Domingo Bello y Espinosa and, besides a flora of Puerto Rico, he published "Apuntes botánicos de las Islas Canarias" (1878) and "Un jardín canario" (1879). Doctor Víctor Pérez González, born on La Palma but settled in Puerto de la Cruz, was a notable expert on vegetation and promoter of some indigenous crops, such as *Chamaecytisus proliferus* ("tagasaste" or tree lucerne), on which he published various works. Manuel de Ossuna y Van den Heede, professor at the University of La Laguna, published several works on nature in the Canaries, several of them on the Anaga peninsula (Tenerife), *e.g.,* his "Noticia sobre la flora y fauna de Anaga" (1898). The pharmacist, writer and painter from Tenerife, Eduardo Rodríguez Núñez, stands out for his botanical knowledge shown in several studies. In 1891 he began to write a Popular Dictionary of the Botany of Tenerife that was never concluded. Doctor Jorge Víctor Pérez Ventoso, from Puerto de la Cruz (Tenerife), devoted himself to the study of the indigenous flora, becoming the first local expert at that time. His work in favor of the "resurrection" of the Canary cedar, a species almost extinct on the islands at that time, is of special interest. That task was continued by his wife Constanza Carnochan. He published numerous studies about the Canary flora in local and European magazines, which to a large extent were compiled by his wife and published in London in 1925 under the title "A Selection of Botanical and other Papers". Agustín Cabrera Díaz, D. Sc. and Professor at the General and Technical Institute of the Canary Islands, was the main Canarian naturalist at the dawn of the twentieth century. He made several study trips to Europe attending congresses and publishing several research papers in the local press, most of them related to applied botany, such as "Plantas extintas de Canarias" (1906).

The twentieth century began with numerous works on the vascular flora of the Islands. The English cleric and botanist Richard Paget Murray tried to write a second Flora of the Canary Islands, which he did not finish. At that time the German Bornmüller visited the islands, and in 1904 published a study on the flora of Madeira and Tenerife. Besides this, he wrote the first important work on the flora of the hitherto forgotten island of El Hierro. Also noteworthy was the work of M. Schiffner, who in 1901 wrote a treatise on the flora of Madeira, Tenerife and Gran Canaria, and in 1902 studied the bryophytes of the Atlantic Islands for Bornmüller. Of great informative value is the work of H. Schenk "Beitrage zur Kenntnis der Vegetation der Canarischen Inseln", published in Jena in 1907. The work of two French botanists, C.J. Pitard and L. Proust, entitled "Les Iles Canaries. Flore de l'Archipel" that saw the light in Paris in 1908, is of paramount importance. It constitutes a comprehensive catalog that brings together all the plants known until then from the Canaries. Between 1910 and 1928, the German physician Oscar Burchard, based in La Orotava, was the leading scholar on the Canarian vascular flora. He traveled all over the islands, discovered several botanical treasures, and published valuable studies on their flora, particularly "Oekologie und Biologie der Canarenpflanzen" (1929). Other European botanists concerned with the Canarian flora and vegetation were Salter, author of "Regional Distribution of the Native Flora in Tenerife" (1917–1918); Thomas Sprague and John Hutchinson, who in 1913 visited the islands and on their return published several articles on its flora, with descriptions of endemic species; H. Knoche, who after his stay in Tenerife published his book Vagandi Mos. Reiseskizzen eines Botanikers. I. Die Kanarischen Inseln (1923); L. Lindinger, who stayed in Tenerife during the First World War and as a result of his research published, among other works, "Beiträge zur Kenntnis von Vegetation und Flora der Kanarischen Inseln" (1926); A. Lloyd Praeger, who studied the succulent flora of the archipelago with special attention to the genera of Crassulaceae, which he published in 1932 as a monograph that remains a model study in its class.

Also at the beginning of the twentieth century, the first monographic publications on the cryptogamic flora of the archipelago began to appear. Thus, C. Sauvageau can be considered as the first phycologist that visited the archipelago and worked "in situ", publishing as a result of his stay "A propos des Cystoseira de Banyuls et de Guéthary" (1912). The aforementioned French botanist C.J. Pitard published an important study on the bryophytes of the islands, increasing their list to 110 species. In the winter of 1920–1921 the phycologist F. Börgesen visited the Canary Islands and produced five research papers compiled under the title "Marine algae from the Canary Islands", published between 1925 and 1930. Geheeb began the first important work on the bryophytic flora of the Atlantic islands that subsequently was continued and completed by his disciple Herzog and published in 1910. Petrak, who did not visit the islands, published in 1929 the first serious work on Canarian micromycetes, recording more than 100 species (26 of them new to science, as well as two genera). He also worked on the imperfecti fungi, recording 12 new species.

In the 1940s, modern research on the flora and vegetation of the Canary Islands proceeded, with a great number of floristic, taxonomic and phytosociological studies. The most important contribution was the huge work of Sventenius (Fig. A2.3), who arrived in the Canary Islands in 1943 as a botanist at the La Orotava Acclimatization Garden. He carried out an intense and extensive exploration throughout the archipelago, publishing more than 20 papers on its flora, among them his "Additamentum ad Flora Canariensem" (1960). He was the founder-director of the Canary Botanic Garden "Viera y Clavijo" of Gran Canaria. He described more than 130 new taxa, including three genera (*Dendriopoterium*, *Marcetella* and *Heywoodiella*). Simultaneously, the forestry engineers Luis Ceballos and Francisco Ortuño published the first great book on Canarian vegetation, "Estudio sobre la Vegetación y la Flora Forestal de las Canarias Occidentales" (1951). K. Lems visited the Canary Islands in 1954 and 1956 and published his "Floristic Botany of the Canary Islands" (1960) and "Structure of Vegetation in the Canary Islands" (1968) and several other studies. Between 1957 and 1968, P. Dansereau published interesting studies on Macaronesian vegetation, with spe-

Fig. A2.3 Eric Ragnor Sventenius

cial emphasis on laurel forest, and proposed an elevational zonation model for the Canaries. From 1954 these islands began to be visited by the Norwegian J. Lid, who published the results of his campaigns in the book “Contributions to the Flora of the Canary Islands” (1967). The Egyptian botanist Boulos made extensive taxonomic studies on the genus *Sonchus* (*Asteraceae*), which he published between 1960 and 1972. Between 1957 and 1967 the Spanish botanist Fernando Esteve Chueca lived in Las Palmas de Gran Canaria and the fruits of this long stay were 11 papers on the flora and vegetation of the islands. Among them “Ensayo fitosociológico de la Crassi-Euphorbietea macaronésica y estudio de los tabaibales y cardonales de Gran Canaria” (1964), written together with Rivas Goday, and “Estudio de las alianzas y asociaciones del orden Cytiso-Pinetalia en las Canarias Orientales” (1969), which constitute two of the first rigorous phytosociological publications on Canarian vegetation. At that time, an interesting phytosociological work by E. Oberdorfer “Pflanzensoziologische Studien auf Teneriffa und Gomera (Kanarische Inseln)” (1965) came out. In the 1970s the first studies of the Norwegian botanist Per Sunding appeared. His major study was “The Vegetation of Gran Canaria” (1972). Then, in collaboration with Eriksson and Hansen, he published the “Check List of Vascular Plants” (1974), for the whole of Macaronesia at island level, which has gone through three revised editions. Günther Kunkel, the most prolific author on Canary Islands botany, with more than 100 pubications, lived on Gran Canaria for 15 years. The difficult and valuable pioneering role played by “Cuadernos de Botánica Canaria”, the first specialized journal on botanical topics published on the islands, was successfully initiated by him. He published in this journal most of his papers. He was the organizer and secretary of the “I Macaronesian Pro-Flora Congress” held in Las Palmas de Gran Canaria. Among his works of interest are “Arboles exóticos de Gran Canaria. I” (1976), “Biogeography and ecology in the Canary Islands” (editor and author, 1976), “Árboles y arbustos de las Islas Canarias” (1981), the four volumes of “Flora de Gran Canaria” (co-authored with his wife, the painter M.A. Kunkel), “Die Kanarischen Inseln und ihre Pflanzenwelt” (1980), “Diccionario Botánico Canario” (1986) and “Flora de las islas Canarias” (1991). Other work important to Canary Islands Botany are the bioclimatic and phytosociological studies of Salvador Rivas-Martínez, and the chorological and phytosociological work of V. Voggenreiter, etc.

In the meantime, the work on cryptogamic botany continued, such as that by K. Mägdefrau, who in 1944 published an in-depth study of the bryophyte vegetation of the Tenerife laurisilva. In 1946, the phycologist J. Feldmann published his study “La Flore marine des Îles Atlantides. Contribution à l’étude du peuplement des îles Atlantides”, which is perhaps the most complete algae study carried out on the Macaronesian archipelagos. In 1952, C.N. Tavares published “Contribution to the lichen flora of Macaronesia”, collecting all available information on the lichen flora of the region, presenting a list of 585 specific taxa for the Canaries. This catalog was later enlarged in later works (1958 and 1964). Jordan de Urries, director of the Botanical Garden of Madrid, worked in the Canary Islands at that time. He studied microscopic parasitic fungi on Gran Canaria, La Palma and La Gomera, producing several works in 1954 and 1957, in which he published many taxa new to science, including 2 genera and 20 species of fungi imperfecti. In 1965 O. Klement published another remarkable contribution to the knowledge of the lichen flora of the Canary Islands, "Zur Kenntnis der Flechtenvegetation der Kanarischen Inseln", increasing the number of Canary lichens to a total of 643 and including the first phytosociological data for this group. In 1967 J. Kohlmeyer published a study recording 21 marine fungi he collected around Tenerife. Also of special interest is the work done by Augier & Noailles on "Les groupements des Bryophytes dans l'île de Tenerife" (1968), in which they described the main bryophytic communities. From the fifties to the present, a group of Norwegian botanists became specialists on the study of Canarian fungi and bryophytes: Jörstad studied the parasitic micromycetes; Störmer worked on bryophytes and published in 1960 "A contribution to the bryology of the Canary Islands. Mosses, chiefly collected by Johannes Lid"; Eckblad studied the gasteromycetes; Gjaerum published several studies between 1969 and 1986 on the Uredinales of Macaronesia, and Ryvarden on the Aphyllophorales of the Canaries describing several endemic species. Between 1966 and 1971 Sunding devoted some works to the bryology of the Eastern Province of the Canaries. The decade of the 1970s was fruitful for lichen research in the Canaries. There were several publications by G. Follmann, a frequent visitor of the archipelago, and C.L. Champion, author of a "Catálogo Preliminar de los líquenes de las Islas Canarias" (1978), in collaboration with L. Sánchez-Pinto, as well as several works on Myxomycetes. The American Richard Korf started to work in the 1970s on a project on the Discomycetes of the Macaronesian Region.

After Sventenius, much of the botanical research on the Canary Islands has been in the hands of another wave of Canarian botanists, located in several research centers: the Botany Department of the University of La Laguna, the Botanic Garden "Viera and Clavijo" at Tafira in Gran Canaria, the University of Las Palmas de Gran Canaria, the Insular Museum of Natural History of Santa Cruz de Tenerife, the Canary Institute for Agricultural Research, etc. But we must not forget the regular important contributions from several foreign universities and institutions, such as Reading, Oslo, Ithaca, New York, Miami, Leiden, Cologne, Göteborg, etc.), whose researchers visit the archipelago frequently.

The creation of the Biology Section within the Faculty of Sciences of the University of La Laguna in 1966 was a real milestone in the study of Natural Sciences in the Canary Islands. The vast majority of botanists who now occupy a position in the Canary research centers studied there. The research work in the Area of Botany dates back to 1967, when Dr. Wolfredo Wildpret de la Torre (Fig. A2.4) was hired to teach botany. Since then there has been a maximum of 18 teachers in the Area, a number that later has been gradually reduced. The initial chair of the Biology Section was transformed into a Botany Department in 1972. In 1982 another chair of Botany was created in the Faculty of Pharmacy, which in 1988 was incorporated into the aforementioned Department. In 1986, in accordance with the guidelines of L.R.U. (University Reform Law), the Department of Botany together with the Area of Plant Physiology became a larger Department and more recently has merged with Ecology into a Department of Botany, Plant Physiology and Ecology. Throughout these years, this center was equipped with modern laboratories, microscopic facilities and a rich library. An important herbarium (TFC) was also established in 1969 that specialized in flora of the Canary Islands, although it also holds plants from other Macaronesian Archipelagos, the Iberian Peninsula and other European countries. Currently it holds some 130,000 vouchers (15,000 algae, 24,000 fungi, 16,000 lichens, 22,000 bryophytes and 53,000 vascular plants). In the Area of Botany, many projects, theses and dissertations have been carried out, as well as several hundred papers and books published in the different research lines founded: Flora and Vascular Vegetation (1967), Marine Botany (1968), Mycology (1969), Taxonomy of Vascular Plants (1970), Lichenology (1976), Bryology (1978), Palynology (1978), Ethnobotany(1980), Plant Cartography (1986), Invasive Plants (1989) and Environmental Education (1999).

Fig. A2.4 Wolfredo Wildpret de la Torre

The Canary Botanical Garden "Viera and Clavijo" at Tafira (Gran Canaria) was founded in 1953 by the Cabildo council of that island on the initiative of Sventenius, who directed it until his tragic death in 1973. From 1974 to 2012 the head of the garden was David Bramwell, a British biologist who had begun to visit the islands since the sixties, doing his doctoral thesis on the flora of the archipelago. Since then he has published floristic, taxonomic, evolutionary and biogeographical works. His most popular work is "Wild Flowers of the Canary Islands" (1974), with several updated editions. Throughout its history the Canary Botanical Garden has been equipped with modern research laboratories and a large specialized library. In order to save endangered plants, cutting-edge technologies have been adopted for in vitro cultures, and germplasm and DNA banks set up for research and preservation of the threatened flora. The Garden contains a magnificent life collection of plants endemic to the Canary Islands. It also has a herbarium (LPA, formed after the merging of the old herbarium of the "Museo Canario" – LPA, and the former JVC herbarium of the Garden), which houses about 65,000 vouchers from the Canaries, Macaronesia and related areas, among which the Sventenius, Kunkel and Bramwell collections are outstanding; it publishes the journal "Botánica Macaronésica". The current lines of research include Taxonomy of Vascular Plants, Palynology, Reproductive Biology, Plant Cytogenetics, In Vitro Cultures, Micropropagation of Endangered Canary Plants and Environmental Education.

In the Department of Ecology and Applied Botany of the "Instituto Canario de Investigaciones Agrarias" (ICIA), located in Valle de Guerra (La Laguna, Tenerife), work has been carried out on the management and study of the collections accumulated over 30 years by the tireless botanist Sventenius. Other work concerns chorological and vegetation maps of the islands, and floristic catalogs. The former Director of the Department, Arnoldo Santos, is the author of several books and papers in scientific journals, referring to studies on the flora and vascular vegetation of the islands, including the book "Flora and Vegetation of La Palma" (1983). The main research lines in botany at this center are Cartography, Taxonomy, Flora and Vascular Vegetation of the Canary Islands, and Forage Plants. The Botanical Garden of Puerto de la Cruz houses a good library and the important herbarium ORT, founded in 1943 by Sventenius. At present, this exceeds 46,000 sheets and the most important collection is that made by its founder. It also contains some flora collections from Sweden, the Iberian Peninsula, Madeira and the Azores, as well as exchange material from Canada, Finland and the Balearic islands. It also includes 500 exsiccata collected by Burchard.

In 1974 within the "Museo Insular de Ciencias Naturales" of Santa Cruz de Tenerife (now included in the "Museo de la Naturaleza y el Hombre"), the Botany Section was founded, at the initiative of its director Professor Wildpret. Two research grants were funded, one for Cryptogamy and other for Phanerogamy. The first occupants were L. Sánchez-Pinto and P.L. Pérez de Paz, respectively. The former remained since then linked to that center until his retirement, in his last stage as botanist curator and finally as director. His main line of research was lichenology on which he published several papers. As for Pérez de Paz, for some years he shared his work in the Museum with teaching at the Faculty of Biology of the University of La Laguna. Later he became fully devoted to the University. He was the founder of the Museum Herbarium (TFMC), currently composed of some 6500 vouchers, including angiosperms and lichens from Macaronesia, SW Morocco and the South of the Iberian Peninsula. After them, the respective scholarships of Cryptogamy and Phanerogamy were occupied by botanists from the University of La Laguna, most of them linked to the Botany Area.

In the Faculties of Sea and Veterinary Sciences of the University of Las Palmas, there are two phycological research groups, respectively led by María Ascención Viera Rodríguez and Ricardo Haroun Tabraue. Also at the "Centro Pesquero de Taliarte" Nieves González Henríquez carries out phycological research. In 1981, the then Institute for Nature Conservation (ICONA) started to develop plans for the recovery of threatened flora from the national parks of the Canary Islands. The botanist in charge of these projects was A. Bañares Baudet. Other centers that have had researchers in the field of Botany are the Higher School of Agricultural Engineering (now a section within the Polytechnic School of Engineering) and the "Viceconsejería" of Environment and Nature Conservation of the Canary Islands Government. At the beginning of the sixties Professor A. González created the prestigious Institute of Natural Products. Under his direction, many phytochemical studies, mainly on the Canary and Latin-American flora were carried out and are presently continued at the "Instituto Universitario de Bio-Orgánica Antonio González" (IUBO) and also in the "Instituto de Productos Naturales y Agrobiología" -IPNA (CSIC).

Some of the research centers of the islands have or have had seed banks, in general linked to their respective herbaria, offering seeds in exchange through their Index Seminum. The Index Seminum of the "Hortus Acclimatationis Plantarum

Arautapae" linked to the Botanical Garden of Puerto de la Cruz was founded by Sventenius in 1944. The Index Seminum of the Canary Botanical Garden Viera y Clavijo began to be published in 1974. The Index Seminum of the Department of Botany of the University of La Laguna was published between 1975 and 1985. The most recent one, the Index Seminum Tenerifensis, founded in 1988 at the School of Agricultural Technical Engineers, lasted only one year.

In this period of intense botanical research in the Canary Islands, there were several specialized journals that began to be published on the islands, and thanks to them, most botanists working in the Canary Islands published papers for the scientific community. The initiative was led by Günther Kunkel with the "Cuadernos de Botánica Canaria" that started to be published in May 1967 in Las Palmas de Gran Canaria. It was the first botanical journal entirely devoted to the Islands, and continued in press until 1977. Its existence was exclusively due to the tenacity of Kunkel, who performed all publishing tasks.

Simultaneously to the "Cuadernos" and complementing them, appeared the "Suplementos de Cuadernos de Botánica Canaria". They appeared irregularly and only four numbers were published. Another journal also published by G. Kunkel, *Monographiae Biologicae Canariensis*, released six numbers between 1970 and 1975 on botanical subjects. It was sponsored by the "Cabildo Insular de Gran Canaria". In the year 1970, Vieraea appeared, a Canarian journal of Biological Sciences (Zoology and Botany). It was founded by Professor W. Wildpret, based in the Botany Department at the University of La Laguna. In 1984 he transferred its rights to the Insular Museum of Natural Sciences of the Cabildo de Tenerife that continues its publication. The journal "Botanica Macaronesica", whose first issue came out in 1976, is sponsored by the Cabildo of Gran Canaria and since its foundation it is based in the Canary Botanical Garden "Viera y Clavijo".

Appendix 3: Ethnobotanical Notes

Phytonymy and Phyto-Toponymy

The origin of the common names of the Canary plants must be sought in the language of the different human groups that mixed to form the current population of these islands. Thus, we find

- Castilian Spanish names (names identical to those used in most of mainland Spain): acebuche (*Olea cerasiformis*), brezo (*Erica arborea*), cardón (*Euphorbia canariensis*), escobón (*Chamaecytisus proliferus*), espino (*Rhamnus crenulata*), madroño (*Arbutus canariensis*), pino (*Pinus canariensis*), retama (*Spartocytisus supranubius*), Sabina (*Juniperus turbinata* subsp. *canariensis*), etc.
- Canary names from Castilian (names in Castilian created in the Canary Islands): bobo o venenero, (*Nicotiana glauca*), cabezote (*Cheirolophus spp.*), estornudera (*Andryala pinnatifida*), fistulera (*Scrophularia glabrata*), leña buena

Fig. A3.1 *Cistus symphytifolius* retains its aboriginal name "amagante". Caldera de Taburiente. La Palma

(*Neochamaelea pulverulenta*), magarza (*Argyranthemu spp.*), palo blanco (*Picconia excelsa*), palomera (*Pericallis lanata*), ratonera (*Forskaolea angustifolia*), tostonera (*Adiantum reniforme*), etc.

- Guanche names (created by the aboriginal population): algaritofe (*Cedronella canariensis*), amagante (Cistus symphytifolius) (Fig. A3.1), balo (*Plocama pendula*), bejeque (*Aeonium spp.*), bicácaro (*Canarina canariensis*), calcosa (*Rumex lunaria*), chajora (*Sideritis spp.*), guaidil (*Convolvulus floridus*), mocán (*Visnea mocanera*), orijama (*Neochamaelea pulverulenta*), tabaiba (*Euphorbia spp.*), tagasaste (*Chamaecytisus proliferus*), tajinaste (*Echium spp.*), taxaigo (*Rubia fruticosa*), verode (*Kleinia neriifolia*), etc.
- Galaico-Portuguese names (from Portugal and Galicia, mostly brought by the Madeiran specialists who worked in sugar mills): acebiño (*Ilex canariensis*), aderno (*Heberdenia escelsa*), follao (*Viburnum rigidum*), balancón (*Traganum moquini*), faya (*Myrica faya*), gibalbera (*Semele androgyna*), hija (*Prunus lusitanica*), malfurada (*Hypericum grandifolium*), malpica (*Carlina salicifolia*), sanguino (*Rhamnus glandulosa*), viñátigo (*Persea indica*), etc.
- Arabic names (Arabic origin used in Castilian Spanish in the Iberian Peninsula): ahulaga (*Launaea arborescens*), almacigo (*Pistacia atlantica*), tarajal (*Tamarix canariensis*), etc.

Many plant names have been used for urban streets, commercial brands, businesses, companies, neighborhood associations, ecologist groups, etc.

The phyto-toponymy records the importance that flora and potential natural vegetation had in the past, their distribution, their uses, etc., indicating the enormous transformations they have undergone. Thus, it is not uncommon to find villages, places, ravines, ridges, mountains, plains, fountains, water galleries (tunnels to intersect the aquifer formations), etc. which owe their name to particular plants. They are spread throughout the islands and, in many cases, can be used to delimit potential natural vegetation. Many are related to *Euphorbia* scrub and shrubland ("tabaibal-cardonal"): Tabaiba (in El Rosario), El Tabaibal (in Hermigua), Tabaibales (in Mogán), El Cardonal (in La Laguna), Cardón (in Pájara), La Cardonera (in San Nicolás de Tolentino). Others to thermo-sclerophyllous forest: La Palma, Las Palmas de Gran Canaria, El Draguillo (in Anaga), Sabina Alta (in Fasnia), Sabinosa (in Frontera). Some are related to willow forest and laurel forest: El Sauzal, Los Sauces, Mocanal (in Valverde), Palo Blanco (in Los Realejos); etc. The pine forest has great prominence in Canarian toponymy, almost all referring to the Canary pine *e.g.* "Pinitos Verdes", "Pinillo", "Pinillos", "Pino Seco", "Pino Gordo", "Pino Gacho", "Siete Pinos", "Pinalete", "Pino Mocho", "Pinitos Nuevos" and "Pino Santo", etc. These names are common in the islands with large pine forests, as are those of associated species (jara (*Cistus spp.*), escobón (*Chamaecytisus proliferus*), codeso (*Adenocarpus viscosus*), etc. There are many names of population nuclei like El Jaral (Guía de Isora), Charco del Pino (Granadilla), El Escobonal (Güímar), Dos Pinos (El Paso), El Pinillo (Tijarafe), El Pinar (El Hierro), Casas del Jaral (Garafía), Pino Santo (Santa Brígida), Barranco del Pinar (Moya), etc.

Toponyms related to old uses of the forest are also found, especially logging: tablados (planks), tableros and tabladitos (boards), aserraderos or serraderos (sawmills), arrastraderos (tracks to drag logs), leña (firewood), carbón (charcoal), hornos and hornitos (pitch furnaces), madera (wood) in a broad sense.

Trees or Other Notable Specimens

Some plants have acquired a certain symbolic value for very different reasons: their longevity, or an outstanding morphological feature (shape), or associated mythical or mystical stories, or special traditional uses, etc., attracted the attention of naturalists who have approached the islands. But sometimes the legendary or historical connotation arose from relictual enclaves of plant formations, like the Doramas forest (Gran Canaria) or Agua García forest (Tenerife), Monte Lentiscal (Gran Canaria) or the palm groves of Betancuria (Fuerteventura).

A large number of ancient, traditional, historic and monumental trees are conserved in Canarian culture. Many of them have already disappeared, although their legendary fame lives on in folk memory. In the lower zone, only a few specimen of large shrubs deserve this consideration, like the Buenavista cardón (*Euphorbia canariensis*), the Guía de Isora balo (*Plocama pendula*), or the tabaiba (*Euphorbia lamarckii*) of Los Baldíos. At mid-elevations, dragon trees and palms are dominant in this respect: the "Palm of the Conquest", which grew in a garden of La Orotava (Tenerife) and was blown down by a "huracán" in 1918; the palm of Paquesito (in Ingenio, Gran Canaria); the celebrated dragon tree (drago) of La Orotava torn apart by a storm in 1819 and demolished in 1867 by another two storms; the "drago" of Icod de los Vinos, in 1990 declared of Cultural Interest by the Canary Government (Fig. A3.2); those at La Laguna Seminary, Tacoronte and Sietefuentes (Los Realejos), in Tenerife; the twin "dragos" of Breña Alta and Puntagorda (La Palma); the centenarian "drago" of Agalán (in Alajeró, La Gomera); those of Meleguinas

Fig. A3.2 The "drago" of Icod de los Vinos, declared in 1990 of Cultural Interest by the Canary Government

and Pino Santo (Santa Brigida) and Luis Verde (Valsequillo), on Gran Gran Canaria. But there are also other trees like the sabinas (Canary juniper) of La Dehesa and Tecorón (El Hierro); the wild olive of El Sauzal (Tenerife); the mastic trees (almácigos) of Arona and Guía de Isora (Tenerife) and Tafira (Gran Canaria), etc. From the laurel forest we cannot forget the legendary Garoé (*Ocotea foetens*) of El Hierro, which supplied water to the aboriginal population, the barbusano (*Apollonias barbujana*) of Cuevas Negras (Los Silos, Tenerife), the mocán (*Visnea mocanera*) of Jinama (Frontera, El Hierro) or the viñátigos (*Persea indica*) of Fuensanta (Agulo, La Gomera) and Aguagarcía (Tacoronte, Tenerife). Many Canary pines have been renowned in the islands for their longevity or their stature. The chronicles say that the church of Los Remedios of La Laguna (present cathedral) was roofed with the wood of a single pine. Also famous were other giant pines that disappeared in more recent times and caused wonder among locals and foreigners, like the "pino de las meriendas" (picnick pine) in the heights of the Orotava Valley, the "pino gordo" (fat pine) of Arico, and the pine of Majágora (Guía de Isora). Many other specimens are noted in history or legend, such as the Pine of the Victory of Acentejo in La Victoria, the pine of Buen Paso (Icod de los Vinos), or the pine of El Calvario in Arafo. Fortunately, several large pines still survive on the southern slope of Tenerife, such as the giants "Pino Gordo" and "Pino de la dos pernadas" above Vilaflor (Fig. A3.3). On El Hierro there are the "pino gordo" and "pino viejo" of El Morcillo. The Pine of the Virgin, the most famous one on La Palma, in El Paso,

Fig. A3.3 The giant pine "pino de las dos pernadas" of Vilaflor, Tenerife (60 m tall, 8.35 m girth, and 2.65 m dbh)

witnessed the conquest of the island, surviving since then thanks to the religious veneration it has always inspired, as also has happened with the "pino gordo" in Puntagorda. On Gran Canaria there was "el pino de Teror" at which appeared the patron virgin of Gran Canaria in 1484, and that fell two centuries later, and the "pino gordo" of Pilancones (San Bartolomé de Tirajana) that survived until 2008. But the "pino del paraguas" (the umbrella pine) persists in Inagua (Mogán), among others. Finally, among the Canary cedars, that of Montaña Rajada (Las Cañadas) is notable, as well as the one in Arico, Tenerife.

Many plants have been chosen as local symbols, and they therefore appear on the respective municipal heraldic shields. This is the case of the Cardón de Buenavista del Norte, the dragon trees on shields of La Orotava, Icod de los Vinos, Alajeró, Breña Alta and Valsequillo, the mastic trees on those of Arona or Guía de Isora, the pines on those of El Paso or La Victoria de Acentejo, the Garoé of El Hierro, the palm of Haría, Santa Úrsula and Valle Gran Rey, the Canary juniper of Frontera, the laurel of Tegueste, the pine of El Rosario, Garafía, Vilaflor and Artenara, etc. In other cases, the plants that give rise to the toponymy appear in the heraldry, such as the palm trees of La Palma and Las Palmas de Gran Canaria, the willow of El Sauzal, and the pine of El Pinar.

Appendix 4: Literature

Acebes Ginovés, J. R., M.C. León Arencibia, M. L. Rodríguez Navarro, M. J. Del Arco Aguilar, A. García Gallo, P. L. Pérez de Paz, O. Rodríguez Delgado, V. E. Martín Osorio & W. Wildpret de la Torre. 2010. *Pteridophyta, Spermatophyta*, pp. 119–172. In: Arechavaleta, M., S. Rodríguez, N. Zurita & A. García (coord.), *Lista de especies silvestres de Canarias. Hongos, plantas y animales terrestres. 2009.* Gobierno de Canarias. Santa Cruz de Tenerife, España.

Afonso Carrillo. J. 1980. Algunas observaciones sobre la distribución vertical de las algas en la isla del Hierro (Canarias). *Vieraea* 10: 3–16.

Afonso-Carrillo, J. 2014. *Lista actualizada de las algas marinas de las islas Canarias 2012.* Elaborada para la Sociedad Española de Ficología (SEF). 64 pp.

Afonso Carrillo, J. & M.C. Gil Rodriguez. 1980. *Cymodocea nodosa* (Ucria) Ascherson (*Zannichelliaceae*) y las praderas submarinas o "sebadales" en el Archipiélago Canario. *Vieraea* 8(2): 365–376.

Afonso-Carrillo J. & M. Sansón. 1999. *Algas, hongos y fanerógamas marinas de las islas Canarias. Clave analítica.* SPULL. 254 pp.

Afonso, L. (ed.). 1997: *Gran Atlas de Canarias.* Ed. Interinsular Canaria. 304 pp.

Afonso, L. (ed.). 1984. *Geografía de Canarias I.* Ed. Interinsular Canaria. S/C de Tenerife. 334 pp.

Álvarez Escobar, A. & O. Rodríguez Delgado. 2008. Contribución al estudio etnobotánico de las especies vegetales del tabaibal-cardonal (*Kleinio neriifoliae-Euphorbietea canariensis*) de la isla de Tenerife. *Anuario del Instituto de Estudios Canarios* 50 (2006): 1–38.

Alonso-Pérez, S., E. Cuevas, C. Pérez, X. Querol, J.M. Baldasano, R. Draxler, J.J. de Bustos. 2011. Trend changes of African airmass intrusions in the marine boundary layer over the subtropical Eastern North Atlantic region in winter. *Tellus B* 63:255–265.

Araña, V. & J.C. Carracedo 1978. *Los Volcanes de las Islas Canarias. Canarian Volcanoes. I. Tenerife.* Departamento de Petrología y Geoquímica. Instituto Lucas Mallada. Consejo Superior de Investigaciones Científicas. 151 pp. Madrid, España.

Arechavaleta, M., N. Zurita, M.C. Marrero & J.L. Martín (eds.). 2005. *Lista preliminar de especies silvestres de Cabo Verde (hongos, plantas y animales terrestres). 2005.* Consejería de Medio Ambiente y Ordenación Territorial. Gobierno de Canarias. 155 pp.

Arechavaleta, M., S. Rodríguez, N. Zurita & A. García (coords.). 2010. *Lista de especies silvestres de Canarias. Hongos, plantas y animales terrestres. 2009.* Gobierno de Canarias. 579 pp.

Bañares Baudet, Á. 2015. *Las plantas suculentas (Crassulaceae) endémicas de las Islas Canarias.* Ediciones Turquesa. Santa Cruz de Tenerife. 280 pp.

Barbour, M.G., M.J. Del Arco, N. Raven & G. García-Baquero. 2012. Anomalous diameter growth and population age structure in mature Canary Islands pine stands. *Phytocoenologia* 42 (1–2): 1–13.

Barres, L., M. Galbany-Casals, A.L. Hipp, J. Molero & R. Vilatersana. 2017. Phylogeography and character evolution of Euphorbia sect. Aphyllis subsect. Macaronesicae (Euphorbiaceae). *Taxon* 66 (2): 324–343.

Beltrán Tejera, E. 2011. Los hongos: notables protagonistas en la biodiversidad canaria. In Afonso Carrillo, J. (ed.): *Biodiversidad: explorando la red vital de la que formamos parte. Actas VI Semana Científica Telesforo Bravo*, pp 29–69. Instituto de Estudios Hispánicos de Canarias.

Beltrán Tejera, E., J. Afonso-Carrillo, A. García Gallo & O. Rodríguez Delgado (eds.). 2009. Homenaje *al Prof. Dr. Wolfredo Wildpret de la Torre*. Instituto de Estudios Canarios. La Laguna, Tenerife. 872 pp. + CD.

Beltrán Tejera, E., W. Wildpret de la Torre, M.C. León Arencibia, A. García Gallo & J. Reyes Hernández. 1999. *Libro Rojo de la Flora Canaria contenida en la Directiva-Hábitats Europea*. Dirección General de Conservación de la Naturaleza. Organismo Autónomo Parques Nacionales. Ministerio de Medio Ambiente. 694 pp.

Biondi, B., M. Allegrezza, F. Taffetani & W. Wildpret de la Torre. 1994. La vegetazione delle coste basse sabbiose delle isole di Fuerteventura e Lanzarote (Isole Canariae, Spagna). *Fitosociologia* 27: 107–121.

Bolmgren, K. & B. Oxelman. 2004. Generic limits in *Rhamnus* L. s.l. (Rhamnaceae) inferred from nuclear and chloroplast DNA sequence phylogenies. *Taxon* 53(2): 383–390.

Borges PAV, Costa A, Cunha R, Gabriel R, Gonçalves V, Martins AF, Melo I, Parente M, Raposeiro P, Rodrigues P, Santos RS, Silva L, Vieira P, Vieira V (eds.). (2010) *A list of the terrestrial and marine biota from the Azores*. Princípia, Cascais, 432 pp.

Borges PAV, Abreu C, Aguiar AMF, Carvalho P, Jardim R, Melo I, Oliveira P, Sérgio C, Serrano ARM, Vieira P (eds.). (2008) *A list of the terrestrial fungi, flora and fauna of Madeira and Selvagens archipelagos*. Direcção Regional do Ambiente da Madeira and Universidade dos Açores, Funchal and Angra do Heroísmo, 440 pp.

Bramwell, D. (ed.). 1979. *Plants and Islands*. Academic Press. London. 459 pp.

Bramwell, D. 1986. Contribución a la biogeografía de las Islas Canarias. *Botánica Macaronésica* 14 (1985): 3–34.

Bramwell, D. & Z. Bramwell. 2001. *Flora Silvestres de las Islas Canarias*. 4ª edición. Editorial Rueda, S.L. Alcorcón (Madrid), España. 437 pp.

Bramwell, D., & I.B.K. Richardson. 1973. Floristic connections between Macaronesia and the east Mediterranean Region. *Monogr. Biol. Canar.* 4: 118–125.

Cáceres Lorenzo, M.T., & M. Salas Pascual. 1995. *Los nombres de las plantas canarias*. Colección "Guagua". Ediciones del Cabildo Insular de Gran Canaria. 162 pp.

Capelo, J. (ed.). 2004. A paisagem vegetal da Ilha da Madeira. *Quercetea* 6: 3–200.

Capelo, J., M. Sequeira, R. Jardim, S. Mesquita & J.C. Costa. 2005. The vegetation of Madeira Island (Portugal). A brief overview and excursión guide. *Quercetea* 7: 95–122.

Capelo, J.H., J.C. Costa, M. Lousã, S. Fontinha, R. Jardim, M. Sequeira & S. Rivas-martínez. 2000. Vegetação da Madeira (Portugal): Aproximação à tipologia fitossociologica. *Silva Lusitanica* 7(2): 257–279.

Carlquist, SH. 1974. Adaptative radiation, Canary Islands and other Macaronesian Islands. In: *Island Biology*. Chapter 5, pp. 180–196. New York.

Carracedo, J.C., F.J. Pérez, E. Ancochea, J. Meco, F. Hernán, C.R. Cubas, R. Casillas, E. Rodriguez & A. Ahijado. 2002. Cenozoic volcanism. II: The Canary Islands. In: Gibbons, W., Moreno, T. (eds.), *The Geology of Spain*, pp. 439–472. The Geological Society of London, London.

Carracedo, J.C., F.J. Pérez Torrado & A. Hansen. 2007. El relieve de las Islas Canarias. The Canary Islands landscape. *Enseñanza de las Ciencias de la Tierra*, 2007 (15.2): 196–205.

Carracedo, J.C., & V.R. Troll. 2016. The Geology of the Canary Islands. Elsevier. 636 pp.

Caujapé-Castells, J., C. García-Verdugo, Á. Marrero-Rodríguez, J.M. Fernández-Palacios, D.J. Crawford & M.E. Mort. 2017. Islands ontogenies, syngameons, and the origins and evolution of genetic diversity in the Canarian endemic flora. *Perspectives in Plant Ecology, Evolution and Systematics* 27 (2017): 9–22.

Ceballos Fernández de Córdoba, L., & F. Ortuño Medina. 1951. *Estudio sobre la Vegetación y la Flora Forestal de las Canarias Occidentales*. Ministerio de Agricultura, Dirección General de Montes, Caza y Pesca fluvial, Instituto Forestal de Investigaciones y Experiencias, Madrid. X + 461 pp. + 4 h. + 57 láms. + 6 mapas.

Ceballos Fernández de Córdoba, L., & F. Ortuño Medina. 1976. *Estudio sobre la Vegetación y Flora Forestal de las Canarias Occidentales*. Reedición corregida. Excmo. Cabildo Insular de Tenerife. Santa Cruz de Tenerife. 433 pp.

Ciferri, R. 1962. La Laurisilva canaria: una paleoflora viviente. *Ricercha Scient*, Ser. 2, 32 (1): 111–134.

Costa, J.C., J.H. Capelo, R. Jardim, M. Sequeira, D. Espírito-Santo, M. Lousã, S. Fontinha, C. Aguiar & S. Rivas-Martínez. 2004. Catálogo sintaxonómico e florístico das comunidades vegetais da Madeira e Porto Santo. *Querceta* 6: 61–185.

Dansereau, P. 1966. Etudes macaronésiennes. III. La zonation altitudinale. *Natur. Canad.* 93: 779–795. Québec.

Dansereau, P. 1968. Végétation de la Macaronésie. *Finisterra* 3: 175–181.

De Castro, M., J. Martín-Vide & S. Alonso. 2005. El clima en España: pasado, presente y escenarios de clima para el siglo XXI. In: Moreno Rodríguez, J.M. (Coord.), *Evaluación preliminar de los impactos en España por efecto del cambio climático*, pp. 1–64. Proyecto ECCE (Informe final), Ministerio de Medio Ambiente.

De Leon Llamazares, A., A. Arriba Balenciaga & M.C. De la Plaza. 1991. *Caracterización agroclimática de la provincia de Santa Cruz de Tenerife*. Dirección General de la Producción Agraria. Ministerio de Agricultura, Pesca y Alimentación. 161 pp. Madrid, España.

Del Arco Aguilar, M.C. 1993. *Recursos vegetales en la prehistoria de Canarias*. Ed. Organismo Autónomo Complejo Insular de Museos y Centros. Cabildo Insular de Tenerife. Santa Cruz de Tenerife. 134 pp.

Del Arco Aguilar, M.J. 1986. Colonización vegetal de las Islas Canarias. In: P.L. Pérez de Paz y F. Valdés Gonzalez (eds.): *Origen y evolución de la vida*. Secr. Public. Universidad de La Laguna. Ser. Informes n° 21: 105–124.

Del Arco Aguilar, M.J. 1989. El origen de la flora canaria. *Quercus* 41: 14–21.

Del Arco Aguilar, M. 2008. Consecuencias del cambio climático sobre la flora y vegetación canaria. In: Méndez Pérez, J.M. & M. Vázquez Abeledo (eds.): *El Cambio Climático en Canarias*, pp. 79–100. Academia Canaria de Ciencias. Serie Monografías N° 1.

Del Arco Aguilar, M.J. & O. Rodríguez Delgado. 1999. Flora y Vegetación. In: *La Enciclopedia Temática e Ilustrada de Canarias*, pp. 62–82. Ed. Centro de la Cultura Popular Canaria.

Del Arco Aguilar, M.J. & O. Rodríguez Delgado. 2000. Canarias. Naturaleza: Flora y Vegetación. En: *Canarias Isla a Isla*, pp. 34–40, 122–124, 182–184, 239–243, 335–339, 431–433, 484–487, 555–557. Ed. Centro de la Cultura Popular Canaria.

Del Arco Aguilar, M.J. & W. Wildpret de la Torre. 1991. Contribución al conocimiento de la vegetación litoral del Archipiélago Canario. I. Las comunidades de *Ruppia maritima, Salsola oppositifolia, Zygophyllum fontanesii* y *Z. gaetulum*. *Homenaje al Profesor Dr. Telesforo Bravo*. Tomo I: 97–116. Secretariado Publicaciones Universidad La Laguna.

Del Arco Aguilar, M.J., & W. Wildpret de la Torre. 1994. Estudio bioclimatológico del Pico Teide. Tenerife (Islas Canarias). *Anuario del Instituto de Estudios Canarios* 38: 7–14.

Del Arco Aguilar M.J., P.L. Pérez de Paz & W. Wildpret de la Torre. 1987. Contribución al conocimiento de los pinares de la isla de Tenerife. *Lazaroa* 7: 67–84.

Del Arco Aguilar, M.J., P.L. Pérez de Paz, W. Wildpret de la Torre, V. Lucía Sauquillo & M. Salas Pascual. 1990. *Atlas cartográfico de los pinares canarios: La Gomera y El Hierro*. Viceconsejería de Medio Ambiente y Conservación de la Naturaleza. Consejería de Política Territorial. Gobierno de Canarias. S/C de Tenerife. 90 pp. + 17 mapas.

Del Arco Aguilar, M.J., P.L. Pérez de Paz, O. Rodríguez Delgado, M. Salas Pascual & W. Wildpret de la Torre. 1992. *Atlas Cartográfico de los Pinares Canarios: II. Tenerife*. Dirección General de Medio Ambiente y Conservación de la Naturaleza, Consejería de Política Territorial, Gobierno de Canarias. 228 pp. + 44 mapas.

Del Arco Aguilar, M.J., J.R. Acebes Ginovés & P.L. Pérez de Paz. 1996. Bioclimatology and climatophilous vegetation of the Island of Hierro (Canary Islands). *Phytocoenologia* 26(4): 445–479.

Del Arco Aguilar, M.J., J.R. Acebes Ginovés, P.L. Pérez de Paz & M.C. Marrero. 1999. Bioclimatology and climatophilous vegetation of Hierro (part 2) and La Palma (Canary Islands). *Phytocoenologia* 29(2): 253–290.

Del Arco Aguilar, M.J., M. Salas, J.R. Acebes, M.C. Marrero, J.A. Reyes-Betancort & P.L. Pérez de Paz. 2002. Bioclimatology and climatophilous vegetation of Gran Canaria (Canary Islands). *Annales Botanici Fennici* 39: 15–41.

Del Arco Aguilar, M., P.L. Pérez de Paz, J.R. Acebes, J.M. González-Mancebo, J.A. Reyes-Betancort, J.A. Bermejo, S. De Armas & R. González-González. 2006. Bioclimatology and climatophilous vegetation of Tenerife (Canary Islands). *Ann. Bot. Fennici* 43 (3): 167–192.

Del Arco Aguilar, M. J., W. Wildpret de la Torre, P. L. Pérez De Paz, O. Rodríguez Delgado, J. R. Acebes Ginovés, A. García Gallo, V. E. Martín Osorio, J. A. Reyes Betancort, M. Salas Pascual, M. A. Díaz, J. A. Bermejo Domínguez, R. González González, M. V. Cabrera Lacalzada & S. García Ávila. 2006. *Mapa de Vegetación de Canarias*. GRAFCAN. 550 pp. + 7 mapas + CD. Santa Cruz de Tenerife, España.

Del Arco Aguilar, M., O. Rodríguez Delgado, J.R. Acebes, A. García Gallo, P.L. Pérez de Paz, J.Mª. González Mancebo, R. González González & V. Garzón Machado. 2009. Bioclimatology and climatophilous vegetation of Gomera (Canary Islands). *Ann. Bot. Fennici* 46: 161–191.

Del Arco Aguilar, M., R. González González, V. Garzón Machado & B. Pizarro Hernández. 2010. Actual and potential natural vegetation on the Canary Islands and its conservation status. *Biodiversity and Conservation* 19: 3089–3140.

De-León, A., A. Arriba & M.C. De-La-Plaza. 1991: *Caracterización agroclimática de la provincia de Las Palmas*. Ministerio de Agricultura, Pesca y Alimentación. Madrid 1990.

Dias, E., C. Mendes, C. Melo, D. Pereira & R. Elias. 2005. The Azores Central Islands Vegetation and Flora. Field Guide. A brief overview and excursión guide. *Quercetea* 7: 123–173.

Emerson, B.C. 2002. Evolution on oceanic islands: molecular phylogenetic approaches to understanding pattern and process. *Molecular Ecology* 11: 951–966.

Esteve Chueca, F. 1968. Datos para el estudio de las clases *Ammophiletea*, *Juncetea* y *Salicornietea* en las Canarias Orientales. *Collect. Bot.* 7 (15): 303–323. Barcelona.

Esteve Chueca, F. 1969. Estudio de las alianzas y asociaciones del orden *Cytiso-Pinetalia* en las Islas Canarias Orientales. *Bol. Real Soc. Esp. Hist. Nat. (Biol.)* 67: 77–104.

Esteve Chueca, F. 1973. Estudio de las asociaciones *Spartocytisetum nubigeni* (Oberd. 1965) emend. y *Sideriti-Pinetum canariensis* ass. nov. en las islas Canarias. *Trab. Dep. Bot. Univ. Granada* 2 (1): 3–9.

Esteve Chueca, F. 1973. Sinopsis de las Alianzas y Asociaciones en la Clase *Cytiso-Pinetea* y Orden *Cytiso-Pinetalia*. *Monogr. Biol. Canar.* 4: 89–92.

Esteve Chueca, F., & O. Socorro Abreu. 1977. Estudio fitosociológico de los prados áridos y otras comunidades vegetales de Lanzarote (Islas Canarias). *Bot. Macar.* 3 (1977): 85–97.

Estévez Domínguez, J. 2004. *Gigantes en las Hespérides. Árboles singulares y monumentales de las Islas Canarias*. La Caja de Canarias. 180 pp.

Euro+Med PlantBase. 2017. http://www.emplantbase.org/home.html

Fernández, A.B. (coord.). 2009. *Parque Nacional de Garajonay. Patrimonio Mundial*. Organismo Autónomo de Parques Nacionales, Ministerio de Medio Ambiente y Medio Rural y Marino. 707 pp.

Fernández Galván, M. 1983. Esquema de la vegetación potencial de la Isla de Gomera. *Comunicações apresentadas ao II Congreso Internacional Pró Flora Macaronésica. Funchal, 19–25 de Junho de 1977*, pp. 269–293. Funchal.

Fernández Galván, M. & A. Santos Guerra. 1984. La vegetación del litoral de Canarias. I. *Arthrocnemetea. Lazaroa* 5 (1983): 143–155.

Fernández-González, F., J. Loidi & J.C. Moreno (aut.); Del Arco, M., Fernández Cancio, A., Font, X., Galán, C., García Mozo, H., Gavilán, R., Penas, A., Pérez Badía, R., Del Río, S., Rivas-Martínez, S., Sardinero, S. y Villar, L. (Contr.). 2005. Impactos sobre la biodiversidad vegetal. In: Moreno Rodríguez, J.M. (coord.), *Evaluación preliminar de los impactos en España por efecto del cambio climático*, pp. 183–248. Proyecto ECCE (Informe final), Ministerio de Medio Ambiente.

Fernández-Palacios, J.M. 2009. Relictualismo en Islas Oceánicas. El caso de la laurisilva macaronésica. In: R. Real & A.L. Márquez, *Biogeografía Scientia Biodiversitatis* : 13–24. Ed. R. Real & A.L.Márquez.

Fernández-Palacios, J.M. & J.L. Martín Esquivel (coord.). 2001. *Naturaleza de las Islas Canarias. Ecología y Conservación*. Ed. Turquesa. 474 pp.

Fernández-Palacios, J.M., R. Otto, J.D. Delgado, J.R. Arévalo, A. Naranjo, F. González Artiles, C. Morici & R. Barone (eds.). 2008. *Los Bosques Termófilos de Canarias*. Proyecto LIFE04/NAT/ES/000064. Excmo. Cabildo Insular de Tenerife. Santa Cruz de Tenerife. 192 pp. + Glosario y Referencias.

Fernández-Palacios, J.M., L. de Nascimento, R. Otto, J.D. Delgado, E. García-del-Rey, J.R. Arévalo & R.J. Whittaker. 2011. A reconstruction of Paleo-Macaronesia, with particular reference to the long-term biogeography of the Atlantic island laurel forests. *Journal of Biogeography (J. Biogeogr.)* 38 (2011): 226–246.

Fernandopulle, D. 1976. Climatic Characteristics of the Canary Islands. In: G. Kunkel (ed.), *Biogeography and ecology in the Canary Islands*, pp. 185 206. La Haya.

Font Tullot, I. 1959. El clima de las Islas Canarias. *An. Est. Atl.* 5: 57–103. Las Palmas de Gran Canaria.

Francisco-Ortega, J., R.K. Jansen & A. Santos-Guerra. 1996. Chloroplast DNA evidence of colonization, adaptative radiation, and hybridization in the evolution of the Macaronesian flora. *Proc. Natl. Acad. Sci. USA* 93: 4085–4090.

Francisco-Ortega, J., A. Santos-Guerra, S-C. Kim & D.J. Crawford. 2000. Planta genetic diversity in the Canary Islands: a conservation perspective. *American Journal of Botany* 87(7): 909–919.

Gaisberg, M. von. 2005. *Die Vegetation der Fußstufe von El Hierro (Kanarische Inseln)*. Dissertationes Botanicae 395, J. Cramer, Berlin Stuttgart, 364 pp.

Gaisberg, M.v. & Ch. Stierstorfer. 2005. The significance of geological traits for speciation of endemic angiosperms on El Hierro (Canary Islands). *Phytocoenologia* 35 (1): 39–52.

García Casanova, J., O. Rodríguez Delgado, W. Wildpret de la Torre. 1996. *Montaña Roja: Naturaleza e Historia de una Reserva Natural y su entorno (El Médano - Granadilla de Abona)*. Ayuntamiento de Granadilla de Abona, Viceconsejería de Medio Ambiente, Centro de la Cultura Popular Canaria. 412 pp.

García Gallo, A. 1988. *Flora y Vegetación del Municipio de La Laguna (Tenerife). Area Central y Meridional*. Exmo. Ayuntamiento de San Cristóbal de La Laguna. 283 pp.

García Gallo, A., O. Rodríguez Delgado, W. Wildpret De La Torre & E. Carqué Álamo. 1992. Contribución al estudio de la Clase *Parietarietea judaicae* Rivas-Martínez *in* Rivas Goday (1955) 1964 *em.nom.* Oberdorfer 1977 en las Islas Canarias. *Documents phytosociologiques* 13 (1991): 239–246.

García Gallo, A., W. Wildpret de la Torre, M.T. Jiménez Felipe & G.E. Vargas Chávez. 1996. Proliferación natural de especies exóticas en el dominio potencial del monte verde de la isla de Tenerife (Canarias). *Anuario del Instituto de Estudios Canarios* 40: 9–16

García Gallo, A., W. Wildpret de la Torre, O. Rodríguez Delgado, P.L. Pérez de Paz, M.C. León Arencibia, C. Suárez Rodríguez & J.A. Reyes-Betancort. 1999. El xenófito *Pennisetum setaceum* en las Islas Canarias (Magnoliophyta, Poaceae). *Vieraea* 27: 133–158.

García Gallo, A., O. Rodríguez Delgado & W. Wildpret de la Torre, 2012. El paisaje vegetal del Espacio Natural Protegido de la Rambla de Castro (Tenerife). *Vieraea* 40: 107–130.

García-Herrera, R., D. Gallego, E. Hernández, L. Gimeno & P. Ribera. 2001. Influence of the North Atlantic oscillation on the Canary Islands precipitation. *Journal of Climate* 14: 3889–3903

García-Herrera, R., A. Macías, D. Gallego, E. Hernández, L. Gimeno & P. Ribera. 2003. Reconstruction of the precipitation in the Canary Islands for the period 1595–1836. *American Meteorolgical Society*: 1037–1039

García-Herrera, R., D. Gallego & A. Macías. 2004. La Oscilación del Atlántico Norte en tiempos históricos. Reconstrucción y validación. In: *La Oscilación del Atlántico Norte y su influencia sobre la Península Ibérica y Canarias*. Aica ediciones, 231 pp.

García-Maroto, F., A. Mañas-Fernández, J.A. Garrido-Cárdenas & D. López Alonso (2009). D6-Desaturase sequence evidence for explosive Pliocene radiations within the adaptive radiation of Macaronesian Echium (Boraginaceae). In: J.L. Guil-Guerrero, B. Guzmán & P. Vargas. *Molecular Phylogenetics and Evolution* 52 (2009): 563–574.

García-Olivares. V., H. López, J. Patiño, N. Alvarez, A. Machado, J.C. Carracedo, V. Soler & B.C. Emerson. 2017. Evidence for mega-landslides as drivers of island colonization. *Journal of Biogeography (J. Biogeogr.)* 44 (2017):1053–1064.

Garzón-Machado, V., P.L. Pérez de Paz & M.J. del Arco. 2009. Degree of habitat conservation and threatened species in the Canary Islands. *Biodiversity Hotspots in the Mediterranean Area: species, communities and landscape level.* Cagliari. 45° International Congress of SISV & FIP. 22 al 29 June 2009. Poster.

Garzón-Machado, V., J.M. González-Mancebo, A. Palomares-Martínez, A. Acebedo-Rodríguez, J.M. Fernández-Palacios, M. Del-Arco-Aguilar & P.L. Pérez-De-Paz. 2010. Strong negative effect of alien herbivores on endemic legumes of the Canary pine forest. *Biological Conservation* 143: 2685–2694.

Garzón Machado, V., M.J. Del Arco Aguilar, F. Valdés & P.L. Pérez de Paz. 2012. Fire as a threatening factor for endemic plants of the Canary Islands. *Biodiversity and Conservation* 21: 2621–2632.

Garzón Machado, V., R. Otto & M.J. del Arco Aguilar. 2014. Bioclimatic and vegetation mapping of a topographically complex oceanic island applying different interpolation techniques. *Int. J. Biometeorol.* 58: 887–889.

Geldmacher, J., P.V.D. Bogaard, K. Hoernle & H.U. Schmincke. 2000. New 40Ar/ 39Ardating of the Madeira archipelago and hotspot track (eastern North Atlantic). Geochem. *Geophys. Geosys.* 1, 1999GC000018.

Gil Rodríguez, M.C., J. Afonso Carrillo & W. Wildpret de la Torre. 1982. Ocurrence de *Halophila decipiens* Ostdenfeld on Tenerife, Canary Islands. *Aquat. Bot.* 12: 205–207.

Gil Rodríguez, M.C., J. Afonso Carrillo & W. Wildpret de la Torre. 1987. Praderas submarinas de *Zostera noltii* (*Zosteraceae*) en las Islas Canarias. *Vieraea* 17: 143–146.

Gómez Campo, C. & col. 1996. *Libro Rojo de Especies Vegetales Amenazadas de las Islas Canarias.* Viceconsejería de Medio Ambiente, Consejería de Política Territorial, Gobierno de Canarias. 672 pp.

González González, R. 2001. *Árboles monumentales, arboledas y flora singular de Tenerife.* Área de Medio Ambiente, Cabildo Insular de Tenerife. 171 pp.

González, R., M.C. León & M. Del Arco. 2002. *Los helechos de la Reserva Natural Integral de El Pijaral.* Consejería de Política Territorial y Medio Ambiente del Gobierno de Canarias. S/C de Tenerife. 194pp.

González-Mancebo, J.M., E. Beltrán Tejera, A. Losada-Lima & L. Sánchez-Pinto. 1996. *La Vida Vegetal en las Lavas Históricas de Canarias. Parque Nacional de Timanfaya.* Organismo Autónomo Parques Nacionales, Ministerio de Medio Ambiente. 255 pp.

Graham, E. 2000. *Temperature and precipitation changes at La Palma, Canary Islands (1971–2000).* Disponible en: http:\\www.eso.org/gen-fac/pubs/astclim/espas/espas-reports

Grehan, J.R. 2016. Biogeographic relationships between Macaronesia and the Americas. *Australian Systematic Botany* 29 (2016): 447–472.

Guadalupe, M.E., M.C. Gil-Rodríguez & M.C. Hernández. 1996. *Flora y Vegetación Marina de Arrecife de Lanzarote*. Fundación César Manrique, Colección Torcusa. Madrid. 269 pp.

Hansen, A., & P. Sunding. 1993. Flora of Macaronesia. Checklist of vascular plants. 4. revised edition. *Sommerfeltia*, 17: 1–295.

Hansen, A., & P. Sunding. 1994. Botanical bibliography of the Canary Islands. *Sommerfeltia*, Supplement 5: 1–116 pp.

Haroun Tabraue, R., M.C. Gil Rodríguez & W. Wildpret de la Torre. 2003. *Plantas Marinas de las Islas Canarias*. Canseco Editores S.L. 319 pp.

Heads, M. 2017. Metapopulation vicarience explains old endemics on young volcanic islands. *Cladistics* 0 (2017): 1–20.

Himstedt, T. 2010. *Untersuchungen zur Ökologie des Monteverde im westlichen Anaga-Gebirge auf Teneriffa, Kanarische Inseln*. Tesis Doctoral inéd. Universität Hannover. 374 pp. + tablas.

Hohenester, A., & W. Welss. 1993. *Exkursionsflora für die Kanarischen Inseln mit Ausblicken auf ganz Makaronesien*. Verlag Eugen Ulmer, Germany. 374 pp.

Höllermann, P. W. 1978. Geoecological aspects of the upper timberline in Tenerife, Canary Islands. *Arctic and Alpine Res.* 10 (2): 365–382.

Holm, P.M., J.R. Wilson, B.P. Christensen, L. Hansen, S.L. Hansen, K.M. Hein, A.K. Mortensen, R. Pedersen, S. Plesner, M.K. Runge. 2006. Sampling the Cape Verde mantle plume: evolution of melt compositions on Santo Antão, Cape Verde Islands. *J. Petrol.* 47: 145–189.

Huetz de Lemps, A. 1969. *Le Climat des Iles Canaries*. Faculté des Lettres et des Sciences Humaines de Paris-Sorbonne. 226 pp.

Humphries, C.J. 1976. Evolution and endemism in *Argyranthemum* Webb *ex* Schultz Bip. (*Compositae: Anthemideae*). *Bot. Macar.* 1: 25–50.

Hüppe, J., R. Pott & W. Wildpret de la Torre. 1996. Standörliche Differenzierungen im subtropischen Sukkulentenbusch der Kanareninsel Teneriffa. *Phytocoenologia* 26 (4): 417–444.

ISTAC. 2017. Instituto de Estadística de Canarias, Gobierno de Canarias, Santa Cruz de Tenerife. Available from www.gobiernodecanarias.org/istac/ (accesed 11/09/2017)

Jardim, R. & M. Menezes de Sequeira. 2008. The vascular plants (Pteridophyta and Spermatophyta) of the Madeira and Selvagens archipelagos. In: Borges, P.A.V., Abreu, C., Aguiar, A.M.F., Carvalho, P., Jardim, R., Melo, I., Oliveira, P., Sérgio, C., Serrano, A.R.M. & Vieira, P. (eds.). *A list of the terrestrialfungi, flora and fauna of Madeira and Selvagens archipelagos*, pp. 157–179. Direcção Regional do Ambiente da Madeira and Universidade dos Açores, Funchal and Angra do Heroísmo. 440 pp.

Kämmer, F. 1974. *Klima und Vegetation auf Tenerife, besonders im Hinblick auf den NebelniedersChlag*. Scripta Geobotanica. Göttingen. 78 pp.

Kunkel, G. (ed.). 1976. *Biogeography and Ecology in the Canary Islands*. Dr. W. Junk, b.v., Publishers. The Hague. 511 pp.

Kunkel, G. 1977. *Endemismos canarios. Inventario de las plantas vasculares endémicas en la provincia de Las Palmas*. ICONA, Ministerio de Agricultura. Serie Monografías 15. Madrid. 436 pp.

KUNKEL, G. 1977. Las plantas vasculares de Fuerteventura (Islas Canarias), con especial interés de las forrajeras. *Naturalia Hispanica* 8: 1–130 pp.

KUNKEL, G. 1986. *Diccionario Botánico Canario. Manual Etimológico.* Edirca, S. L. Las Palmas de Gran Canaria, España. 273 pp.

KUNKEL, G. 1991. *Flora y vegetación del Archipiélago Canario. Tratado florístico.* 2ª parte. Dicotiledóneas. Edirca. Madrid. 312 pp.

KUNKEL, G. (coord.). 1992. *Flora y vegetación del Archipiélago Canario. Tratado florístico.* 1ª parte. Edirca. Madrid. 295 pp.

LA ROCHE, F. & J.C. RODRÍGUEZ-PIÑERO. 1994. Aproximación al número de taxones de la flora vascular silvestre de los Archipiélagos Macaronésicos. *Rev. Acad. Canar. Cienc.* 6: 77–98.

LEMS, K. 1984. *Phytogeographic study of the Canary Islands.* Thesis. 1958. University of Michigan. Authorized facsimile printed by microfilm/xerography on acid-free paper by University microfilms International. Ann Arbor, Michigan, U.S.A. vol 1, 204 pp.; vol. 2, 144 pp.

LENS, F., N. DAVIN, E. SMET, M. DEL ARCO. 2013. Insular woodiness on the Canary Islands: a remarckable case of convergent evolution. *Int. J. Plant Sci.* 174(7): 992–1013.

LI, L., J. LI, J.G. ROWNER, H. VAN DER WERFF, Z-H. WANG & H-W. LI. 2011. Molecular phylogenetic analysis of the *Persea* group (Lauraceae) and its biogeographic implications on the evolution of tropical and subtropical Amphi-Paciphic disjunctions. *American Journal of Botany* 98(9): 1520–1536.

LIN, HO-Y. 1989. *Systematics of Aeonium (Crassulaceae).* Special Publication No 3, National Museum of Natural Science. Taiwan. 102 pp.

LLEDÓ, M.D., M.B. CRESPO, M.F. FAY & M.W. CHASE. 2005. Molecular phylogenetics of *Limonium* and related genera (Plumbaginaceae): biogeographical and systematic implications. *American Journal of Botany* 92(7): 1189–1198.

LOBIN, W. 1982 - *Untersuchung über Flora, Vegetation und biogeographische Beziehungen der Kapverdischen Inseln.* CFS 53. 112 p. Frankfurt, M.

LOBIN, W., T. LEYENS, A. SANTOS, H. COSTA NEVES & I. GOMES. 2005. The genus Sideroxylon (Sapotaceae) on the Madeira, Canary Islands and Cape Verde Archipelagos. *Vieraea* 33: 119–144.

LOHMEYER, W., & W. TRAUTMANN. 1970. Zur Kenntnis der Vegetation der Kanarischen Inseln La Palma. *Schriftenreiche für Vegetationskunde* 5: 209–236.

LOSADA LIMA, A. 1988. Algunas consideraciones sobre la brioflora macaronésica. *Lazaroa* 10: 213–218.

LU, P-L. & C.W. MORDEN. 2014. Phylogenetic relationships among Dracaenoid genera (Asparagaceae: Nolinoideae) inferred from chloroplast DNA loci. *Systematic Botany* 39(1): 90–104.

LÜPNITZ, D. 1971. Zur physiognomie des Kanarischen Sukkulentenbusches. *Mz. Naturweis. Arch.* 10: 133–148.

LÜPNITZ, D. 1995. Beitrag zur phytogeographischen Stellung der Kanarischen Inseln. *Mainzer naturwiss. Archiv* 33: 83–98.

MACHADO, A., & M. MORERA (coords.). 2005. *Nombres comunes de las plantas y los animales de Canarias.* Academia Canaria de la Lengua, Islas Canarias, España. 277 pp.

Marrero Gómez, M.C., O. Rodríguez Delgado & W. Wildpret De La Torre. 2001. Contribución al estudio bioclimático y corológico de los tabaibales dulces de Tenerife (Islas Canarias). *Vieraea* 29: 135–158.

Marrero Gómez, M.C., O. Rodríguez Delgado & W. Wildpret de la Torre. 2003. *Plocametum pendulae* ("balera") nueva asociación de las Canarias Occidentales. *Vieraea* 31: 377–390.

Marrero Gómez, M.V., O. Rodríguez Delgado & W. Wildpret de la Torre. 1992. Contribución al estudio fitocorológico de los restos de sabinares y otras comunidades termófilas del Sur de Tenerife (Islas Canarias). *Rev. Acad. Canar. Cienc.* 3 (4): 25–44.

Marrero, A. 1991. La flora y vegetación del Parque Natural de "Los Islotes del Norte de Lanzarote y Riscos de Famara". Su situación actual. *Comunicaçôes apresentadas nas 1ª Jornadas Atlânticas de Protecçâo do Meio Ambiente. Açores, Madeira, Canarias e Cabo Verde (Angra do Heroísmo, 25 Junho - 1 Fevereiro de 1988)*: 195–211.

Marrero Rodríguez, A., C. Suárez Rodríguez & J.D. Rodrigo. 1989. Distribución de especies significativas para la comprensión de las formaciones boscosas en Gran Canaria (Islas Canarias). II. *Bot. Macar.* 18 (1989): 27–46.

Marrero, A., R.S. Almeida & M. González Martín. 1998. A new species of the wild dragon tree, Dracaena (Dracaenaceae) from Gran Canaria and its taxonomic and biogeographic implications. *Botanical Journal of the Linnean Society* 128: 291–314.

Martín Esquivel, J.L. 2010. *Atlas de biodiversidad de Canarias*. Gobierno de Canarias. Publicaciones Turquesa S.L. 287 pp.

Martín Esquivel, J.L., H. García Court, C.E. Redondo Rojas, I. García Fernández & I. Carralero Jaime. 1995. *La Red Canaria de Espacios Naturales Protegidos*. Viceconsejería de Medio Ambiente, Consejería de Política Territorial, Gobierno de Canarias. 412 pp.

Martín Esquivel, J.L., M. Arechavaleta Hernández, P.A.V. Borges & B.F. Faria (eds.). 2008. *Top 100. Las cien especies amenazadas prioritarias de gestión en la región europea biogeográfica de la Macaronesia*. Interreg III-B Bionatura, Dirección General del Medio Natural, Gobierno de Canarias. ARENA, Governo Regional dos Açores. Direcçâo Regional do Ambiente, Governo Regional da Madeira. 500 pp.

Martín Esquivel, J.L., B. Santana, N. Nazco & B. López. 2013. *Evaluación preliminar de la vulnerabilidad ante el cambio climático en las Islas Canarias.* Proyecto Climaimpacto (Mac/3/C159). Informe inédito. Viceconsejería de Medio Ambiente del Gobierno de Canarias.

Martín Esquivel, J.L., M.V. Marrero, M. del Arco & V. Garzón. 2015. Aspectos clave para un plan de adaptación de la biodiversidad terrestre de Canarias al cambio climático. In: Herrero, A. & M.A. Zavala (eds.). *Los Bosques y la Biodiversidad frente al Cambio Climático: Impactos, Vulnerabilidad y Adaptación en España*. Capítulo 53: 573–580. Ministerio de Agricultura, Alimentación y Medio Ambiente, Madrid.

Martín Osorio, V.E., W. Wildpret de la Torre & S. Scholz. 2011. Relict ecosystems of thermophilous and laurel forest as biodiversity hotspots in Fuerteventura, Canary Islands. *Plant Biosystems* 145 (Supplement): 180–185.

Martín Osorio, V.E., W. Wildpret de la Torre, M. Del Arco Aguilar, P. L. Pérez de Paz, B. Hernández Bolaños, O. Rodríguez, J. R. Acebes & A. García Gallo. 2007. Estudio Bioclimático y Fitocenótico comparativo de la alta cumbre canaria: Tenerife-La Palma. Islas Canarias. *Phytocoenologia* 37 (3–4): 663–697.

Martín Osorio, V.E., S. Scholz & W. Wildpret de la Torre. 2013. La clase fitosociológica Greenovio-Aeonietea en Jandía, Fuerteventura, con descripción de dos nuevas asociaciones. *Vieraea* 41: 155–162.

Martín-Vide, J. & D. Fernández Belmonte. 2001. El índice NAO y la precipitación mensual en la España peninsular. *Investigaciones Geográficas* 26: 41–58

Médail, F., & P. Quézel. 1999. The phytogeographical significance of S.W. Morocco compared to the Canary Islands. *Plant Ecology* 140: 221–244.

Mester, A. 1987. Estudio fitosociológico de las comunidades de la clase *Pruno-Lauretea azoricae* en La Gomera (Islas Canarias). *Vieraea* 17 (1–2): 409–428.

Mies, B. 1995. On the comparison of the flora and vegetation of the island groups of Socotra and Macaronesia. *Bol. Mus. Mun. Funchal, Supl.* 4: 455–471.

Montelongo Parada, V., J.D. Rodrigo Pérez & D. Bramwell. 1986. Sobre la vegetación de Gran Canaria. *Bot. Macar.* 12–13 (1984): 17–50.

Moreno, J.C. (coord.) 2008. *Lista Roja 2008 de la flora vascular española.* Dirección General de Medio Natural y Política Forestal (Ministerio de Medio Ambiente, y Medio Rural y Marino, y Sociedad Española de Biología de la Conservación de Plantas), Madrid. 86 pp.

Moro, L., J.L. Martín, M.J. Garrrido & I. Izquierdo (eds.), 2003. Lista de especies marinas de Canarias (algas, hongos, plantas y animales) 2003. Consejería de Política Territorial y Medio Ambiente, Gobierno de Canarias. 250 pp.

Nakamura, U., W. Wildpret de la Torre, M.J. Del Arco Aguilar & J.A. Reyes Betancort. 2000. A phytosociological study on Mediterranean laurel foresta area of Tenerife, Canary Islands – in comparison with Japanese laurel forest landscape area of Izu, Central Japan. *Phytocoenologia* 30 (3–4): 613–632.

Naranjo Cigala, A., M. Salas Pascual & R. Almeida Pérez. 2001. La vegetación del Paisaje protegido de Las Cumbres (Gran Canaria, Islas Canarias): El matorral de leguminosas (Telino-Adenocarpion, Andryalo-Ericetalia, Pruno-Lauretea). *Vieraea* 29: 59–70

Nezadal & Wells. 2009. Aportaciones al conocimiento del bosque termófilo en el noroeste de Tenerife (Islas Canarias). In: BELTRÁN TEJERA, E., J. AFONSO-CARRILLO, A. GARCÍA GALLO & O. RODRÍGUEZ DELGADO (eds.). 2009. Homenaje al Prof. Dr. Wolfredo Wildpret de la Torre. Instituto de Estudios Canarios. La Laguna, Tenerife Publ. Inst. Est. Canarios, Monografía 78: 229–244.

Oberdorfer, E. 1965. Pflanzensoziologische Studien auf Teneriffa und Gomera (Kanarische Inseln). *Beitr. Naturk. Forsch. SW-Deutschl.*, 24: 47–104. Karlsruhe.

Ohsawa, M., W. Wildpret & M. del Arco (eds.). 1999. *Anaga Cloud Forest.* Chiba University, Chiba.

Palacios, C.J., & D. Trujillo. 2009. *Árboles y Arboledas Singulares de Canarias. Memorias Vivas de la Tierra.* Gobierno de Canarias. Publicaciones Turques SL. 263 pp.

Pérez de Paz, P.L. 1982. Perspectiva histórica de los últimos 50 años (1932–1982) de la Botánica en Canarias. In: *Instituto de Estudios Canarios 50 Aniversario (1932–1982)*, tomo 1: 295–340. Aula de Cultura del Cabildo Insular de Tenerife.

Pérez de Paz, P.L. (ed.). 1990. *Parque Nacional de Garajonay, Patrimonio Mundial.* ICONA y Excmo. Cabildo Insular de La Gomera. 349 pp.

Pérez de Paz, P.L. & J.R. Acebes Ginovés. 1978. Las Islas Salvajes: Contribución al conocimiento de su Flora y vegetación. In: *Contribución al estudio de la Historia Natural de las Islas Salvajes*, pp. 79–105. Ed. act/Museo Insular de Ciencias Naturales. Excmo. Cabildo Insular de Tenerife, Santa Cruz de Tenerife.

Pérez de Paz, P.L., & C.E. Hernández Padrón, 1999. *Plantas medicinales o útiles en la Flora Canaria. Aplicaciones populares.* Francisco Lemus, Editor. La Laguna. 386 pp.

Pérez de Paz, P.L., M. del Arco Aguilar, J.R. Acebes Ginovés & W. Wildpret de la Torre. 1986. *Leguminosas forrajeras de Canarias.* Excmo. Cabildo Insular de Tenerife. Act. Serie Publ. Científicas. Subser. Mus. Ins. de Cienc. Nat. 2. 157 pp.

Pérez de Paz, P.L., M. Del Arco Aguilar & W. Wildpret de la Torre. 1987. Contribución al conocimiento de la vegetación hidrofítica de Canarias. In: M.J. Del Arco Aguilar & W. Wildpret de la Torre (eds.): *V Jornadas de Fitosociología. Vegetación de Riberas de Agua dulce II*, pp. 11–34. Secretariado de Publicaciones, Universidad de La Laguna. Serie Informes, 22.

Pérez de Paz, P.L., M. del Arco Aguilar & W. Wildpret de la Torre. 1990. Contribución al conocimiento de los matorrales de sustitución del Archipiélago Canario. Nuevas comunidades para El Hierro y La Palma. *Vieraea* 19: 53–62.

Pérez de Paz, P.L., M.J. Del Arco Aguilar, O. Rodríguez Delgado, J.R. Acebes Ginovés, M.V. Marrero Gómez & W. Wildpret de la Torre. 1994. *Atlas Cartográfico de los Pinares Canarios III: La Palma.* Viceconsejería de Medio Ambiente, Consejería de Política Territorial, Gobierno de Canarias. Santa Cruz de Tenerife. 160 pp. + 7 mapas.

Pérez de Paz, P.L., M. Salas Pascual, O. Rodríguez Delgado, J.R. Acebes Ginovés, M.J. Del Arco Aguilar & W. Wildpret de la Torre. 1994. *Atlas Cartográfico de los Pinares Canarios IV: Gran Canaria y plantaciones de Fuerteventura y Lanzarote.* Viceconsejería de Medio Ambiente, Consejería de Política Territorial, Gobierno de Canarias. Santa Cruz de Tenerife. 199 pp. + 22 mapas.

Pérez de Paz, P.L., W. Wildpret de la Torre, M.J. del Arco Aguilar, O. Rodríguez Delgado, J.R. Acebes Ginovés & A. García Gallo. 1995. *Inventariación de los tipos de habitats incluidos en el Anexo I de la Directiva 92/43/CEE. Canarias.* Memoria no publicada. ICONA. 117 pp. + 33 mapas.

Pitard, J., & L. Proust. 1908. *Les Iles Canaries. Flore de l'Archipel.* Librairie des Sciences Naturelles, Paris. 502 pp.

Press, J.R. & M.J. Short. 1994. *Flora of Madeira.* London. 574 pp.

Puppo, P., M. Curto & H. Meimberg. 2016. Genetic structure of Micromeria (Lamiaceae) in Tenerife, the imprint of geological history and hybridization on within-island diversification. *Ecology and Evolution* 6 (11): 3443–3460.

Quézel, P. 1979. Analysis of the Flora of Mediterranean and Saharan Africa. *Ann. Missouri Bot. Gard.* 65: 479–534 (1978).

Quintana Vega, G., M. Salas Pascual & E. Fernández Negrín. 206. Contribución al estudio de las comunidades rupícolas de la vertiente norte de Gran Canaria (Islas Canarias). *Lazaroa* 27: 89–102.

Reyes Betancort, J.A., M.C. León Arencibia, W. Wildpret de la Torre & M.M. Medina Pérez. 2000. *Estado de Conservación de la Flora Silvestre Amenazada de Lanzarote (Islas Canarias).* Viceconsejería de Medio Ambiente. Gobierno de Canarias. 177 pp.

Reyes-Betancort, J.A., W. Wildpret de la Torre & M.C. León Arencibia. 2001. The vegetation of Lanzarote (Canary Islands). *Phytocoenologia* 31 (2): 185–247.

Rivas-Martínez, S. 2007: Mapas de series, geoseries y geopermaseries de vegetación de España [Memoria del mapa de vegetación potencial de España], Parte I. - *Itinera Geobotanica* 17: 5–435.

Rivas-Martínez, S., W. Wildpret de la Torre, T.E. Díaz González, P.L. Pérez de Paz, M. Del Arco Aguilar & O. Rodríguez Delgado. 1993. Excursion guide. Outline vegetation of Tenerife Island (Canary Islands). Sinopsis de la vegetación de la Isla de Tenerife (Islas Canarias): Guía de la excursión. *Itinera Geobotanica* 7: 5–167.

Rivas-Martínez, S., W. Wildpret de la Torre, M. del Arco Aguilar, O. Rodríguez Delgado, P.L. Pérez de Paz, A. García Gallo, J.R. Acebes Ginovés, T.E. Díaz González & F. Fernández González. 1993. Las comunidades vegetales de la Isla de Tenerife (Islas Canarias). *Itinera Geobotanica* 7: 169–374.

Rivas-Martínez, S., F. Fernández González, J. Loidi, M. Lousa & A. Penas. 2001. Syntaxonomical checklist of vascular plant communities of Spain and Portugal to association level. *Itinera Geobotanica* 14: 5–341.

Rivas-Martínez, S., T.E. Díaz, F. Fernández González, J. Izco, J. Loidi, M. Lousã & A. Penas. 2002. Vascular plant communities of Spain and Portugal. Addenda to the syntaxonomical checklist of 2001. *Itinera Geobotanica* 15(1): 1–432, 15(2): 433–922.

Rivas-Martínez, S., S. Rivas Sáenz & A. Penas. 2011. Worldwide bioclimatic classification system. *Global Geobotany* 1: 1–634 + 4 maps.

Rivas-Martínez, S. & coautores. 2011: Mapa de series, geoseries y geopermaseries de vegetación de España [Memoria del Mapa de Vegetación Potencial de España, 2011]. Parte II. *Itinera Geobotanica* 18 (1): 5–424; 18 (2): 425–800.

Rivas-Martínez, S., A. Penas, T.E. Díaz González, L. Herrero y S. del Río (eds.). 2014. Mapas de vegetación de las series, geoseries y geopermaseries de España. 1:250.000 [Maps of vegetation series, geoseries and geopermaseries from Spain. 1:250.000]. *Global Geobotany* 3: 1–634 pp. + mapas.

Rodrigo Pérez, J., & V. Montelongo Parada. 1986. Distribución de especies significativas para la comprensión de las formaciones boscosas en Gran Canaria (Islas Canarias).I. *Bot. Macar.* 12–13 (1984): 3–16.

Rodríguez Delgado, O. 1990. *Flora y Vegetación de las Bandas del Sur de Tenerife. La Comarca de Agache (Güímar).* Secretariado de Publicaciones, Universidad de La Laguna. Resumen Tesis Doctoral. 54 pp.

Rodríguez Delgado, O. 1993. Bibliografía geobotánica canaria. *Itinera Geobotanica* 7: 437–509.

Rodríguez Delgado, O. (ed.). 1999. *Homenaje a Webb en el II centenario de su nacimiento.* Instituto de Estudios Canarios, Departamento de Biología Vegetal (Botánica), Universidad de La Laguna. La Laguna de Tenerife. 119 pp.

Rodríguez Delgado, O. (ed.). 2003. *Apuntes sobre Flora y Vegetación de Gran Canaria (Guía de la excursión geobotánica de las XIX Jornadas de Fitosociología y Simposio Internacional de la FIP 2003).* Cabildo de Gran Canaria, Medio Ambiente y Aguas. 271 pp.

Rodríguez Delgado, O. (ed.). 2005. *Patrimonio Natural de la isla de Fuerteventura.* Cabildo de Fuerteventura, Dirección General del Medio Natural del Gobierno de Canarias, Centro de la Cultura Popular Canaria. 460 pp.

Rodríguez Delgado, O., & E. Beltrán Tejera. 1995. Botánica. In: *Gran Enciclopedia Canaria.* Tomo III (Bar-Can), pp. 639–641. Ediciones Canarias.

Rodríguez Delgado, O., & M.J. Del Arco Aguilar. 2007. Los pinares canarios. Pasado y presente. *Rincones del Atlántico* 4: 148–159.

Rodríguez Delgado, O., & R. Elena-Rosselló (eds.). 2005. *Evolución del paisaje vegetal del Parque Nacional del Teide.* Organismo Autónomo Parques Nacionales, Ministerio de Medio Ambiente. 364 pp.

Rodríguez Delgado, O., & M.V. Marrero Gómez. 1991. Evolución y aprovechamiento de los bosques termófilos (los "montes bajos") de la Isla de Tenerife. *Anuario de Estudios Atlánticos* 36 (1990): 595–630. Madrid, Las Palmas.

Rodríguez Delgado, O., & A. Naranjo Cigala. 2005. *El azúcar y su cultura en las Islas Atlánticas. Paisajes, geografía, etnobotánica y geobotánica. Canarias, siglos XVI al XX.* Proyecto "Atlántica". Consejería de Educación, Cultura y deportes del Gobierno de Canarias, Dirección General de Patrimonio Histórico, Ayuntamiento de Los Llanos de Aridane. 192 pp.

Rodríguez Delgado, O., W. Wildpret de la Torre, M.J. del Arco Aguilar & P.L. Pérez de Paz. 1990. Contribución al estudio fitosociológico de los restos de sabinares y otras comunidades termófilas de la Isla de Tenerife (Canarias). *Rev. Acad. Canar. Cienc.*, 2: 121–142.

Rodríguez Delgado, O., M.J. Del Arco Aguilar, A. García Gallo, J.R. Acebes Ginovés, P.L. Pérez de Paz & W. Wildpret de la Torre. 1998. *Catálogo sintaxonómico de las comunidades vegetales de plantas vasculares de la Subregión Canaria: Islas Canarias e Islas Salvajes. Syntaxonomic checklist for the vascular plant communities in the Canarian Subregion: Canary Islands and Salvage Islands.* Versión Español / Inglés. Materiales Didácticos Universitarios. Ser. Biología 1. Servicio de Publicaciones. Universidad de La Laguna. 130 pp.

Rodríguez Delgado, O., A. García Gallo & J. A. Reyes Betancort. 2000. Estudio fitosociológico de la vegetación actual de Fuerteventura (Islas Canarias). *Vieraea* 28: 61–98.

Rodríguez Losada, J.A. 2000. Las Islas Canarias y el origen y clasificación de las rocas ígneas. Universidad de La Laguna. Tenerife. http://www.gobiernodecanarias.org/geotecnia/canarias_clasif_roc_rguez-losada.pdf

Rodríguez Luengo, J.L., J. García Casanova, G. Díaz Reyes & A. Delgado (dirs.). 2003. *Fauna y Flora de Canarias en el Catálogo Nacional de Especies Amenazadas*. Consejería de Política Territorial y Medio Ambiente, Gobierno de Canarias. 204 pp.

Rodríguez, J.C., M. del Arco Aguilar & W. Wildpret de la Torre. 1986. Contribución al estudio fitosociológico de los sauzales canarios. *Rubo-Salicetum canariensis* asociación nueva. *Doc. Phytosociol. N.S.*, 10(1): 379–388.

Rüdiger, O., J.M. Fernández-Palacios & B.O. Krüsi. 2001. Variation species composition and vegetation structure of succulent scrub on Tenerife in relation to environmental variation. *Journal of Vegetation Science* 12: 237–248.

Salas Pascual, M., M.J. Del Arco Aguilar & P.L. Pérez de Paz. 1998. Contribución al estudio fitosociológico del pinar grancanario (Islas Canarias). *Lazaroa* 19: 99–117.

Salas, M., M. Del Arco, J.R. Acebes, A. Naranjo & R. González. 2003. Los retamares de *Retama rhodorhizoides* Webb & Berthel. en Gran Canaria: *Echio decaisnei-Retametum rhodorhizoidis* ass. nov. *Vieraea* 31: 201–206.

Sánchez-Pinto, L., Rodríguez, M.L., Rodríguez, S., Martín, K., Cabrera, A. & Marrero, M.C. 2005. *Pteridophyta, Spermatophyta.* In: *Lista preliminar de especies silvestres de Cabo Verde (hongos, plantas y animales terrestres)*, pp. 38–57. Consejereía de Medio Ambiente y Ordenación Terrotorial, Gobierno de Canarias.

Santos Guerra, A. 1980. *Contribución al conocimiento de la flora y vegetación de la isla de Hierro (I. Canarias)*. Fundación Juan March, Serie Universitaria, 114. Madrid. 51 pp.

Santos Guerra, A. 1983. *Vegetación y Flora de La Palma*. Ed. Interinsular Canaria S.A. Cabildo Insular de La Palma. Santa Cruz de Tenerife. 348 pp.

Santos Guerra, A. 1990. *Bosques de Laurisilva en la Región Macaronésica*. Consejo de Europa. Colección Naturaleza y Medio Ambiente 49. Estrasburgo. 79 pp.

Santos Guerra, A. 1999. Origen y evolución de la flora canaria. In: Fernández-Palacios, J.M., J.J. Bacallado & J.A. Belmonte (eds.), *Ecología y cultura en Canarias*, pp. 107–129.

Schaefer, H. 2003. Chorology and Diversity of the Azorean Flora. *Diss. Bot.* 374: 1–670.

Schaefer, H. 2005. *Flora of the Azores. A Fieldguide*. 2nd ed. Margraf Publishers/ Backhuys Publishers, Weikersheim. 346 pp.

Schmidt-Lebuhn, A.N. 2008. Monophyly and phylogenetic relationships of Minthostachys (Labiatae, Nepetoideae) examined using morphological and nrITS data. *Plant Systematics and Evolution* 270: 25–38.

SCHOENFELDER, P. & I. SCHOENFELDER. 2005. *Die Kosmos-Kanarenflora*. 2nd ed. Franckh-Kosmos, Stuttgart. 319 pp.

SCHÖNFELDER, P., & V. VOGGENREITER. 1994. Zur Abgrenzung und Gliederung der Klassen *Spartocytisetea supranubii* cl. nov. und *Cytiso-Pinetea canariensis* auf Tenerife/Kanarische Inseln. *Phytocoenologia*, 24: 461–493.

SHMIDA, A. & M.J.A. WERGER. 1992. Growthform diversity on the Canary Islands. Vegetatio 102: 173–181.

SILVA, L., M. MOURA, H. SCHAEFER, F.RUMSEY, E. DIAS. 2008. List of vascular plants (Tracheobionta). *In*: Borges, P.A.V., A. Costa, R. Cunha, R. Gabriel, V. Gonçalves, A.F. Martins, I. Melo, M. Parente, P. Raposeiro, P. Rodrigues, R.S. Santos, L. Silva, P. Vieira & V. Vieira (eds.), *A list of the terrestrial and marine biota from the Azores*, pp. 117–146, Princípia, Cascais, 432 pp.

SILVA, L., E. OJEDA LAND & J.L. RODRÍGUEZ LUENGO (eds.). 2008. *Flora y Fauna Terrestre Invasora en la Macaronesia. TOP 100 en Azores, Madeira y Canarias*. ARENA, Ponta Delgada. 546 pp.

SOSA HENRÍQUEZ, P., A. NARANJO CIGALA, M. MÁRQUEZ GARCÍA, A. ESCANDELL BERMÚDEZ & M.Á. GONZÁLEZ PÉREZ. 2007. *Atlas de los palmerales de Gran Canaria*. Obra social La Caja de Canarias. 197 pp.

STIERSTORFER, C. 2005. *The Vascular Plant Vegetation in the Forest Belt of El Hierro (Canary Islands)*. Diss. Bot. 393. J. Cramer. Stuttgart. 375 pp. + láms.

SUÁREZ BETANCOR, J. (dir.). 1997. *La toponimia de Gran Canaria*. Ediciones del Cabildo Insular de Gran Canaria, Gestur. Las Palmas de Gran Canaria. Dos volúmenes, 332 + 333 pp. + un CD-Rom.

SUÁREZ RODRÍGUEZ, C. 1994. *Estudio de los relictos actuales del monteverde en Gran Canaria*. Ediciones del Cabildo Insular de Gran Canaria, Consejería de Política Territorial del Gobierno de Canarias. 617 pp.

SUN, Y., Y. LI, C.F. VARGAS-MENDOZA, F. WANG & F. XING. 2016. Colonization and diversification of the Euphorbia species (sect. Aphyllis subsect. Macaronesicae) on the Canary Islands. *Scientific Reports* 6: 34454, DOI: 10.1038/srep34454.

SUNDING, P. 1970. Bibliographia Phytosociologica. The Canary Islands. *Excerpta Botánica* (Sec. B.) 10: 257–268.

SUNDING, P., 1972. *The vegetation of Gran Canaria*. Scrifter utgit av det Norske Vindenskaps-Akademi i Oslo. I.Mat.-Naturv. Klase. NY. Serie N° 29. (Universitetsforlaget) Oslo. 186 pp. + LIII láms.

TAKHTAJAN, A. 1986. *Floristic Regions of the World*. University of California Press. 522 pp.

VAN DEN BOGAARD, P. 2013. The origin of the Canary island Seamount Province – New ages of old seamounts. *Scientific Reports* 3: 1–7.

VERA GALVÁN, M.Á., C. SAMARÍN BELLO, G. DELGADO CASTRO & G. VIERA RUIZ. 2010. *Natura 2000 en Macaronesia. Azores, Madeira, Salvajes y Canarias. Guía técnica*. Consejería de Medio Ambiente y Ordenación Territorial, Gobierno de Canarias. 567 pp.

VIERA Y CLAVIJO, J. DE. 2005. *Diccionario de Historia Natural de las Islas Canarias o Indice alfabético descriptivo de sus tres reinos: animal, vegetal y mineral*. Nivaria Ediciones. La Laguna, Tenerife. 639 pp.

Vitales, D., T. Garnatje, J. Pellicer, J. Vallès. A. Santos-Guerra & I. Sanmartín. 2014. The explosive radiation of *Cheirolophus* (Asteraceae, Cardueae) in Macaronesia. *BMC Evolutionary Biology* 14: 118. https://doi.org/10.1186/1471-2148-14-118.

Voggenreiter, V. 1974. *Geobotanische Untersuchungen an der natürlichen Vegetation der Kanareninsel Tenerife (Anhang: Vergleiche mit La Palma und Gran Canaria) als Grundlage für den Naturschutz*. Dissertationes Botanicae 26, J. Cramer, Lehre, 718 pp.

Voggenreiter, V. 1976. *Euphorbio canariensis-Pinetum canariensis* ass. nov. y límite inferior del Pinar Canario en Gran Canaria. *Vieraea* 6(1): 3–16.

Voggenreiter, V. 1995. *Atlas Fitocorológico Ilustrado de Tenerife en Cuadrícula U.T.M. 5 km x 5 km (Puntos de Repartición para Cuadrados de 25 Km²) -un Prodromus-*. Edición manuscrita. Bonn. 399 pp.

Weber, H.E., Moravec, J. & J.P. Theurillat. 2000. International code of phytosociological nomenclature. 3rd ed. J. Veg. Sci. 11: 739–768. [http://www.iavs.org/pdf/code.pdf]

Welss, W., & R. Lindacher. 1994. Beiträgue zur Chorologie und Florenstatistik der Kanarischen Inseln. *Hoppea, Denkschr. Regensb. Bot. Ges.* 55 (Hohenester-Festschrift): 845–857.

Wildpret de la Torre, W. & M. Del Arco Aguilar. 1987. España Insular: Las Canarias. In: Peinado, M. & S. Rivas-Martínez (eds.), *La Vegetación de España*, pp. 517–544. Universidad de Alcalá de Henares, Secretariado de Publicaciones. Colección "Aula Abierta" N° 3. Madrid.

Wildpret de la Torre, W., W. Greuter & B. Zimmer. 1987. *Guide to Excursion No. 35. The phanerogamic and cryptogamic flora and vegetation of Tenerife (Canary Islands).* 56 pp. Berlin.

Wildpret de la Torre, W., M.J. del Arco Aguilar & A. García Gallo. 1988. Contribución al estudio de la clase *Onopordetea acanthii* Br.-Bl. 1964 em. Riv.-Mart. inéd. en las Islas Canarias. *Scolymo maculati-Cynaretum ferocissimae ass. nov. Documents phytosociologiques* N.S. 11:153–158.

Wildpret de la Torre, W., P.L. Pérez de Paz, M.J. del Arco Aguilar & A. García Gallo. 1988. Contribución al estudio de la clase *Polygono-Poetea annuae* Rivas-Martínez 1975 en las Islas Canarias. *Acta Bot. Barc.*, 37: 355–361.

Wildpret de la Torre, W., A. García Gallo & E. Carqué Álamo. 1996. Crasuláceas endémicas macaronésicas en las comunidades pioneras de tejados y muros de huertas en Canarias. *Documents phytosociologiques N.S.* 16: 59–68.

Wildpret de la Torre, W., E. Beltrán Tejera & M.C. León Arencibia. 1997. Flora and vascular vegetation of the Islet of Montaña Clara (Canary Islands). *Proceeding Book IAVS Symposium Tenerife 12–16 April 1993*: 237–246.

Wildpret de la Torre, W., A. García Gallo, I. Pérez Vargas & J.S. Socorro Hernández. 2005. *Flora ornamental del casco histórico de La Laguna*. Excmo. Ayuntamiento de San Cristóbal de La Laguna. 407 pp.

Wolff, P. & M. Rosinski. 1999. Nuevas e interesantes plantas vasculares de las Islas Canarias. *Vieraea* 11–21.

Syntaxonomic Index

M. J. del Arco Aguilar, O. Rodríguez Delgado, *Vegetation of the Canary Islands*, Plant and Vegetation 16, https://doi.org/10.1007/978-3-319-77255-4

R

S

T

U

V

Taxonomic Index

M. J. del Arco Aguilar, O. Rodríguez Delgado, *Vegetation of the Canary Islands*, Plant and Vegetation 16, https://doi.org/10.1007/978-3-319-77255-4

M